Electoral Geography

Igor Okunev

Electoral Geography

Bruxelles - Berlin - Chennai - Lausanne - New York - Oxford

Library of Congress Cataloging-in-Publication Data
A CIP catalog record for this book has been applied for at the Library of Congress.

Bibliographic Information published by the Deutsche Nationalbibliothek
The Deutsche Nationalbibliothek lists this publication in the Deutsche Nationalbibliografie; detailed bibliographic data is available online at http://dnb.d-nb.de.

Translated from Russian by Philip Taylor

ISBN 978-3-0343-5033-4 (Print)
E-ISBN 978-3-0343-5034-1 (E-PDF)
E-ISBN 978-3-0343-5035-8 (E-PUB)
DOI 10.3726/b22026
D/2024/5678/28

© 2024 Peter Lang Group AG, Lausanne
Published by Peter Lang Éditions Scientifiques Internationales - P.I.E. SA, Brussels, Belgium

info@peterlang.com - www.peterlang.com

All rights reserved.

All parts of this publication are protected by copyright. Any utilisation outside the strict limits of the copyright law, without the permission of the publisher, is forbidden and liable to prosecution. This applies in particular to reproductions, translations, microfilming, and storage and processing in electronic retrieval systems.

Table of Contents

Chapter 1. Introduction to Electoral Geography 11

    1.1 Electoral Geography: Object of Study and Methods 12
    1.2 Stages in the Development of Electoral Geography 18
    1.3 Geography of Elections Around the World 25
    1.4 Elections as Part of the Political Process 31
    1.5 Electoral Statistics and Electoral Sociology 35
    1.6 Applied Electoral Geography 38
    1.7 International Electoral Geography 41

Chapter 2. Territorial Differentiation of Electoral Systems 47

    2.1 Public Choice Theory. The Condorcet and Arrow Paradoxes 48
    2.2 Electoral and Voting Districts 52
    2.3 Typology of Electoral Systems 54
    2.4 Majority Electoral Systems 60
    2.5 Semi-Proportional Electoral Systems 64
    2.6 Proportional Electoral Systems 66
    2.7 Mixed Electoral Systems 71
    2.8 Methods of Distributing Seats 75
    2.9 Weighted Electoral Systems 81
    2.10 Geographic Favouritism in Electoral Systems 85

Chapter 3. Territorial Differentiation of Party Systems 97

    3.1 Parties and Their Territorial Differentiation 98
    3.2 Regional and Regionalist Parties 104

3.3 Spatial Typology of Party Systems 105
3.4 Effective Number of Parties 108
3.5 Nationalization and Regionalization of Party Systems 110
3.6 The Ideological Spectrum of the Electoral Field 114
3.7 Cleavages in the Ideological and Political Space 118

Chapter 4. Spatial Models of Voting 131
4.1 Theories of Electoral Behaviour 132
4.2 The Hotelling–Downs One-Dimensional Model .. 133
4.3 The Enelow–Hinich Linear Model 139
4.4 The Granberg–Brown Parabolic Model 142
4.5 The Rabinowitz–MacDonald Vector Model 144

Chapter 5. The Spatial Effects of Voting 151
5.1 The Scalar Effects of Voting 152
5.2 The Vector Effects of Voting 155
5.3 Malapportionment 160
5.4 Gerrymandering 162
5.5 The Efficiency Gap 165
5.6 Compactness of Electoral Districts 169

Chapter 6. Spatial Analysis in Electoral Geography 177
6.1 Electoral-Geographic Maps 177
6.2 Exploratory Data Analysis 185
6.3 Exploratory Spatial Data Analysis 192
6.4 Centrography 196
6.5 Spatial Neighbourhood Weights 202
6.6 Spatial Lag 212
6.7 Spatial Autocorrection 217
6.8 Local Spatial Autocorrelation 228
6.9 Spatio-Temporal Autocorrelation 234
6.10 Spatial Regression 240

Appendix. Spatial Statistical Analysis in Political and Electoral Geography: Methodological Guide 253

Bibliography 419

List of Tables and Figures 459

Index of Terms 463

Chapter 1
Introduction to Electoral Geography

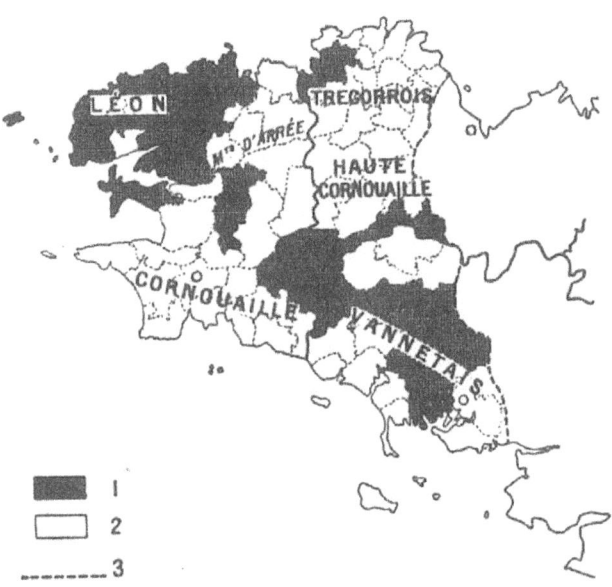

FIG. 56. — CARTE POLITIQUE (PAR CANTONS) DE LA BRETAGNE BRETONNANTE AUX ÉLECTIONS LÉGISLATIVES DE 1910
1. Cantons ayant donné une majorité à la coalition de droite
2. — — — — . gauche
3. Limite orientale des dialectes bretons

Source: Siegfried, A. (1913). *Tableau politique de la France de l'ouest sous la 3e Republique*. Armand Colin.

- What does electoral geography study? What place does it occupy in the geographic and social sciences?
- What aspects of elections have a spatial dimension?
- What are the ways in which power can be transferred? Where do elections fit among these methods?

- How are electoral systems, on the one hand, and forms of government and political regimes, on the other, related?
- How is knowledge of electoral geography valuable in practice?

1.1 Electoral Geography: Object of Study and Methods

Electoral geography is the study of the spatial dimension of the electoral process.

Our location contains a lot of data about us – not only about the physical coordinates of where we are, but also about our interests, values, and even political preferences. For example, information about the average cost of housing in the area where we live, or data about the stores we frequent, may say a lot about how much we earn, and therefore about our social status and likely political orientation. This kind of information is everywhere in modern society: we check in when we travel and when we stay at hotels; we create geotags when we post something to the internet; and we constantly transfer anonymous information about the location of our personal devices. Processing arrays of this information makes it possible to analyse and even predict trends in the political process. It turns out that just as the terrain can vary, that is, it can be spatially heterogenous, so too can the voting habits of the people living on it. And it also turns out that a person's political choice is influenced, among other things, by his or her physical location. The differentiation of space based on the electoral preferences of residents is what forms the electoral space – a special layer of the earth's surface that is the object of study of the discipline we call electoral geography.

Electoral geography stands at the intersection of the geographic and social sciences, namely: social (political) geography on the one hand, and political science, electoral sociology and election law on the other. The interdisciplinary field related to the study of electoral systems and processes is sometimes called *psephology*, in which case electoral geography can be seen as a science that applies geographical methods to the object of study of psephology. Just as elections are the focus of political science, electoral geography can be seen as the focus of political geography.

The key concept of geography – space – is the basic form of the existence of matter and is characterized by its physical dimensions and volume. This means that the analysis of an object in space allows us, first of all, to identify its location relative to other objects or phenomena and, secondly, to identify how this position in space affects the properties of the object itself. This method of studying an object is at the heart of the methodology of geographic science.

The object of study in electoral geography is the electoral process – the system of interactions between voters and elected representatives before, during and after elections to government bodies. The particular focus of electoral geography is the spatial dimension of the electoral process.

The spatial dimension of the electoral process is studied in one of the types of electoral space (Table 1.1):

1. Electoral landscape – a projection of *physical (absolute) space*, the coordinate system of which corresponds to the Earth's surface and is measured in degrees and other physical units (metres, feet, etc.);
2. Electoral field – a projection of *formal (relative) space*, which is a mathematical abstraction, the coordinate system of which is determined by the structure of interactions of elements in space and their topological properties;
3. Electoral worldview – a projection of *perceptual (cognitive) space*, the coordinate system of which is determined by the subjective perception of elements in space and the relationships between them. The electoral worldview is thus a spatially organized system of images, narratives, discourses, perceptions, myths, identities and other subjective points of view on the electoral process.

Table 1.1. Types of electoral space

Type of electoral space	Type of space	Map example	Area of electoral geography
electoral landscape	physical (absolute) – natural	a geographic map	voting geography
electoral field	formal (relative) – social	a subway map	spatial modelling of voting
electoral worldview	perceptual (cognitive) – individual	a travel map/ itinerary	critical electoral geography

The three types of space were first conceptualized by the German–American philosopher Rudolf Carnap in his famous early work *Der Raum (Space)* (1922) [235]. Physical (absolute) space is thus a mathematical matrix in which each object is given exact geographic coordinates (latitude, longitude and elevation). In formal (relative) space, it is not the exact coordinates that are important, but the location of objects relative to each other and the relationships that result from such an arrangement. Perceptual (cognitive) space is the reflection of an object's location in our memory and imagination. Each of these types of space represents a fundamental perspective on who shapes the world around us. Physical space is created by nature – that is, by an external force – and is thus not within our control and can only act as an explanatory force in social processes. Relative space is created by society, it is essentially the projection of society on a surface, which means that its properties are inseparable from the essence of the social order. Finally, cognitive space is a product of the individual's subjective perception, it is unique and cannot be generalized.

To better understand this, let us compare how a subway map would look in the three types of space. In the physical (absolute) space, we see the subway lines superimposed on top of a map of the city; they are tied to specific geographic coordinates. In the formal (relative) space, the subway lines straighten out and the distance between them is standardized, since we are only concerned about how many stops we need to travel and where we need to change trains – this is the kind of map that is usually posted at subway stations and within train cars. In the perceptual (cognitive) space, subway stations are associated with local sites, our own memories of life events that took place in different parts of the city, and an idea of the distance between them. Distance is the basic parameter of analysis in space and, as such, it too can be absolute, relative and cognitive. Absolute distance is measured in universal units of measurement (degrees, metres, miles, and so on); relative distance may be measured in the number of stops, train changes, money or time spent on the journey, etc.; and cognitive distance may be measured, for example, in the number of songs you listened to or the number of pages of your book you read during your journey

The basic hypothesis of electoral geography is that the structure of space has an independent influence (that is, separate from other factors) on the electoral process and voting results.

Modern electoral geography is made up of the following key areas and subdisciplines:

Electoral Geography: Object of Study and Methods 15

- geography of voting identifies the factors and patterns that underlie long-standing territorial differences in the political activity of voters and their voting habits by administrative and territorial unit, constituency and district, as well as the geographic favouritism and disproportionality of electoral systems – that is, the proclivity of electoral procedures to the territorial differentiation of election results [253, 480];
- geography of representation deals with the level of representation of territories in the executive and legislative bodies of power at various levels and the negative experience of unequal and unfair distribution (*malapportionment*) [274, 394, 428, 446];
- electoral limology systematizes the influence of constituency and district boundaries on election results, primarily by analysing the negative experience of moving the boundaries of electoral districts (gerrymandering), and identifies transparent and fair strategies for changing such boundaries (redistricting) [211, 224, 381];
- geography of social cleavages deals with the territorial dimension of ideological and political splits in society as a factor in the electoral behaviour of citizens [373, 403];
- electoral geography of parties and party systems focuses on issues of the nationalization and regionalization of parties and party systems [234];
- electoral geography of political campaigns studies patterns in the conduct of election campaigns [335];
- spatial modelling of voting concerns the mathematical modelling of the electoral process in an abstract space [320];
- urban electoral geography explains the territorial patterns of the electoral process at the city and municipal level [252, 319];
- international electoral geography is concerned with geographical patterns in the voting of countries in supranational structures (the United Nations, the World Trade Organization, the International Labour Organization, the European Union, the Eurovision Song Contest, etc.) [204, 371];
- critical electoral geography focuses on the factor of identity in the territorial differentiation of the electoral process among different social groups in the post-positivist paradigm (for example, feminist electoral geography [382, 452], emotional electoral geography [453], racial electoral geography, etc.).

Despite the fact that electoral geography today is divided into several independent areas of research, the key goal of the discipline is to understand the factors and patterns behind the stable electoral and geographic heterogeneity of space. Within this area of focus, the following main stages in the analysis of the electoral differentiation of space can be singled out:

1. assessment and flattening out of geographic favouritism within the political system;
2. quantitative and qualitative analysis of the structure of the party system, the number of segments that make it up and their relative weight;
3. assessment of the degree of nationalization/regionalization of the party system and the proclivity for territorial differentiation of the parties that make up that system;
4. identification of the ideological platform of the segments of the party system and linking them to the long-standing ideological and political splits in society;
5. construction of a spatial voting model that reflects the configuration of political forces on the ideological spectrum and the electoral field;
6. operationalization of a comparison matrix of candidates and parties based on stable segments of the electoral field (in the case of studies that cover more than one country or time period);
7. assessment of the overall level of electoral-geographic segmentation of space and identification of stable clusters and anomalous territories among them;
8. electoral-geographic zoning of space through the identification of spatial clusters that demonstrate a similar type of electoral behaviour;
9. identification and verification of potential factors of electoral behaviour, determination of spatial differentiation of the level of their validity;
10. assessment of the effects of electoral geography and the degree of electoral-geographic engineering.

It is well known that voting preferences undergo a number of systemic transformations during the electoral process, which means that the end result may differ markedly from the original intention. The key task of electoral research is thus to assess the degree to which voter intention is deformed by a system of interrelated filters: (1) deeply ingrained ideological and political divisions and the structure of the party system they mediate; (2) the course of the election campaign (including scalar effects of electoral geography); (3) the neighbourhood effect (i.e. the vector effect of electoral geography); and (4) the institutional favouritism of the electoral system (including geographic favouritism). An important task of electoral geographic analysis in this context is to explain the nature of this deformation of voter intention.

In the course of research, electoral geography relies primarily on special geographic methods: participant observation and interviews during field trips, mapping, zoning, modelling and spatial analysis.

Electoral-geographic field trips: this includes observation, collecting, digitizing, geocoding and processing primary data on election results and voting behaviour, and interviews with people involved in the electoral process (including expert interviews, focus groups, etc.).

Electoral-geographic mapping: this involves comparing voting results with the position in space of various objects (residential areas, campaigning locations, election headquarters, the homes of candidates, etc.) and phenomena (protest activity, conflicts, etc.). Plotting the objects of study on maps allows us to present the location of objects relative to each other. Then, a search is carried out to find the nearest objects to the one we are studying and patterns in their positional relationship are analysed. Finally, superimposing various thematic layers of objects on top of one of other and changing the scale of the map gives us new information about the patterns of territorial differences in voting behaviour.

Electoral-geographic zoning: the spatial differentiation of the territory according to a given feature, phenomenon or condition (usually the degree of political activity or the level of support for a particular party or candidate), its severity, and how compatible or incompatible it is with other features. The resulting parts of space (regions), or the space-time continuum (*chorions*), allow us to formulate certain conclusions about the structure of society and the electoral processes that take place within it.

Electoral-geographic modelling: the search for spatial links that explain the properties of objects and the nature of electoral processes. This method involves: (1) identifying key and secondary objects of the space that is being analysed, spatial dependency vectors and spatial barriers that prevent dependency; and (2) modelling schemes of interdependence of elements and levels of spatial organization.

Spatial analysis in electoral geography: assessment of global and local spatial correlation in voting behaviour; evaluation of the level of electoral-geographic segmentation of space; identifying stable spatially continual statistical clusters of similar voting behaviour; qualitative and quantitative assessment of spatial factors in the distribution of electoral phenomena.

While it is true that electoral geography leans heavily on the positivist theoretical paradigm and relies on statistical methods of processing data, in its modern interpretation it also follows a constructivist approach that opens up new ways of understanding electoral processes.

1.2 Stages in the Development of Electoral Geography

The history of electoral-geographic research dates back a little over a century and can be divided into three stages: the descriptive stage (1910s–1960s), the analytical stage (1970s–2000s), and the synthetic stage (from the 2010s onwards).

The *descriptive stage* of the development of electoral geography (1910s–1960s) involved the collection and systematization of empirical knowledge about the territorial differentiation of electoral processes. The main achievement of this stage was formation of electoral geography as a new subject area that combines elements of social geography, political science, electoral sociology and election law to describe the influence that geography has on elections.

The French geographer André Siegfried (1875–1959) is credited with having founded and named the discipline of electoral geography (Fig. 1.1). In his 1913 book *Tableau politique de la France de l'Ouest sous la Troisième République* (Political Picture of Western France under the Third Republic) [461], Siegfried analyses the territorial distribution of elections in 14 departments for the period 1871–1910 and concludes that geological morphology is, indirectly, a key factor in voting behaviour in rural areas (this thesis was confirmed in a similar study carried out in 1949 in the department of Ardèche in the south of France [460]). The area of the country that is on granitic soil is sparsely populated, and most of the residents are large landowning farmers. Additionally, the role of the church as an institution of intra-communal communication is strong here. This leaning towards large landowners and the church made the local electorate more conservative and monarchist. Conversely, people in the limestone soil areas in the river basin valleys live more closely together, with a large concentration of small and medium-sized farms. Here, competition and egalitarianism develop, dependence on the church decreases and, consequently, the electorate is more like to vote liberal and republican. While many reduce Siegfried's work to the formula "*le granite vote à droite, le calcaire vote à gauche*" ("granite votes right, limestone votes left"), it is noteworthy for the pioneering conceptualization of what the author called the influence of the environmental factor on political sensibilities [55]. Siegfried's research formed the basis of the highly influential French school of electoral geography. His work would be continued (and critiqued) by such thinkers as François Goguel [303–304], Jean Billet [212], Jacqueline Beaujeu-Garnier [209], Albert Brimo [219], Lucien Gachon [294], Claude Leleu [368], and Marie-Thérèse and Alain Lancelot [361].

Fig. 1.1. André Siegfried – the founder of electoral geography

Like most political geographers of the first wave, Siegfried was accused of *geographic determinism* – an affinity for the theoretical and methodological paradigm which implies that location in physical space is the most important (if not the only) factor explaining the properties of the object of analysis. Most critics of Siegfried's work belonged to the second school of political geographers that adhered to the opposite view, or *geographic nihilism* (or *indeterminism*), which states that geography has little effect on social processes in the modern world. As is often the case, the truth is likely to be found somewhere in between, in *geographic possibilism*, where modern electoral and geographic research are quite at home: physical space creates windows of opportunity, probabilistic scenarios for the properties of an object to form, but it cannot unambiguously determine the nature of that object.

Two other seminal works on the issue appeared around the same time that Siegfried was developing his theory: "Geographic Influences in British Elections" by British scholar Edward Krehbiel (1916) [359] and "Geography and the Gerrymander" by American researcher Carl Ortwin Sauer (1918) [448]. These works arguably laid the foundation for the Anglo-American school of electoral geography, which dominates to this day. Other influential works of this time period include those of Vera Dean [256], Valdimer Key Jr. [347–349], George Kish [354], John Kirtland Wright [499], and David Walter [493]. As for the prominent works of other national schools of thought, we cannot ignore Swedish political scientist Herbert Tingsten's *Political Behavior: Studies in Election Statistics* (1937) [484], which put the victory of the socialists in Stockholm down to the high numbers of manual labourers living in the suburbs of the city.

Gradually, research into electoral geography began to stretch beyond descriptions of individual cases, which raised the question of its status as a discipline in its own right with its own subject of study and a theoretical and ideological basis. It also became clear that simply mapping election results (the dominant method of the time) was not enough to understand the essence of electoral and geographic processes. This is precisely how British–Australian political geographer John Prescott framed the question in his article "The Function and Methods of Electoral Geography" (1959) [427], which effectively summed up the first stage in the development of the subject and laid out directions for future development.

A watershed moment that allowed electoral geography the move into the analytical stage of its development (1960s–2000s) was the appearance of the theory of ideological and political divisions put forward by Seymour Lipset and Stein Rokkan in 1967 [373]. The British–American researcher Kevin R. Cox [253], who actively collaborated with Rokkan, was instrumental in translating these ideas into the language of electoral geographer. Using the work of Lipset, Rokkan and Cox as a starting point, the English political geographers Peter J. Taylor [476–480] and Ron Johnston [331–339] (Fig. 1.2) published a monograph entitled *Geography of Elections* [480] in 1979, which is considered a seminal work in the theory of electoral geography. Among other things, the work puts forward universal patterns of the influence of space on the electoral process that have been termed the "effects of electoral geography" [500].

Fig. 1.2. Kevin R. Cox, Peter J. Taylor and Ron Johnston (left to right) – founders of the theory of electoral geography

From this moment on, the number of works on electoral geography started to grow exponentially. The evolution of the discipline is marked by two milestone works – a collection of essays entitled *Developments in Electoral Geography* published in 1990 [339] and another collection called *Revitalizing Electoral Geography* released in 2011 [494] – both of which served as a bookend to the intermediate results of the development of electoral geography. The most prominent names in the advancement of the field at this stage include Phil Dunham [267], Daniel Kahneman [234], Gary King [351–352], Nicholas Quinton [365], Jonathan Leib [365, 494], William Miller [389], Charles Pattie [337–338, 340, 376, 421], David Reynolds [345–436], Lynn Staeheli [319], Fred Shelley [184, 199, 339], Bernie Warf [494], Gerald Webster [496], Colin Flint [413], Benjamin Forest [287], James Forrest [288], David Hodge [319] and last but certainly not least John A. Agnew [184–187, 459].

Today, systematic mapping and electoral-geographic analysis of election results is common in most countries, and many have developed their own schools of electoral-geographic analysis. The United States, United Kingdom and France continue to pave the way in this respect, although well-established national schools of electoral geography also exist in Canada [214, 313, 390], Mexico [246, 270], Brazil [377], Chile [254],

South Africa, Portugal, Spain [502], Italy [222–223, 259], Germany [228, 276, 316], Switzerland [325, 482, 488], Belgium [255, 465], the Netherlands [492], Poland [356–357, 444], Hungary [355], the Czech Republic, Slovakia [346, 505], Romania [216], Lithuania [485], Russia, Ukraine [317], Turkey [210, 501], Iran [391], Israel, India [188, 192, 239, 260–263, 329, 384, 463–464], China [375, 455], Japan [241, 473], Australia, and New Zealand [383]. In a sense, one of the main criteria for assessing the level of maturity of a national school is the publication of an electoral-geographic atlas of a country or macro-region, and new works of this kind are appearing all the time (see, for example, the recent atlas of Poland edited by Mariusz Kowalski and Przemysław Śleszyński [358], and the atlas of Romania edited by Ionel Boamfa [215]). All this notwithstanding, we still cannot claim that electoral geography has attained the status of a full-fledged academic discipline in its own right: specialized departments and research centres, journals, and sections in scientific societies are a rarity.

A watershed moment in the development of the Russian school of electoral geography was the publication in 1990 of the book *The Spring of '89. Geography and Anatomy of Parliamentary Elections* [35], edited by V.A. Kolosov, N.V. Petrov and L.V. Smirnyagin, which offered a comprehensive spatial analysis of the first democratic elections in the Soviet Union. In the years that followed, the Laboratory of Geopolitical Research at the Institute of Geography of the Russian Academy of Science (V.A. Kolosov [26, 75–81, 144, 423], O.I. Vednina [77]), A.B. Sebentsov, N.L. Turov), the Laboratory of Georesources and Political Geography at the V.B. Sochava Institute of Geography of the Siberian Branch of the Russian Academy of Sciences (A.N. Fartyshev [165], A.A. Cherenev [65, 135–137, 167–169], Y.S. Razmakhnina [139–140], P.L. Popov [65, 135–137], V.G. Saraev [135–137], D.A. Gales [135–137]), the Laboratory of Regional Analysis and Political Geography at the Department of Economic and Social Geography of Russia of Moscow State University (V.E. Shuvalov, A.V. Berezkin [22, 26], A.P. Zhidkin [56], A.Y. Zimokha [62–63], A.A. Sidorenko [179]), the Department of Regional Policy and Political Geography at St. Petersburg State University (K.E. Aksenov [5–6], A.S. Zinovyev [6, 54]), and the Laboratory for Regional Political Studies of the National Research University Higher School of Economics (R.F. Turovsky [78–81, 156–163], Y.O. Gayvoronsky, E.M. Korneeva [86]), established themselves as the leading centres for research into electoral geography in Russia. Other researchers who made major contributions to the development of the Russian school of electoral geography at that stage include F.T. Aleskerova [1–4], A.S. Akhremenko [14–21], M.N. Arbatskaya [8–12], A.V. Baranov [23–25], A.A. Bunina [29], I.M. Busygina, A.Y. Buzina [30], V.Y. Gelman [36, 38, 129, 155], G.V. Golosov [36, 41–43, 129, 305], N.V. Grishin [45–51], B.A. Isaev [67–68], A.L. Kireev [71], Y.G. Korgunyuk [82–85], M.I. Krishtal [88, 150], A.V. Kynev [90–92], A.E. Lyubarev [98–101], B.I. Makarenko [128], A.G. Manakov [102], E.Y. Meleshkina [36, 104–106, 129], E.N. Minchenko, I.E. Mintusov, O.S. Morozova [109–111], D.B. Oreshkin [116–118], P.V. Panov [125–127], N.V. Petrov [26, 94, 131, 424], S.B. Radkevich, I.N. Tarasov [150–151], A.S. Titkov [152–153], and A.N. Zhuravlev [58–59].

Electoral geography reached the peak of its popularity in the 1990s, both in Russia and around the world. First, the overly positivist electoral geography was difficult to fit into the constructivist turn in political geography. Second, the descriptive statistics and mapping used in numerous studies did not allow for theoretical generalizations beyond a specific election cycle or country. At the same time, computer software was not powerful enough to process large amounts of electoral or sociological data and use complex spatial analysis algorithms.

Recent years have seen a transition to the *synthetic stage* in the development of electoral geography (starting in the 2010s). The theoretical and methodological base of the discipline has been significantly enriched thanks to the synthesis of conclusions obtained from the analysis of individual election campaigns and national (and, less frequently, regional) electoral traditions. The main reason behind this leap to a qualitatively new level of development in electoral geography is the widespread use of spatial statistical analysis methods, which combine computer modelling in geographic information systems (GIS) and mathematical statistics (specifically spatial econometrics). This is partly due to the fact that social scientists (who are not professional cartographers) have increasingly been turning to geographic information systems for political research (what Michael Shin calls, as an extension of Andrew Turner's notion of general neogeography, "political neogeography" [458]). In addition to the statistical methods traditionally used in electoral geography, the discipline is being synthesized with other related branches of knowledge (social anthropology, psychology, semiotics, etc.), which, among other things, leads to the adoption of new qualitative analytical methods.

The foundations of theoretic geography, as opposed to descriptive geography based on mathematical methods and spatial analysis, can be traced back to the works in William Bunge in the early 1960s [225]. The basic axiom of geography about the influence of space on the properties of an object (the foundation of spatial analysis) was formulated in 1970 by American–geographer Waldo R. Tobler and would become known as the "first law of geography": "Everything is related to everything else, but near things are more related than distant things." Tobler who later clarify this definition by pointing out that proximity in geography can be measured in different ways (not only in degrees and metres, but also, for example, in minutes, money, turns, stops, etc.) [196]. In 2004, Tobler formulated a *second law of geography*: "The phenomenon external to an area of interest affects what goes on inside," which can be interpreted as a statement that a phenomenon is influenced not only by the properties of the region in which it is located, but also by phenomena that are characteristic of other regions (for example, neighbouring regions, as proximity can also be measured in different ways). This provision is even more important for testing the hypothesis of electoral geography, as it generalizes the spatial effects of voting. Tobler's laws thus combine the two main principles of spatial influence on a phenomenon or process: the principle of proximity and the principle of neighbourhood [470].

Fig. 1.3. Patrick Moran, Keith Ord and Luc Anselin (left to right) – founders of spatial statistical analysis

However, the statistical foundation for these ideas was laid somewhat later by Keith Ord [242–244, 301, 417–418] and Luc Anselin [193–196, 307, 413], the founders of spatial statistical analysis, in the 1970s–1980s (Fig. 1.3), who took the measures of spatial autocorrelation proposed by the Australian statistician Patrick Moran [395] and the Irish mathematician Roy C. Geary [298] in the early 1950s as a starting point. It is this recourse to the school of spatial econometrics and the ability to easily carry out big data modelling in geoinformation systems today that opens up opportunities for electoral geography to empirically confirm the theory. General approaches to the use of this methodology in electoral geography have been formulated by John O'Loughlin [372, 413–415], Andrew M. Linke [372] and Michael Shin [458–459].

In Russia, developments in this area come from the Center for Spatial Analysis in International Relations of the Institute for International Studies at the Moscow State University of International Relations (MGIMO) (I.Y. Okunev [114, 119–122, 410–412], M.N. Shestakova [176], E.A. Zakharova [61], L.S. Zhirnov [57]), as well as from the Centre for Spatial Econometrics in Applied Macroeconomic Research at the Faculty of Economics of the National Research University Higher School of Economics (O.A. Demidova, E.A. Podkolzina, L.E. Kuletskaya [89, 132]).

1.3 Geography of Elections Around the World

Elections have always been part and parcel of politics, and are routine in most countries these days. In this section, we will study the political map of elections around the world and identify the dominant trends in the forms and rules of electoral procedures.

All power has both a source and a bearer. The *source of power* is not elected; it is the person to whom it belongs in accordance with political tradition. The source of power in monarchies is a single person and his or her dynasty. In republics, it is all the citizens of the country. The source of power can rule independently (as in the case of an absolute monarchy or a republic with a direct democracy), that is, it can be both the source and bearer of power, or divide power among itself and the citizens or bodies of power elected by them as a bearer of power (as in the case of a dualistic monarchy). However, most often, the source of power transfers those powers to an elected holder who directly exercises rule in the country (as in the case of a limited monarchy or a republic with representative democracy) (Table 1.2).

Table 1.2. Elected bodies of power in different forms of government

Source of power	Bearer of power	Form of government
Monarch	Monarch	Absolute monarchy
Monarch	Monarch and elected bodies of power	Dualistic monarchy with representative democracy
Monarch	Elected bodies of power	Limited monarchy or republic with representative democracy
Citizens	Elected bodies of power	Republic with representative democracy
Citizens	Citizens	Republic with direct democracy

Thus, there is no need for elected bodies of power in absolute monarchies (with the exception of individual consultative bodies of power), where the head of state is both the source and bearer of power, or in republics with direct democracy, where citizens themselves make all the political decisions during referendums (or *plebiscites*). Direct democracy no longer exists as a form of government, although the share of direct participation of citizens in the political process in democracies is gradually increasing (the most interesting examples of this are Iceland and Switzerland). Absolute monarchies exist in Brunei, Oman, Eswatini, Qatar, Saudi Arabia, the United Arab Emirates and Vatican City. Dualistic monarchies operate in Bahrain, Bhutan, Jordan, Kuwait, Liechtenstein, Morocco, Monaco, and Tonga. And limited monarchies exist in Andorra, Belgium, Cambodia, Denmark, Spain, Japan, Lesotho, Luxembourg, Malaysia, Netherlands, Norway, Sweden, Thailand and the United Kingdom.

Unusual forms of elections exist in some monarchical states. Traditionally, Andorra has been co-ruled by the President of France and the Roman Catholic Diocese of Urgell in Spain. Thus, one representative of the source of power is elected by the citizens of another country, while the other is appointed by the head of a third state (the Pope). In Eswatini, a council of elders called Liqoqo elects a queen mother (or, more precisely, an "elephant queen"), whose son becomes the absolute ruler ("lion king"). In the United Arab Emirates, the absolute ruler of the state is elected by the heads of regions (emirs), who in turn are absolute monarchs in their emirates. A similar model exists in Malaysia, with the difference being that some regions have a monarchical form of government, while others have a republican form of government. All of these cases include elements of an aristocratic form of government, when the source of power is not concentrated in a single monarch, but rather in a narrow group of people.

There are three main forms of legitimate transfer of power: by rule (through inheritance or by title/position), by appointment, or through elections (direct or indirect). The transfer of power by rule implies that power is transferred automatically by law or unwritten tradition. For example, in most monarchies, the principle of inheriting power through the family line is written into law. Parliamentary powers can also be inherited. This was the case in the United Kingdom, for example, before the reform of the House of Lords, where hereditary Peers were part of the nobility with titles not lower than that of baron/baroness. In Lesotho, two thirds of the seats in the upper house of parliament continue to be held by hereditary tribal chiefs. Similar to heredity is the practice of obtaining deputy's seat by title (or position). For example, 26 Lords Spiritual sit in the House of Lords in the United Kingdom by sole virtue of their ecclesiastical offices in the Anglican Church. In Belgium, the children of the king, or, if there are none, the closest relatives in the royal family's descending line automatically become senators at the age of 18. In Botswana, the president and attorney general are *ex officio* deputies, while only the attorney general is in Dominica, Saint Kitts and Nevis, Saint Vincent and the Grenadines and Tanzania. In Russia, a former president can become a senator if he or she so desires.

Transfer of power by appointment implies that a person is given power by decision of a non-representative political institution. For example, a monarch can be appointed by the previous crowned ruler. In Norway, for instance, if there is no male heir to the throne, the king can propose a candidate of his own choosing to the parliament (the Storting). In the Commonwealth of Nations (Canada, Australia, New Zealand, etc.), governors general are appointed by the British monarch. The principle of appointment is used in presidential republics, where the president decides who will be prime minister and government ministers. Appointment of members of parliament is also possible: in Antigua and Barbuda, the Bahamas and Canada, all senators are appointed by the governor general or head of state at the suggestion of the prime minister (and occasionally the opposition leader). Members of the Senate of Canada (similar to the House of Lords in the United Kingdom) used to exercise their powers for life, and now they sit until they reach the age of 75. In Algeria and Kazakhstan, the president appoints one third of the senators; in Bhutan, the monarch appoints one fifth of senators; and in Myanmar, the commander of the armed forces appoints a quarter of both houses of parliament. In Botswana, the Gambia, Namibia, Singapore and Tanzania, a small number (between three and eleven) of deputies in the lower (or only) house are appointed by the president. This right is afforded to the governor general in Saint Kitts and Nevis and Saint Vincent and the Grenadines, and to the prime minister (Taoiseach) in Ireland.

The most common form of the transfer of power is elections, which involve the use of a competitive procedure. Elections can be direct and indirect. *Direct elections*, where people vote directly for candidates to office or their representatives in the legislature, are by far the most common. This is how three quarters of the heads of republics and the vast majority of deputies of legislative bodies of all levels are elected: from municipal to supranational. Israel had a unique experience of electing its prime minister by popular vote.

Indirect (or multi-stage) elections are carried out by intermediates (electors) or special institutions, including legislatures. In contrast to appointment, the decision here is made by a representative political institution whose members have been elected in direct elections. The main types of indirect elections are:

1. Deputies elect the head of state (in parliamentary and socialist republics, while two captains regent are elected in San Marino) or prime minister (in all parliamentary republics; in Italy, both houses of parliament take part in the elections).
2. Deputies from parliaments of various levels elect the head of state (Germany, Italy, India and Pakistan).
3. Deputies from parliaments of various levels elect deputies to the national parliament (the French Senate).
4. A special college of electors voted in through popular elections elect the head of state (the United States).
5. Members of parliament elect other members of the same parliament (this is how a certain number of women deputies are elected in some Muslim countries).
6. Election by appointed persons (the Pope is elected by a conclave of cardinals of the Roman Catholic Church who themselves were appointed by previous pontiffs, while he gets his power from God;

the head of Afghanistan is elected by the Leadership Council, which is made up of political, military and religious leaders).
7. Elections of members of the upper chambers of parliament by the lower parliament (Austria, India, China, the Comoros, half of the senators of Russia) or regional governors (Germany, the other half of the senators of Russia).
8. Election to parliament of representatives from public organizations or communities of deputies of representatives (in Kazakhstan, nine deputies are elected from the Assembly of the People of Kazakhstan representing ethnic minorities; in Cyprus, three deputies are elected by religious communities; in Rwanda, two deputies are elected by the National Youth Council, and one is elected by the Federation of the Associations of the Disabled).

In addition to elections proper, some countries hold *primary* (intra-party) *elections* to determine candidates for elections. The presidential primaries are big business in the United States. An alternative way to nominate candidates is through voting at party conferences (*caucuses*).

Elected officials and bodies typically have a fixed term of office (*tenure*), which in practice varies – from six months (San Marino) to seven years (Israel) for a president, and from two (Micronesia) to nine (Liberia) years for members of parliament. It is also common practice to set limits on the number of terms that the current head of state (*incumbent*) can serve. Two terms are most common, although one term (in Armenia and Mexico, for example) and three terms (Cameroon and Kiribati) are possible. The time period between two *regular* (or *general*) elections – that is, elections held in accordance with the established terms of office, is called an *electoral cycle*. In the event that the parliament terminates its powers mid-cycle (if parliament is dissolved, for example), *early* (*snap*) *elections* are held. If individual members of parliament need to be replaced, then *by-elections* may be organized. By-elections usually use the majority rule, while under the principle of proportionality the mandate is transferred either by decision of the party or by the results of a previous vote. The political culture of the United States also uses the term *midterm elections* to describe elections to Congress and local governments that take place two years after the presidential election, that is, in the middle of the president's term.

An important metric for evaluating the electoral process is voter turnout, which shows the level of political participation and, by extension, the degree of legitimacy of the elections. Low turnout reflects electoral *absenteeism*, or the desire to evade participation in elections. Certain legal instruments have been put in place to help raise voter turnout, including compulsory, early, absentee and remote voting. *Compulsory voting* makes voting an obligation, rather than a right. A few dozen countries have implemented such a system, although their number is dwindling because most realize that this measure is ineffective. Some countries (Belgium, Bolivia, Greece, Luxembourg, Mexico, Thailand and Turkey) that have compulsory voting do not impose any penalties for violating this requirement, while others levy fines (Argentina, Australia, Brazil, Ecuador, Nauru, Peru and Uruguay). *Early voting* guarantees that voters can take part in elections in the event that circumstances may prevent them from doing so on election day. *Absentee voting* allows voters to cast their vote in a different precinct to which they are normally allocated. To do this, they must give advance notice, although in some cases it is not required (for example, if a person is voting at an embassy when abroad). *Remote voting* is when the person does not physically visit the polling station and casts their vote by post, telephone or electronically (online). Estonia is a world leader and pioneer in remote voting. In indirect elections, members of parliament may use *proxy voting*, where they transfer their right to vote to another deputy in the event that they cannot vote themselves.

Today, most agree that elections should be universal, that is, that all *electoral qualifications* imposing restrictions on *active suffrage* (the right to participate in elections as a voter) – for example, restrictions based on gender, race and property, which were commonly enforced in the past – are considered illegitimate. Saudi Arabia was the last country to give women the vote, doing so in 2015. However, a number of electoral qualifications still apply in the modern world. The most notable among these is voting age: in most countries, people must be 18 years old in order to vote, although there are examples where the voting age is lower (16 in Austria, Brazil, Cuba, Nicaragua and Somalia; 17 in East Timor, Indonesia, North Korea and Sudan), and higher (20 in Cameroon, Japan, Liechtenstein, Morocco, Nauru, South Korea and Tunisia; 21 in Bahrain, Fiji Gabon, Kuwait, Lebanon, Malaysia, the Maldives, Pakistan, Samoa, Singapore and Tonga). Another electoral qualification that is commonly used is the naturalization qualification, which means that only citizens of a state can vote in elections, although there are countries where all legal residents can take part (residents in Hungary, Denmark, Spain, the Netherlands, Norway, Finland and Sweden can take part in local elections, while residents in New Zealand can take part in national elections). As regards *passive suffrage* (the right to be elected), the age qualification is typically even higher, and the naturalization qualification stricter. In addition, candidates are often subjected to harsher rules regarding settled residence, criminal convictions (criminal records must be clean) and education.

1.4 Elections as Part of the Political Process

Elections are a key social institution, and their place in the political process is best understood using the "black box" model developed by David Easton [268]. Easton proposed looking at state institutions of power as a "black box" that organizes the political process in its interactions with society. Demands and support (*input*) are entered into the black box by society and then transformed into a reaction (*output*) – decisions, policies, regulations, laws, etc. Society evaluates the degree to which the reaction of the political system meets its demands, creating feedback in the form of new demands. The political system thus becomes cyclical. A political system is considered synchronized and effective if the demands match the decisions. Elections are nothing more than a part of the political system's feedback loop with society. They are a means of entering demands into the "black box" and can be used to assess how adequately society believes the "black box" is responding to its needs.

To put it even simpler, imagine that the political system is like a washing machine. You put dirty laundry in, select the cycle (form a demand), and expect the machine to perform the task you have set it (to take the required action). You may decide once the cycle is complete that the clothes are not completely clean, so you start the process again – this is an expression of your feedback. How annoying would it be if the washing machine carried out a drying cycle when you specifically selected a delicate wash? In this case, you would have to get it repaired or replaced.

The political process is unique in that demands are formed and decisions are evaluated differently depending on the context. And the context is determined primarily by two parameters – political culture and state regime.

Political culture is the degree to which a society has traditionally participated in the political process. Gabriel Almond and Sidney Verba identified three types of political culture: parochial, subject, and participant (Table 1.3) [189, 190]. Their conclusions were based on their observations of the political culture in the United States, which suggests – and this is important – that political culture is not a characteristic of a country or society as a whole, but of an individual or social group. Different types of political culture exist within every country, the question is: What share do they occupy and how prevalent are they in public life?

Table 1.3. Types of political culture

Political culture	Input	Output	Political socialization
Parochial	0	0	None
Subject	0	+	Collective
Participant	+	+	Individual

A prime example of *parochial political culture* is the rural America of the past, where the only form of social activity was attending Sunday service in a church parish (hence the name). Here, society is not interested in politics, does not create demands and does not evaluate reactions. Logically, it might follow from this that those in power in such a situation are free to do as they please, but this is not the case. The fact that the people do not make demands does not mean that there are no demands whatsoever. This kind of society is also called traditional: by not demanding change, the people are in fact demanding that nothing be changed at all. It is extremely difficult to carry out any kind of transformation in such a society. The opposite of parochial political culture is *participatory political culture* – the culture of participation. This is where the people are interested in politics and actively take part in it.

The most interesting type of political culture is in between these two extremes: *subjective political culture*. As political socialization – the transition from a parochial to a participatory culture – takes shape, the individual is initially interested in the decisions of the political system, but is not yet ready to make demands of that system. This leads to a situation where elections are little more than a beauty contest between politicians, a competition to see who can offer the most seductive promises, because at this stage it is more important to convince the people of the bright future you are offering than to bog them down with details of how this will actually be achieved. In order to formulate demands of the political system, the individual first tries to identify with the people around them: relatives, friends, colleagues. In other words, political socialization first occurs at the collective level through a group of interests, and only then at the individual level. Consequently, a key part of the electoral process is working with interest groups: if you can win over the leaders of these groups, then you will have also won the votes of their members.

What happens when the feedback within the political system suggests that the decisions do not satisfy the demands, that is, when the system is desynchronized? This is where the political system falls into crisis. A comparative analysis of this phenomenon was carried out by the Stanford Political Development Workshop. The biggest problem in such a crisis is that the natural process of finding a compromise in society is disrupted, because people's views are radicalized, and their positions polarized. Consequently, it is extremely difficult in such an ideologically fragmented society to form a coalition of forces that would represent the views of the majority of voters.

The second key parameter of the external environment is the state regime. Political competitions in elections necessarily involves bringing different views on the country's development into the political system – that is, there must be a pluralism of opinions, the hallmark of a democratic regime. Pluralism is supported by numerous institutions: guarantees of human rights and freedoms; the freedom of speech; the independence of the courts and the media; the presence of civil society; accountability of the authorities; transparent and competitive elections, etc. To (over) generalize, a democratic regime is characterized by public control over the political sphere and state institutions. Regimes that have problems when it comes to pluralism of opinions are called undemocratic, although they differ markedly from one another. There are regimes where there is only one point of view in society – a *monism* of opinions, as it were. This is typically observed in traditional societies with a parochial type of political culture, where there is no interest in political changes and, accordingly, the unanimity of views is focused on maintaining tradition and social order. Such a regime is called *sultanistic*. The rapid spread of radio at the beginning of the 20th century, which allowed the poorer classes to be included in politics, led to a unique phenomenon, namely, totalitarian regimes. In a totalitarian regime, monism is supported not by a lack of interest in politics, but, paradoxically, by mass interest in politics, which is achieved by extreme ideologization and mobilization of society. This is achieved primarily through such institutions as a single state ideology, public violence, the creation of an atmosphere of fear in society, a cult of personality, etc. In its purest form, totalitarianism is the antithesis of democracy – maximum state control over society. Totalitarian regimes are thus similar to sultanistic regimes in that they are both monistic, but they differ in terms of the level of participation of society. Regimes that hover somewhere between democratic and non-democratic are considered transitional (unstable) and called authoritarian (hybrid).

As we can see, the political process hinges on the existence of a constant point of rivalry in society, an internal conflict (in the positive sense of the word), which determines the dynamics of the struggle of the forces of influence to determine the country's development strategy. This conflict may come in various forms, but it is invariably determined by the electoral process. Let us demonstrate this using the typology of the relationship between electoral and political systems developed by Michael Wallerstein (not to be confused with his cousin Immanuel, a renowned scholar in his own right) (Table 1.4).

Table 1.4. Relationship between electoral and political systems

	Majoritarian formula and a two-party system	Proportional formula and a multi-party system
Parliamentary system	Conflict between the ruling party and the opposition (United Kingdom)	Conflict as a result of the need to constantly build coalition governments (Germany)
Presidential system	Conflict between the legislative and executive branches of government (United States)	Conflict of the "Kremlin towers" (Russia)

Let us divide political systems into presidential, in which the president is elected by the people and forms the executive power, and parliamentary, where the parliament forms the executive power (including in limited monarchies). As for electoral systems, they can be divided into majoritarian, which, as we know, leads to a stable two-party system, and proportional, which support a multi-party system. So, we have a dependent executive in parliamentary systems and an independent executive in presidential systems, and two strong parties in majoritarian systems and numerous weaker parties in proportional systems. The combination of a parliamentary system and majoritarianism (as is the case of the Westminster system in the United Kingdom) leads to the main political conflict in the country unfolding through the confrontation between the two largest parties and the corresponding ruling and shadow cabinets.

In essence, there is no separation of powers. There are also two strong parties in the United States, but the combination of majoritarianism and the presidential system makes the separation of powers the main point of confrontation within the system – the confrontation between the legislative and executive branches of government (between a strong president and strong parties). In the absence of a strong president and strong parties that are able to form a government by themselves – a combination of proportional and party systems (as in the case of Germany) – the main driver of the political process is the formation and maintenance of ruling coalitions. Finally, the main conflict within political systems that combine a strong and independent executive power with a multi-party, presidential and proportional system (as in the case of Russia) is the rivalry of institutions within the executive power (the president and the prime minister; the presidential administration and the cabinet of ministers; various ministries and the groupings with each other, etc.). This kind of confrontation mostly takes place outside of the public's gaze and resembles, in the words of Winston Churchill, a "bulldog fight under a rug," or a "conflict of the Kremlin towers."

To sum up, we can say that the electoral process influences the work of the political system at the most significant of levels, since, in a sense, it sets the rhythm and nature of its functioning.

1.5 Electoral Statistics and Electoral Sociology

The main types of empirical data used for electoral-geographic analysis are electoral statistics and electoral sociology.

Electoral statistics is aggregated data on the frequency distribution of political will. The basic unit of measurement used in absolute terms is the number of votes cast, calculated on the basis of data from electoral ballots, which contain a list of items corresponding to the available electoral choices. In addition, electoral statistics aggregates data on the number of voters and ballots (including those used in special electoral procedures, such as early, absentee or remote voting). As a rule, relative indicators are calculated from the actual voter turnout, although for the purposes of certain tasks they can also be determined from the number of registered voters (for example, to calculate turnout percentage).

Table 1.5. **Electoral statistics and electoral sociology**

Data characteristic	Electoral statistics	Electoral sociology	Exit polls
Type of data	secondary	primary or secondary	primary or secondary
Size of data set	includes all data	selected data	selected data
Individual information	no	yes	yes
Linked to territory	yes	not typically	yes
Timeframe	medium term	short term	medium term

The data used in electoral statistics is completely different to that used in electoral sociology (Table 1.5). First of all, the researcher can conduct surveys him/herself (primary data), or they can use the results of research carried out by someone else (secondary data). That said, electoral statistics are almost never primary and are usually collected by special election commissions. Second, electoral statistics is the result of processing data on the entire population of the contingent of citizens being analysed, whereas sociological studies are based on a sample survey of the electorate. Third, for reasons of anonymity, electoral data does not contain any additional information on the individual (age, gender, social status, etc.), something that is central to determining the correctness of the sample and developing basic hypotheses in electoral sociology. Fourth, in electoral statistics, data on votes is tied to polling stations – that is, they have a spatial dimension. Meanwhile, in electoral sociology, there is rarely a need to display verifiable territorial distribution of respondents' answers.

And, finally, electoral statistics allows us to carry out cross-temporal analysis for the medium term only (i.e. by electoral cycles), while electoral sociology allows us to take measurements in the short term as well, for example, during a single election campaign [14].

Although most studies on electoral geography primarily use electoral statistics, a full picture is obtained when combining such data with regular opinion polls. In this case, measuring electoral support presents certain difficulties, since opinion polls allow us to identify, in addition to candidate ratings, such parameters as recognizability, voter trust, disapproval rating, etc., as well as to link them to the social and economic profiles of respondents [183].

As we mentioned earlier, the key criterion for the validity of sociological data is a carefully selected sample of respondents. Experience tells us that the results of opinion polls conducted using a representative sample – that is, a sample that reflects the proportions, correlations and linkages between parts of the electorate in the general population – do not differ significantly from those obtained from studying the entire population. A confidence interval is used to assess the accuracy of the sample. If an opinion poll shows that a candidate enjoys a 50 % rating and the recommended confidence interval is 3 %, then support among the general population should be in the range of 47–53 %. Additional parameters (gender, age, class) are used to ensure that the sample is representative, and the quotas of each parameter in the sample should be close to their share in the general population (i.e. within a certain confidence interval). Sociologists generally recommend a sample size of roughly 350–400. However, if the general population is less than 3,000 people, the sample size will be slightly less. In such cases, a typical sample is recommended, that is, a small group of respondents whose qualities reflect the average in the general population. With a general population of more than 100,000 people, the size of the sample will not increase significantly. However, if individual social parameters need to be analysed – for example, in electoral geography, we need to study the distribution of the electorate across territories – then each social group in the sample should be represented by 50–100 people, proportionally to the region's share in the country. So, a sample of 400 people will require an analysis of spatial dispersion for approximately four to eight territorial units. National polls that monitor two to four social parameters should use a sample of 1,600 [146].

Another commonly used form of empirical data is exit polls – surveys taken when exiting polling stations that combine features of electoral statistics and electoral sociology. Exit polls involve two-stage sampling – first, the polling station is selected, and then the voters themselves. For example, 800 polling stations were used to conduct exit polls in Russia (of approximately 100,000 across the country), each of which polled 1,000 voters. Probability sampling is also used to select voting districts. For example, a voting district that has twice the number of registered voters is twice as likely to be selected as a place for exit polls. Finally, the number of voters polled at the exit at any given time should correspond to the average proportion of the electorate visiting the polling station at that time of day. One advantage of this type of data for electoral geography is that an analysis of exit poll results using a competent sample allows us, on the one hand, to connect data to specific territories, and, on the other, to identify socio-demographic parameters of the electorate as a whole.

1.6 Applied Electoral Geography

Electoral geography is more than simply research. In practice, it is the key to the application of political technologies, since knowledge of the patterns and trends in candidate ratings on the ground is crucial to a successful election campaign at any level. Knowledge of long-standing geographical voting patterns makes it possible to devise a regional election campaign strategy, including, among other things, opinion polls, field work and legal support for the campaign.

Election campaigns are primarily focused on attracting undecided voters (*changeable electorate*): spending significant resources on one's traditional supporters (*core electorate*) or people with established alternative views (*opposition electorate*) is considered ineffective. So, knowing exactly where the changeable electorate lives informs campaign decisions – these are the regions the candidate should visit personally to meet with voters, or the issues that concern voters there are those that should be addressed in the campaign programme and its main ideas (messages).

The proportion of voters who make up the changeable electorate is usually calculated as the difference between the party's best and worst performances (or between candidates with a similar ideological background) over a given number of electoral cycles, or based on a number of opinion polls asking who the residents of a given region expect to vote for in the next elections. Another method is to determine the difference between the number of votes the party won in the previous electoral cycle in several parallel elections (for example, in national, regional and municipal elections, or in elections held using proportional representation with majoritarian representation for single-member districts).

It thus follows that the core electorate is defined as the party's baseline in terms of worst electoral performance (or of candidates with a similar ideological background) over a given number of electoral cycles, or based on a number of opinion polls asking who the residents of a given region expect to vote for in the next elections. Again, an alternative method is to determine the lowest level of party support during the previous electoral cycle in several parallel elections.

Finally, the share of opposition electorate can be calculated as the sum of the lowest values of alternative parties (or of candidates with a similar ideological background) over a given number of electoral cycles, or based on a number of opinion polls asking who the residents of a given region expect to vote for in the next elections. Another method is to calculate the sum of the worst performances of alternative parties in the previous electoral cycles in several parallel elections.

Further, to estimate the absolute numbers of core, opposition and changeable voters, the resulting share should be multiplied by the number of voters in the given district or territory, and by the voter turnout in the previous elections.

Table 1.6. Political technology strategies in elections

	Always go to the polls	Sometimes go to the polls	Never go to the polls
Core electorate	Maintaining traditional channels of communication with voters (A)	Main focus group for motivating voter turnout (B)	Secondary group for motivating voter turnout (C)
Changeable electorate	Main focus group for election campaign (D)	Secondary focus group for election campaign and motivating voter turnout (E)	No strategy (F)
Opposition electorate	Main focus group for discouraging voter turnout (G)	Secondary focus group for discouraging voter turnout (H)	No strategy (I)

The chosen political technology strategy thus depends both on the type of electorate and on its readiness to come to the polls (Table 1.6). With a core electorate that always votes (Category A), it is enough to maintain existing channels of communication with voters to reassure them that their candidate is taking part in the elections and has not switched parties. However, the campaign focuses on different groups. The main focus will always be the politically active changeable electorate (Category D), with the goal being to attract them to your programme through your ideology. The core electorate who occasionally go to the polls (Category B) may also need to be persuaded to actually cast their vote. Attention is also paid to the opposition electorate who regularly vote (Category G), in relation to whom political technologists apply the so-called "voter suppression" strategy, that is, they implement a campaign aimed at dissuading them from turning out to vote. Secondary focus is placed on the semi-active changeable electorate (Category E), which needs to be both interested in the election programme and motivated to come to the polling station; the core electorate who never vote (Category C) and thus need a little pushing to fulfil their civic duty; and occasionally the opposition electorate (H), which you would prefer to not go to the polls at all. Finally, campaigns are almost never targeted at the changeable and oppositional electorates who never vote (Category F and Category I, respectively).

A recent trend that has taken hold in political technology in recent years is the use of data on where potential voters are located in order to target the distribution of campaign materials. This is particularly relevant with regard to political advertising on the internet, where data on the physical location of voters is obtainable via information they have provided about their place of residence, geotags on their posts, the IP addresses or Wi-Fi access points they use, and the GPS coordinates of their mobile devices. *Geotargeting* involves placing campaign materials exclusively in areas where potential voters of a candidate live, or showing them as ads on the connected devices of people located in these areas. Meanwhile, *geofencing* (*geozoning*) is the targeting of political advertising to people located in buffer zones near certain points. For example, placing a billboard of a candidate who is against the proposed closure of a power plant next to the power plant itself will send a clear message to the workers there about that candidate's position on the matter. Geofencing can also be used on the internet. In this case, a political advertisement of the candidate who opposed the shutting down of the power plant will appear on the connected devices of people who happen to be in the vicinity.

Let us stress once again that geotargeting identifies potential voters through their belonging to a certain position or area, while geofencing targets voters through their connection with or proximity to a certain position or area. This is a subtle but important difference between the two. In practice, geotargeting is used for smaller-scale campaigns, while geofencing is used on larger campaigns. There is another, more advanced tactic that combines geotargeting and geofencing, and that is *geoframing*, which involves two steps: (1) identifying groups of potential voters based on proximity to a given object; and (2) targeting advertising to them regardless of their current location. Combining geotargeting, geofencing and geoframing with other data about the user (for example, stated preferences, search history, sites visited, etc.) can make political advertising on the internet even more precise.

1.7 International Electoral Geography

A separate area of research in electoral geography is the study of geographic patterns in elections at the supranational level. The most comprehensive of these works analyse voting patterns at the UN General Assembly and the Eurovision Song Contest, although some look at voting in the EU, WTO and ILO bodies and in other international organizations.

Researchers have been analysing voting patterns in the United Nations for as long as the organization has existed. Seminal works in this area include Margaret Ball's 1951 study [204] and Arend Leiphart's 1963 investigation into voting in the General Assembly [371]. In 1958, Waldo Chamberlain [238] produced a pioneering study on the relationship between regional integration and behaviour in the UN General Assembly, in which he demonstrated that the future NATO countries displayed such solidarity that there was really no need to create the alliance in the first place. Studies these days typically use the updated database of votes in the UN General Assembly since 1946 compiled under the direction of Georgetown University professor Erik Voeten and available on the Harvard University website [491].[1] In most studies, voting in the General Assembly is used as a dependent or independent variable to look for statistical correlation with other factors, such as the likelihood of inter-state conflicts, the provision of international humanitarian aid, or the colonial past, although some focus on the spatial nature of voting. Two topics dominate the latter: (1) spatial modelling of voting in the General Assembly; and (2) the analysis of the degree of regionalization in the world based on national expressions of will in the United Nations.

---

[1] Voeten E., Strezhnev A., and Bailey M. United Nations General Assembly Voting Data. (Cambridge, MA: Harvard Dataverse, Harvard University, 2009–2021). https://doi.org/10.7910/DVN/LEJUQZ.

The most commonly used method of assessing similarities in the voting patterns of countries (calculated as the Euclidean distance between them) is Signorino and Ritter's formula [462]:

$$S_{ab} = 1 - \frac{\sum |Y_{an} - Y_{bn}|}{V}$$

where $n = 1, ...V$ of counted votes, a and b are the two countries in question, and Y is the result of a vote with three options: $Y = 1$ – "for"; $Y = 2$ – "abstain" (or was not present); and $Y = 3$ – "against." Thus, $S_{ab} = 1$ if the countries vote identically on all issues, and $S_{ab} = -1$ if they vote differently on all issues. Figure 1.4 is an example of a chorochromatic map created using the foreign policy similarity index based on voting results in the UN General Assembly [312].

Yet another area of international electoral geography is the study of voting patterns at the supranational level, specifically with regard to the annual Eurovision Song Contest [257, 469]. It would seem that a contest where juries and TV viewers vote for what they think the best song is should have nothing to do with politics. However, it turns out that the opposite is true, as the votes are a clear reflection of the geographic and geopolitical patterns that make up Europe, where people tendentiously vote for their neighbours, ethnically close peoples and foreign policy partners. Grouping the results of the voting suggests the existence of geographical blocs: a post-Soviet bloc (which is further split into pro- and anti-Russian mini-blocs), a Baltic bloc, a Scandinavian bloc, a Yugoslav bloc, and an Eastern Mediterranean bloc. Generally speaking, academic research has confirmed the significance of the proximity factor in determining the propensity to vote for a given song in the competition for the entire continent, especially Eastern Europe.

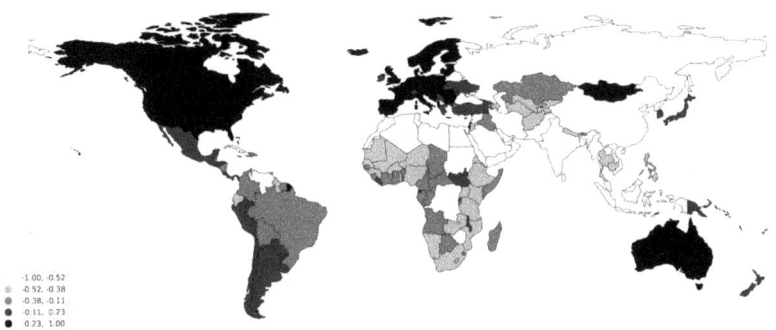

Fig. 1.4. Chorochromatic map of the foreign policy similarity index (based on voting in the UNGA)
Source: Häge, F.M. "Choice or Circumstance? Adjusting Measures of Foreign Policy Similarity for Chance Agreement." *Political Analysis*, Vol. 19, No. 3 (2011), 287–305.

Key Terms

- electoral geography, psephology, geography of voting, geography of representation, geography of social divisions, electoral limology, electoral geography of parties and party systems, electoral geography of political campaigns, spatial modelling of voting, international electoral geography, urban electoral geography, critical electoral geography
- electoral space, electoral landscape, electoral field, electoral worldview
- electoral-geographic field work, electoral-geographic mapping, electoral-geographic zoning, electoral-geographic modelling, spatial analysis in electoral geography
- stages in the development of electoral geography: descriptive, analytical, synthetic
- geographic determinism, nihilism and possibilism
- source and bearer of power, electoral cycle
- direct elections, indirect elections, general elections, preliminary elections, regular elections, snap elections, early elections, midterm elections and by-elections
- early voting, remote voting, absentee voting, mandatory voting, proxy voting
- active and passive suffrage, electoral qualification

- absenteeism, incumbent, cadence, caucus, plebiscite, primaries, referendum, exit poll
- political culture: parochial, subject, participant
- monism and pluralism of opinions
- core, oppositional and changeable electorate
- geotargeting, geofencing, geoframing

Questions and Exercises

1. Electoral geography is sometimes accused of being overly positivist in its research. Give arguments for and against this position. Which do you agree with most?
2. Match the name of the researcher to the stage in the development of electoral geography: Luc Anselin, Ron Johnston, André Siegfried, Valdimer Key Jr., Kevin R. Cox, Vladimir Kolosov, John O'Loughlin, John Prescott, Peter Taylor, Rostislav Turovsky, John Agnew.
3. How do indirect elections differ from appointment and proxy voting? Why are these procedures preferred over direct elections?
4. Give examples of general, preliminary, regular, snap, early, midterm and by-elections in the history of a single country.
5. How are early elections different from early voting? Can early voting be used in early elections?
6. Discuss the advantages and disadvantages of reinstating the property qualification in democratic elections.
7. Conduct mini-interviews with members of your family and determine their respective types of political culture. What electoral strategy would you recommend to win over their votes? Will there be just one strategy?
8. Can a sovereign in a country whose society has no interest in politics rule as he or she pleases?
9. Which of the types of correlation between electoral and political systems described by Michael Wallerstein seems to be the least stable?
10. Determine the absolute numbers of core, changeable and oppositional electorates of your favourite party in all regions adjacent to the one in which you live and identify the optimal strategy for applying electoral technologies in the next election.

11. Take one election. Study its results, as well as the preliminary results of a public opinion poll and exit poll data. How well do they match up? How can you explain the discrepancies?
12. Analyse the voting results of the last Eurovision Song Contest. Which countries tended to vote for their neighbours and why?

Chapter 2
Territorial Differentiation of Electoral Systems

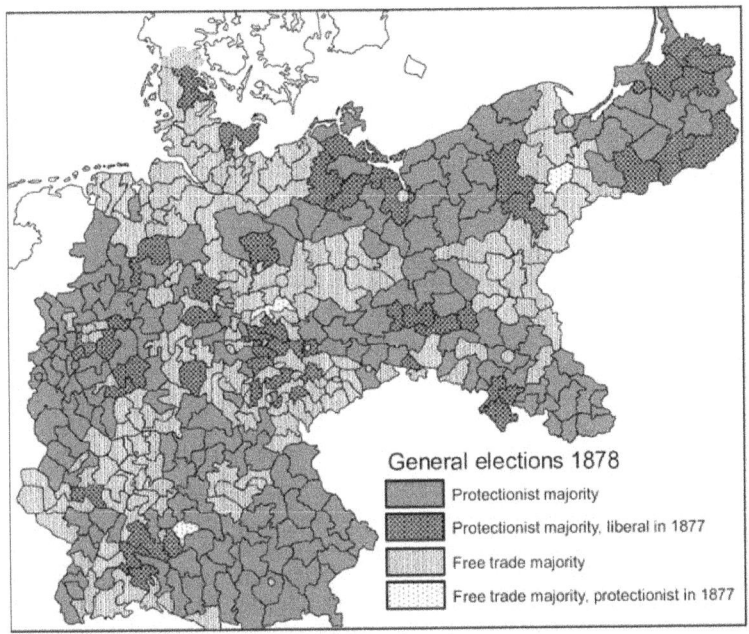

Note: the map shows the results of the general election for all constituencies of the German Kaiserreich in

Source: Lehmann, S.H. "The German Elections in the 1870s: Why Germany Turned from Liberalism to Protectionism," *The Journal of Economic History*, Vol. 70, No. 1 (2010), p. 151.

- What criteria should the perfect electoral system meet? Is such a system possible?
- How do electoral systems differ in terms of how they are organized in different territories and the electoral formula used?
- Which electoral systems contribute to greater personalization, and which contribute to greater choice? How can the advantages of both be combined?

- How can electoral systems distort the territorial distribution of support for political parties?
- What methods exist for distributing seats in parliament between regions and parties?

2.1 Public Choice Theory. The Condorcet and Arrow Paradoxes

Before territorial factors can be considered, it is first necessary to identify and define more general patterns of public (collective) choice. These can be placed within the framework of a separate economic (and partly mathematical) discipline called *public choice theory*.

The ability to make a choice (and not only a political choice) is a key process that is characteristic of any society. Every day, we make individual decisions, choosing from any number of options the one we consider to be the best – that is, the one that corresponds to our personal needs, interests and beliefs. The totality of individual choices is a collective or public choice. But how can we tell whether the collective choice is also the best one for each member of society, that it corresponds to the needs, interests and beliefs of every individual? The only way to know is if the collective decision is taken unanimously (by *consensus*). That said, the essence of human nature and the organization of society is such that the optimal strategy for one person may be in conflict with the optimal strategy for other members of society. This means that unanimous decisions are, as a rule, not possible. Hence the need to create rules for making collective decisions based on differing and conflicting opinions of members of society. In a broad sense, this mechanism is called an electoral system (or electoral rule).

The key question posed by the mathematical theory of elections is thus whether an ideal electoral system – one that makes optimal decisions for society having taken the various opinions of members of society into account – is even possible. Such a system is necessary for each member of society to consider the collective decision legitimate (fair) and thus abide by it, even if it goes against his or her personal convictions. In other words, it is necessary to agree on the criteria that an electoral system must meet in order to be considered optimal. Researchers have come up with a list of criteria that an optimal electoral system must meet [72]:

1. *Anonymity.* All voters are equal – that is, if any two voters exchange ballots, then the outcome of the election will remain the same. Dictatorships, for example, do not meet this criterion, as it is the decision of one member of society (the dictator) that always prevails, meaning that voters are not equal.
2. *Neutrality.* All candidates are equal – that is, if all voters change their choice (i.e. cast their vote for another candidate), the result of the election will change accordingly: the winner will lose, and the loser will win. Electoral systems that use mandatory voting, for example, do not meet this criterion, as decisions are made based on the opinions of people outside the community, thus making the position of candidates *a priori* unequal.
3. *Monotonicity.* This means that the winning candidate cannot lose by receiving additional votes, and the loser cannot win by losing votes. Electoral systems that use minoritarianism – where the candidate who receives the fewest votes wins – do not meet this criterion.

In 1952, the American mathematician Kenneth O. May proved that the only system that can meet all three criteria (assuming that there are two candidates and an unequal number of voters) is a majority electoral system in which the candidate who receives more than 50 % of the votes is the winner (*May's theorem*).

Thus, majority voting is not only the optimal, but also the only possible social choice function when unanimity is impossible. The very presence of a *simple (absolute) majority* (50 % + 1 vote) among voters changes the electoral strategy of candidates dramatically. Until this threshold of support is passed, it is theoretically possible that another candidate may emerge who can offer a more popular programme. This forces candidates to make concessions and adjust their programme in order to win a simple majority. At the same time, it often happens that once this threshold is achieved, the winning candidate is no longer interested in finding new supporters because, theoretically, no one can offer a more popular programme.

In political science terminology, there is also the concept of *qualified majority* (usually ⅔ or ¾ of votes), but such unanimity is only required in rare cases (for example, when making amendments to the country's constitution, which is why this is also called a *constitutional majority*). It has also been empirically proven that a candidate who receives 45 % of votes in the first round of a multi-round election, or even 40 % if he or she is more than 10 % ahead of the candidate in second place, is unlikely to lose. This situation is called a *mixed majority*. The types of majority are summarized in Table 2.1.

Table 2.1. Type of majority

Type of majority	Definition
Relative	more than the other candidates
Mixed	approximately 40–50 %
Simple (absolute)	50 % + 1
Qualified (constitutional)	typically ⅔ or ¾
Consensus	100 %

In some exceptional cases, the principle of majority is limited to the right of *veto*, which means that a decision cannot be made (or a candidate elected) if the holder of the right of veto opposes it.

Are simple majority systems ideal? Does the candidate who enjoys the support of the majority of the community always win? The problem is that, in reality, we are rarely given just two choices. When there are more than two candidates, a *relative majority* (the most votes among all candidates), rather than a simple majority, is needed. Orders of preference now appear in electoral behaviour: Candidate B may be more liked than Candidate C, for example, but less liked than Candidate A. At the same time, if Candidate A receives more support than Candidate B, and Candidate B has more support than Candidate C, then Candidate A will also have more support than Candidate C, that is, it is *transitive*.

$$A \succ B \succ C$$

As the number of candidates increases, so too does the likelihood of unexpected complications appearing in public choice. Let us take a look at one such example. Imagine the following order of preferences among voters for candidates A, B and C.

Place	% of votes			
	35	28	20	17
1	A	B	C	C
2	B	A	A	B
3	C	C	B	A

In a majority electoral system, Candidate C would win, with 37 % of the vote. At the same time, it is worth noting that the majority of voters (63 %) had Candidate C as their third choice, that is, the majority of voters would prefer either Candidate A or Candidate B to Candidate C. Moreover, in a one-on-one, Candidate C would lose to both Candidate A and Candidate B

$$A(63\%) \succ C(37\%), B(63\%) \succ C(37\%), A(55\%) \succ B(45\%)$$

In other words, the order of preference for majority rule and pairwise voting will be different: $C \succ A \succ B$ and $A \succ B \succ C$.

Therefore, when there are more than two candidates (and, consequently, an order of preferences), the list of criteria for an optimal electoral system necessarily grows [420].

4. *Universality.* Every possible set of ballots with a transitivity of preferences must lead to a transitive order of preferences. Electoral systems should not impose any condition other than transitivity on how voters should order candidates in an election.

5. *Independence from outside initiatives.* Public preferences between any two candidates depend exclusively on individual preferences between these candidates.

Let us now imagine the following order of preferences among voters X, Y and Z with regard to candidates A, B and C.

	X	Y	Z
1	A	B	C
2	B	C	A
3	C	A	B

Note that if the election was between candidates A and B, then A would win; if it was between B and C, then B would win; and if it was between C and A, then C would win.

$$A \succ B, B \succ C, C \succ A$$

$$A \succ B \succ C \succ A$$

It turns out that a situation can arise in public choice theory that is theoretically impossible from the point of view of mathematical logic. This situation is called the Condorcet paradox after the 18th-century mathematician Marquis de Condorcet [247]. If $A \succ B$ and $B \succ C$, but at the same time $C \succ A$, then we say the system is nontransitive.

In 1951, the American mathematician Kenneth Arrow proved that the Condorcet paradox (the possible non-transitivity of collective choice) meant that it is impossible in principle to create an electoral system with more than two candidates that would meet the criteria given earlier, that is, one that would be optimal (*Arrow's impossibility theorem*) [201]. His work in this area earned him the Nobel Prize in Economics in 1972.

Let us meditate further on what the Condorcet and Arrow paradoxes actually mean. According to them, by definition, no electoral system in the world can be considered optimal, that is, as fairly reflecting the needs and expectations of the people. In other words, selecting and finetuning an electoral system for a given society is an analytical task that aims to implement a system in such a way that the shortcomings of electoral procedures do not prove fatal for society. As we will see below, the key parameter of the electoral system is its territorial organization, and the functioning of the entire system thus depends on the calibration of this parameter.

2.2 Electoral and Voting Districts

While political geography focuses on such elements of the territorial organization of society as states and their administrative units, electoral geography deals with special types of territorial units, namely *electoral districts* and *voting districts*. It is important to note here that these are entirely different things: an electoral district is a unit of the territorial organization of the electoral system, whereas a voting district is a unit of the territorial organization of the state (or region) itself.

The need to divide the electoral system into *electoral districts* arises when several representatives are to be elected (for example, to parliament). In this case, the country is divided into territorial units (electoral districts), each of which elects one or several representatives. The need to divide the country into *voting districts* has nothing to do with the intricacies of the electoral process, but rather stems from the physical impossibility of having every single person cast their vote in one

Electoral and Voting Districts

location. In this case, the country is divided into voting districts, to which voters are assigned according to their place of residence. To further ensure the smooth running of elections, a hierarchy can be created whereby the process is overseen by a central election commission, as well as by regional, territorial, local and other authorities, down to the lowest level – voting districts. Voting districts may also be assigned votes in extraterritorial units (for example, in embassies abroad, military bases, polar stations, etc.).

The size of an electoral district is thus measured by the number of candidates elected, while the size of a voting district is measured by the number of voters. Electoral districts may be *single-member* (when voters elect one representative) or *multi-member* (when two, three, four, five, etc., are elected) if several representatives are elected at once. It is not uncommon to see huge multi-member districts (where 11, 21, etc., representatives are elected), but they are usually limited to five or six mandates. *Single* (or national) districts are a special case. They can be either single-member (used in presidential elections, for example, when the entire country elects a single representative), or multi-member (when the entire country elects the members of parliament, and the total number of seats is distributed among the parties in proportion to the total number of votes cast). Voters are invariably assigned to the same voting district (typically the one nearest to their place of residence). However, upon arrival at the polling station, it may turn out that they belong to more than one electoral district: for example, a single-member district when voting for the president, and a multi-member district when voting for members of parliament (where the district borders coincide with the region's borders).

This dichotomy in the choice of electoral district size is particularly prominent in proportional systems. The larger the district, the less distorted the proportionality of the number of seats distributed. However, the smaller the district, the stronger the connection between the candidate and the voter. This problem is partly solved by using multi-level electoral districts, as is the case in Austria, Bulgaria, Denmark and Sweden, for example. In these countries, the majority of seats are distributed in ordinary multi-member districts, while the rest are distributed in a single multi-member district.

The patterns that emerge in relation to the territorial heterogeneity of the electoral space can also be studied in two planes: through the differentiation of voting results by electoral or by voting district. If there are additional levels in the hierarchy of the electoral territorial division of a country (between the centre and the electoral districts – regional level, territorial level, etc.), then a comparative analysis of the spatial distribution of electoral preferences at these levels is also possible. In this case, the following limitations should be taken into account. First, comparison between different electoral districts can only be based on the party (or ideological) affiliation of candidates, as there will be different candidates in different electoral districts. And, second, when you have a single district, territorial heterogeneity can only be studied at the level of electoral territorial division.

Electoral districts do not always have a territorial link. For example, it is not uncommon for electoral districts to be allocated for expatriates, ethnic minorities, and even representatives of legal persons (for example, functional districts for business representatives in Hong Kong, or university districts for graduates of two of the top universities in Ireland).

2.3 Typology of Electoral Systems

An electoral system is a way of determining the results of elections. The following criteria are used to classify the different types of electoral system [99]:

1. Electoral formula (the method of converting votes into seats). This is the main criterion, and, as such, it is used to determine whether an electoral system is *majoritarian* (where mandates are distributed on the basis of the majority principle) or *proportional* (where mandates are distributed in proportion to the votes cast). Majoritarian systems are further divided into relative (*plural*) or simple (absolute) majority.
2. Method of voting: *categorical* (where voters choose one candidate from the list only); *approval* (where voters can select several candidates without expressing a preference for any one of them); *disapproval* (where voters express, along with their approval for certain candidates, disapproval of others); *cumulative* (where voters have a set number of votes and can cast them among the candidates as they please); and *preferential* (where voters rank the candidates in order of preference).

3. The size of the electoral district: single-member, in which one representative is elected (*uninominal system*); or multi-member, in which several representatives are elected (*plurinominal systems*).

We can now use these basic criteria to classify the main electoral systems that exist around the world (Table 2.2.).

Table 2.2. Types of electoral system

Electoral System	Electoral formula	Voting method	District size
Single-member plurality (informally referred to as first-past-the-post) – SMP/FPTP	M(P)	Cat.	Single
Two-round system – TRS	M	Cat.	Single
Instant-runoff voting or Alternative vote – IRV/AV	M	Pref.	Single
Party block voting – PBV	M(P)	Cat.	Multi
Block voting – BV	M(P)	Appr.	Multi
Single non-transferable vote – SNTV	M(P)	Cat.	Multi
Fixed ratio vote	M(P)	Cat.	Multi
Single transferable vote – STV	P	Pref.	Multi
Party-list proportional representation – List PR	P	Cat./ Appr./ Pref.	Multi

Key: electoral formula (M – majoritarian (simple majority), M(P) – majoritarian plurality (relative majority), P – plurality (proportional)); voting method (Cat. – categorical, Appr. – approval, Pref. – preferential); district size (Single – single-member, Multi – multi-member).

Table 2.3 and Figures 2.1–2.2 show the distribution of electoral systems for elections to the upper and lower houses of parliament among UN member states in 2022. In addition to the electoral systems described above, the table also includes multi-component (mixed and hybrid) systems that combine elements of the basic majority and proportional systems. In cases where a country uses a multi-component system but one component clearly dominates (for example, an additional component may consist of a very limited number of seats being up for grabs), it has been classified according to the main element only.

Table 2.3. Electoral systems used in parliamentary elections around the world (as of 2022)

Electoral system	Upper house	Lower (or only) house
Single-member plurality	Bhutan, Grenada, Haiti, Kenya, Myanmar, Nigeria, Poland, United States	Antigua and Barbuda, Azerbaijan, Bahamas, Bangladesh, Barbados, Belarus, Belize, Bhutan, Botswana, Canada, Dominica, Eritrea, Gambia, Ghana, Grenada, India, Jamaica, Kenya, Kiribati, Liberia, Malawi, Malaysia, Maldives, Micronesia, Myanmar, Nigeria, Pakistan, Palau, Qatar, Saint Kitts and Nevis, Saint Vincent and the Grenadines, Saint Lucia, Samoa, Sierra Leone, Solomon Islands, Tonga, Trinidad and Tobago, Turkmenistan, Tuvalu, Uganda, United Kingdom, United States, Yemen, Zambia
Two-round system	Czech Republic, Switzerland	Bahrein, Central African Republic, Comoro Islands, Congo, France, Gabon, Haiti, Iran, Mali, North Korea, Uzbekistan, Vietnam
Instant-runoff voting		Australia, Kiribati, Nauru, Papua New Guinea
Party block voting		Côte d'Ivoire, Singapore
Block voting	Brazil, Liberia, Palau, Philippines, Spain	Ethiopia, Laos, Mauritius, Mongolia
Single non-transferable vote	Indonesia, Japan	Afghanistan, Iraq, Kuwait, Oman, Vanuatu
Fixed ratio vote	Argentina, Mexico	
Single transferable vote	Australia	Ireland, Malta

Table 2.3. Continued

Electoral system	Upper house	Lower (or only) house
Party-list proportional representation	Bolivia, Chile, Colombia, Dominican Republic, Equatorial Guinea, Namibia, Paraguay, Romania, Uruguay, Zimbabwe	Albania, Algeria, Angola, Argentina, Armenia, Austria, Belgium, Benin, Bulgaria, Bosnia and Herzegovina, Brazil, Burkina Faso, Burundi, Cambodia, Cape Verde, Chile, Colombia, Costa Rica, Croatia, Cyprus, Czech Republic, Denmark, Dominican Republic, East Timor, Ecuador, El Salvador, Estonia, Equatorial Guinea, Fiji, Finland, Greece, Guatemala, Guinea-Bissau, Honduras, Iceland, Indonesia, Israel, Kazakhstan, Kyrgyzstan, Latvia, Lebanon, Liechtenstein, Luxembourg, Montenegro, Morocco, Moldova, Mozambique, Namibia, Netherlands, Nicaragua, Niger, North Macedonia, Norway, Paraguay, Peru, Poland, Portugal, Romania, Rwanda, San Marino, São Tomé and Príncipe, Serbia, Slovakia, Sri Lanka, Slovenia, South Africa, Spain, Suriname, Sweden, Switzerland, Togo, Tunisia, Turkey, Uruguay
Mixed with no linkage	Egypt, Italy	Andorra, Cameroon, Chad, Democratic Republic of the Congo, Egypt, Georgia, Guinea, Japan, Jordan, Italy, Lithuania, Mauritania, Mexico, Monaco, Nepal, Panama, Philippines, Russia, Senegal, Seychelles, Tajikistan, Tanzania, Thailand, Ukraine, Venezuela, Zimbabwe
Mixed with seat/vote linkage		Bolivia, Djibouti, Germany, Hungary, Lesotho, New Zealand, South Korea
Hybrid		Marshall Islands

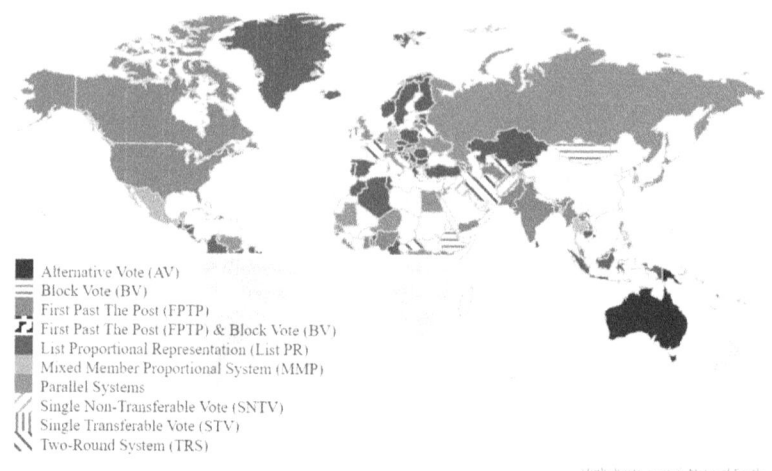

Fig. 2.1. Electoral systems used to elect candidates to the lower (or only) house of parliament
Source: Inter-Parliamentary Union, 2022.

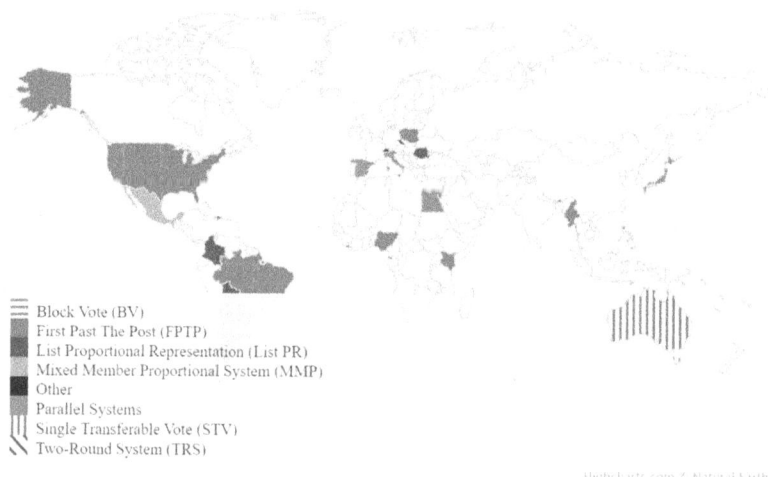

Fig. 2.2. Electoral systems used to elect candidates to the upper house of parliament
Source: Inter-Parliamentary Union, 2022.

Typology of Electoral Systems

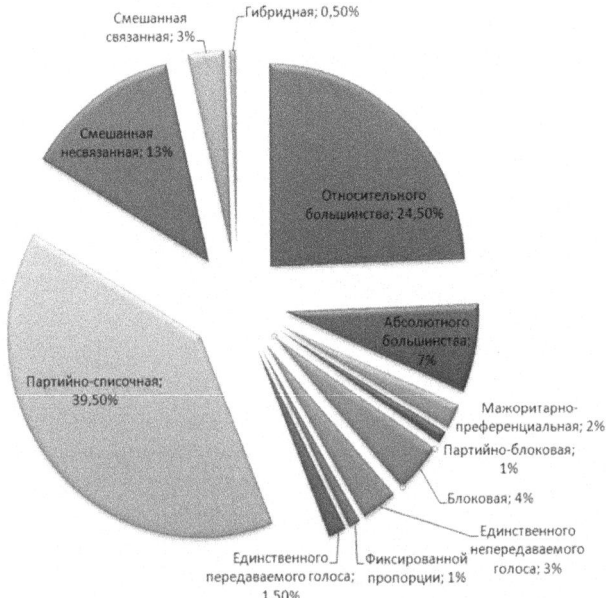

Fig. 2.3. Pie chart of electoral systems used in parliamentary elections around the world

Относительного большинства; 24,50 %	Single-member plurality; 24.5 %
Абсолютного большинства; 7 %	Two-round system; 7 %
Мажоритарно-Преференциальная; 2 %	Instant-runoff voting; 2 %
Партийно-Блоковая; 1 %	Party block voting; 1 %
Блоковая; 4 %	Block voting; 4 %
Единственного непередаваемого голоса; 3 %	Single non-transferable vote; 3 %
Фиксированной пропорции; 1 %	Fixed ratio vote; 1 %
Единственного передаваемого голоса; 1,50 %	Single transferable vote; 1.5 %

Партийно-Списочная; 39,50 %	Party-list proportional representation; 39.5 %
Смешанная несвязанная; 13 %	Mixed with no linkage; 13 %
Смешанная связанная; 3 %	Mixed with seat/vote linkage; 3 %
Гибридная; 0,50 %	Hybrid; 0.5 %

Electoral systems are extremely unevenly distributed around the world. A comparison of their share in elections to both houses of national parliaments would provide the clearest picture of such distribution (Fig. 1.3). Majoritarian systems account for 42.5 % of all electoral systems around the world (including 8 % that are semi-proportional systems that essentially use the majority principle too), followed by proportional systems (41 %), and then mixed and hybrid systems (16.5 %). At the same time, proportional lists are used in 55.5 % of all elections (in their pure form in 39.5 %, plus another 16 % in a mixed form when combined with a majoritarian system). Majoritarian systems mostly use a plurality (relative majority) formula (one in four elections use this), while two-round voting and absolute majority are used in 7 % of cases. The remaining systems are little more than exotic exceptions.

2.4 Majority Electoral Systems

In *majoritarian* systems, parliamentary seats are distributed according to the majority principle. The simplest such system is the *plural* system [42] (also known as a *single-member plurality*). In this set-up, the candidate who receives the most votes is deemed the winner. Single-member electoral districts are typically used when electing one person (usually the head of state or head of the executive branch of government), while multiple single-member districts (equal to the number of deputies being elected) are used to elect candidates to collective bodies (most often a legislative body such as the parliament), with each district electing one deputy.

Approximately one in every five electoral systems in the world today is plural, with heads of state elected in this manner. The plural system is also used to elect candidates to the lower houses of parliament in the United Kingdom, Canada, Pakistan and Ethiopia, and to elect members of the upper houses in Nigeria and Poland. The main advantage of single-member plurality is its simplicity. Second is the personification of choice, and third is the greater dependence of candidates on the electorate in their district, which makes them more accountable and thus forces them to focus on the needs of the people. This latter point in particular allows candidates to tailor their approaches to different kinds of specialized regions (for example, agrarian regions) and makes the plurality system preferable in local elections.

Single-member plurality leads to several significant transformations of the political process in a country. The most well-known transformation is known as Duverger's law, which holds that single-member plurality systems favour the development of a two-party system [53]. In a situation where only the winning party receives a mandate, voters understand that there is no point supporting parties that have no chance of winning and thus start to vote *strategically* (*tactical voting*) – that is, they will vote for the party whose programme best reflects their ideological preferences if it means preventing another leading party with opposing views from winning. Over time, this leads to the electoral landscape being dominated by two major parties and the establishment of a two-party system. These parties tend to be centrist or broad-based (and therefore less consistent and strict), as they are the ones that manage to accumulate the largest numbers of strategic voters. On the one hand, this means that the interests of society cannot be fully represented, since elections are decided not by a loyal electorate, but by strategic voting. On the other hand, this almost completely eliminates the possibility of extremist parties entering parliament by democratic means. Duverger's law also leads to a one-party (and thus stable) government in plural systems and, as a natural consequence, a consolidated and strong one-party opposition. In addition, plurality systems – because they encourage negative voting – contribute to the formation of broad coalitions of political forces, which is particularly important for multi-component or democratizing societies.

Another consequence of the single-member plurality system is the so-called *spoiler effect*. It turns out that the optimal strategy in such a system is to put forward a spoiler candidate or party that will draw votes from a major candidate or party with similar policies. Drawing even a small share of the votes will see the ideological opponent emerge victorious. We should note here that Duverger's law is not actually a law at all, but rather a tendency, and there are examples that refute it (Canada, India, and recently, United Kingdom).

Table 2.4. Electoral systems for heads of state around the world (as of 2022)

Direct		Indirect
Single-round	Two-round	
Angola, Azerbaijan, Bosnia and Herzegovina, Central African Republic, Democratic Republic of the Congo, Equatorial Guinea, Gabon, Gambia, Honduras, Iceland, Ireland, Kenya, Kiribati, Malawi, Mexico, Nicaragua, Panama, Paraguay, Philippines, Rwanda, Seychelles, Singapore, Solomon Islands, South Korea, Sri Lanka, Tajikistan, Uganda, Venezuela	Afghanistan, Algeria, Argentina, Austria, Belarus, Benin, Bolivia, Brazil, Bulgaria, Burkina Faso, Burundi, Cameroon, Cape Verde, Chad, Chile, Colombia, Comoro Islands, Congo, Costa Rica, Côte d'Ivoire, Croatia, Cyprus, Czech Republic, Djibouti, Dominican Republic, East Timor, Ecuador, El Salvador, Egypt, Finland, France, Georgia, Ghana, Guatemala, Guinea, Guinea-Bissau, Haiti, Indonesia, Iran, Kazakhstan, Kyrgyzstan, Liberia, Lithuania, Madagascar, Maldives, Mali, Mauritania, Moldova, Mongolia, Montenegro, Mozambique, Namibia, Niger, Nigeria, North Macedonia, Palau, Peru, Poland, Portugal, Romania, Russia, São Tomé and Príncipe, Senegal, Serbia, Sierra Leone, Slovakia, Slovenia, Syria, Tanzania, Togo, Tunisia, Turkey, Turkmenistan, Ukraine, Uruguay, Uzbekistan, Yemen, Zambia, Zimbabwe	Albania, Armenia, Bangladesh, Barbados, Botswana, China, Cuba, Dominica, Eritrea, Estonia, Ethiopia, Fiji, Germany, Greece, Guyana, Hungary, India, Iraq, Israel, Italy, Laos, Latvia, Lebanon, Malta, Marshall Islands, Mauritius, Micronesia, Myanmar, Nauru, Nepal, North Korea, Pakistan, Samoa, San Marino, South Africa, Suriname, Switzerland, Trinidad and Tobago, United States, Vanuatu, Vietnam

The *absolute majority* (or *two-round*) *system* is similar to the plurality formula, but does away with some of its shortcomings. In such a set-up, the winning candidate must receive more than 50 % of the votes. If there is no winner, then a second round (*re-ballot*) is often held between the two candidates who received the largest number of votes. But there are exceptions here too: in France, candidates in the parliamentary elections need 12.5 % of the votes in order to go through to the second round. Using a second round of voting to achieve an absolute majority generally reduces the degree of unrepresentativeness that occurs with a plurality system. At the same time, the system can be manipulated, which is seen when the parties or candidates that lost in the previous election (typically the opposition) unite against the winner in the first round.

The two-round system is the most popular for electing heads of state (Table 2.4): 148 of the 193 UN member states elect their leaders in this way (the head of state in the remaining countries is usually a monarch). Of these, 28 countries (19 %) use a single-member plurality system, 79 (53 %) use a two-round majoritarian system, and 41 (28 %) use a system whereby the head of state is elected indirectly (by parliament or an electoral college). The absolute majority system is also the preferred method of electing members of parliament, in particular in the elections to the upper houses in the Czech Republic and Switzerland, and to the lower houses in Vietnam, Iran, France and Uzbekistan.

Mixed-member proportional representation is a sub-species of the two-round system. As we noted earlier, a candidate who wins more than 45 % of votes in the first round of a multi-round election, or even 40 % if he or she is more than 10 % ahead of the candidate in second place, is unlikely to lose. For this reason, some countries (Argentina, Bolivia, Costa Rica and Ecuador) have lowered the requirements for victory in the first round. Specifically, a mixed majority, rather than a simple majority (50 %), is needed. That is, candidates need 40–45 % of the votes, depending on the country, and sometimes with the additional condition that the candidate in question has a lead of at least 10 %, in order to be declared the winner after the first round.

The main problem with the two-round system is that it is an organizationally complex process: holding a second round requires additional time and money. *Instant-runoff voting* is intended to address this issue, while remaining committed to the principle of absolute majority. This system has been instituted, for example, in Irish presidential elections, and is also used to elect deputies to the lower houses of parliament in Australia, Papua New Guinea and, in a slightly modified form, Kiribati. Here, preferential voting is used (as opposed to categorical voting), where candidates are ranked in order of preference. If a candidate gains an absolute majority in the preference lists, then he or she is declared the winner. But even if there is no absolute majority, there is no need for a re-ballot. What happens is that the candidate with the fewest first-choice votes is removed from the calculation, and the votes cast for them are redistributed in accordance with the preference lists to those candidates who were listed second. If there is still no winner by absolute majority, then this procedure is repeated until a winner is found. It turns out that the possibility of a second ballot is baked into the instant-runoff system; the only difference is that the winner does not necessarily have to be from among the leading two candidates (as is the case with the two-round system), and could very well have been third, fourth, or even lower, on the original preference lists.

The version of the instant-runoff system used in Nauru is rather unique and is based on a *points system* (also known as a *Borda count*). In this system, candidates receive points based on their ranking in voting lists. The variant used in Nauru is called the Dowdall System (named after the island's former secretary for justice): one point is awarded for first place, half a point for second place, one third of a point for third place, and one quarter of a point for fourth place. The result is then determined using a plurality formula that is based not on the percentage of support for candidates, but on the total number of points accumulated.

Another subtype of instant-runoff voting is the *conditional vote system*, which was used, in particular in the first half of the 20th century in Alabama in the United States and Queensland in Australia. Here, the voting procedure is the same, with all the candidates ranked, but if one of them gains an absolute majority during counting, then he or she is declared the winner. If no clear winner emerges, instead of eliminating candidates one by one as in the traditional instant-runoff system, the first two candidates receive the votes from all other preference lists simultaneously. In other words, this system *de facto* repeats the logic of the two-round system, but without a second round. Sri Lanka uses this system in its presidential elections, although voters only rank the first three candidates (rather than all of them).

Another development of the relative majority system used in parliamentary elections is *party block voting*. This is where there are multi-member districts, but voters have one vote that is cast for a block of candidates from a single party. The party with a relative majority thus receives all mandates that are contested within a given district. As a result, this system transfers and multiplies the consequences of the disproportionality of the plurality formula to the party level. This kind of set-up serves to strengthen the dominant party and is most often used in non-democratic states – in Côte d'Ivoire and Singapore, for example.

2.5 Semi-Proportional Electoral Systems

Electoral systems in which the majority principle operates in multi-member districts are called *semi-proportional* (or *pre-proportional*), since this can lead to a proportional result. Electing several candidates from a single district allows for the interests of a wider number of voters to be taken into account, which makes such systems similar to proportional systems. However, these systems are still majoritarian, as they nevertheless use the majority principle to determine the winners. An exception to this is the party block system, since its plurinominality cannot lead even to partial proportionality. Semi-proportional systems used at the national level today include the block voting system (with subtypes: unlimited, limited and cumulative voting), the single non-transferable voting system, and the fixed ratio voting system.

The *block voting system* (its classical version being the *unlimited block approval voting*) works on the principle that voters in multi-member districts have the same number of votes as there are seats up for grabs. They can cast all their votes for representatives of a single party (who implore the electorate to vote for them as a block, hence the name), or they can split their votes among representatives of different parties. This model is used to elect members of the upper houses of parliament in Brazil, Liberia, Palau and the Philippines, and members of the lower houses in Laos, Ethiopia, and Mauritius, as well as Mongolia, which instituted this system relatively recently. This method may have its advantages when the competition is limited, i.e. when there are at most twice as many candidates as mandates, although it can lead to situations where the party that received an absolute majority of votes actually loses. The block voting system only leads to proportionality when the voters are well-acquainted with all the candidates. If they are not, the reverse will happen: relatively unknown political figures will get into power on the back of a popular candidate from their party.

A subset of the block voting system used, for example, in elections to the upper house of the Spanish parliament, is *limited block approval voting*. This also involves approval voting in multi-member districts, but here the number of votes per voter is less than the number of seats being contested. Another subset of block voting is *cumulative voting*, where a cumulative system is used instead of an approval system. In practice, this means that voters have several votes and can cast them as they please among the candidates. Here too the number of votes is equal to the number of mandates, but voters are free to give all their votes to a single candidate or distribute them among several. This system is famously used in the state of Illinois in the United States.

The *single non-transferable vote system* is characterized by categorical voting in multi-member districts. In other words, several candidates are elected per district, but voters have only one vote. Japan famously uses this system, but it has also been instituted in Afghanistan, Indonesia, Iraq, Kuwait, Oman and Vanuatu. Its primary function is to reduce the role of the dominant party, forcing it to count on popular support for several candidates.

Fixed ratio voting is a specific kind of semi-proportional system used for partisan elections. Here, the results of some parties are determined by the majoritarian principle, while others are determined by proportionality. For example, elections to the senates in Argentina and Mexico involve a system of bonuses: in three-member districts, for example, the winning party takes two seats, while the party that came second receives one. A version of this principle was also implemented during the rule of Augusto Pinochet in Chile, called the *binomial system*: in two-member districts, the winning party received both seats in the senate if it received more than two thirds of the votes, and one seat if it received less. In both its versions, this system supports the dominant and second most popular parties and cuts off small, more radical forces.

2.6 Proportional Electoral Systems

In *proportional systems*, seats are allocated proportionally to the number of votes cast. This principle was first implemented in the form of a single transferable vote, and then became part of a whole family of party-list systems.

The *single transferable vote* (or *proportional-preferential*) *system* uses a version of preferential voting in multi-member districts. This rather rare formula proved its effectiveness in the elections to the upper house of Australia and the lower houses of Ireland and Malta. As far as the voter is concerned, the system resembles instant-runoff voting: candidates are ranked in order of preference. The difference is that in instant-runoff voting an absolute majority is needed to win, whereas single transferable voting requires a fixed share of votes, which is more or less equal to the ratio of the number of voters to the number of mandates being contested. After the first count, the candidate who has passed this threshold is declared the winner. This is followed by a recount: just as with instant-runoff system, the candidate who finished last is eliminated and his or her votes are redistributed to those candidates who were placed second on the preference lists of those voters who voted for that candidate. But, unlike instant-runoff voting, votes cast for the winning candidate (who has already passed the fixed share threshold) are also excluded from counting and go to candidates listed second on the preference lists of those voters who voted for that candidate. This procedure is repeated until all the seats in a district are filled. The system combines the advantages of the proportional and majoritarian approaches: it is a good reflection of the range of opinions in society, yet it also ensures that the connection between voters and candidates remains intact. At the same time, it is not particularly voter-friendly, as it requires considerable political maturity from them. Nor does it eliminate tactical voting: parties conspire with regard to which of them will receive second-place votes, and coalition-building becomes a part of election campaigns.

Party-list representation is the classical example of a proportional system. It works as follows: voters select a list of candidates, and seats are allocated in proportion to the percentage of votes for the list. This is the most popular electoral system in the world, and more than one third of all national elections use some form of it. Party-list systems can be open-list, closed-list or free-list.

In *closed-list* systems, voters vote for the party that nominated the list of candidates and have no say as to who will represent their district in parliament. This model is used, for example, in Bulgaria, Israel, Portugal, Romania, Turkey and South Africa. Meanwhile, in some Latin American countries, voters can cast their vote for individual factions (lemmas) within parties, supporting a part of the general party list. And in Italy, votes for individual parties are collated when calculating the final distribution of seats in party coalitions. These methods lead to intra-party (or coalition) fighting, and small radical parties are cut off. Closed lists can be *regional* (or local), where the party determines the lists of candidates in different regions, and when the seats are allocated in multi-member districts, candidates from regional lists in those regions where the party received the majority of votes have an advantage.

With *open-list* voting, the voter influences which candidates win seats within the quota that the party receives as a result of proportional voting. While a single multi-member district is suitable for closed-list voting, open-list voting typically involves dividing the country into several multi-member districts, and seats are allocated proportionally to the number of votes cast in each.

The main types of allocating seats with open-list voting are as follows:

1. categorical voting – where voters select a candidate from the party list (Austria, Belgium, Cyprus, Denmark, Finland, Poland, Sweden);
2. approval voting – where voters select a number of candidates from the party list that is equal to (or occasionally less than) the number of seats that are up for grabs (Czech Republic, Greece, Slovakia);
3. disapproval voting – where voters both select preferred candidates and express formal disapproval of other candidates from the party list (Latvia);
4. cumulative voting – where voters can cast their votes among several candidates in unequal proportion from the party list;
5. preferential voting – where voters rank candidates from the party list (Finland used to have this system).

There is another version of the proportional list system that is also the most democratic – *free* lists, which is also called *panachage* (from the French for "mixture"). Here, voters are not limited to a single party list and can cast their votes in multimember districts among candidates from different lists. At the same time, unlike the majoritarian block system, seats are initially allocated proportionally among the parties, and then within parties based on the number of votes cast for specific candidates. This principle is used at the national level in Switzerland, Liechtenstein and Luxembourg.

In a sense, party-list proportional representation is the exact opposite of the plurality system, which negatively highlights both its advantages and disadvantages. On the one hand, it fixes the main issues of the plurality formula – the fact that it is not representative and a large number of votes end up being wasted. On the other hand, it surrenders the key advantage of uninominality – that is, the close connection between members of parliament and those who elected them. This leads to the depersonalization of elections, when seats in the legislature are decided not so much by the candidate's election campaign as by his or her lobbying connections within the party leadership.

Many also believe that the proportional system favours smaller parties, which can lead to the excessive fragmentation of society and the erosion of the majority necessary for the stable functioning of the government. To overcome these shortcomings in party-list systems, so-called majority bonuses and electoral thresholds are put in place. A *majority bonus* is when the winning party is given extra seats above its share as determined by proportional distribution. Greece, for example, awards an additional 50 seats in parliament to the winning party as part of its majority bonus system. Meanwhile, San Marino uses a jackpot system, where the winner is guaranteed at least 35 of the 60 seats in parliament. Cameroon and Chad have implemented a system that is somewhere in between the party-list and party block paradigms: if the winning party in a multi-member district receives more than half of the votes, then it gets all the seats; if it wins less than half of the votes, it gets 50 % of the seats, and the rest are divvied up among the other parties on a proportional basis. Majority bonuses have a long tradition in Italy, where they have been used in one form or another since 1923.

An artificial *electoral threshold* is the minimum share of the votes that a candidate or party has to receive in order to be allowed into the legislature. In most countries, the threshold is set at 3–5 % (see Table 2.5). This is not in line with the recommendation of the Parliamentary Assembly of the Council of Europe (PACE), which states that democratic representation can only be ensured if the threshold does not exceed 3 %. Countries with extremely low thresholds do not actually have any formal rules regarding such entrance barriers, which means that the electoral threshold corresponds to the natural quota – in order to get into parliament, candidates (parties) have to receive the requisite number of votes to get at least one seat. In the Netherlands, this is 0.67 %. Other countries that have or had low thresholds include Uruguay (1 % in order to win a seat in the lower house), Israel (1 % until 1992), and Brazil (1.5 % until 2022). The highest electoral thresholds today are found in the microstates of Europe (7.14 % in Andorra, 8 % in Liechtenstein) and some Turkic countries (7 % in Kazakhstan, 9 % in Kyrgyzstan, and 10 % in Turkey). Sweden appears to have the highest threshold, where 12 % of votes are required in district elections (although 4 % is enough on the national level). Non-party alliances often face even higher thresholds (by around 2 %) in order to get opportunistic coalitions to unite into formal parties.

Table 2.5. Electoral thresholds in parliamentary elections around the world (as of 2022)

Electoral threshold	Countries where it is used
No threshold	Finland, North Macedonia, Portugal, South Africa
0.67 %	Netherlands
1 %	Uruguay
2 %	Brazil, Burundi, Denmark, Philippines
3 %	Albania, Argentina, Bolivia, Bosnia and Herzegovina, Colombia, Georgia, Greece, Italy, Mexico, Montenegro, Nepal, Serbia, South Korea, Spain
3.25 %	Israel
3.6 %	Cyprus
4 %	Austria, Bulgaria, East Timor, Indonesia, Norway, Slovenia
5 %	Armenia, Belgium, Croatia, Czech Republic, Estonia, Fiji, Germany, Hungary, Iceland, Latvia, Lithuania, Mozambique, Moldova, Monaco, New Zealand, Peru, Poland, Romania, Russia, Rwanda, San Marino, Slovakia, Tajikistan, Ukraine
6 %	Moldova
7 %	Kazakhstan
7.14 %	Andorra
8 %	Liechtenstein
9 %	Kyrgyzstan
10 %	Turkey
12 %	Sweden

Introducing an electoral threshold limits the number of factions in parliament and ultimately leads to an increase in the number of deputies in those factions. It also helps weed out unrepresentative (and sometimes extremist) parties from the political process, which means that it encourages these parties to build coalitions and move away from their radical positions. All this is considered a boon, since it increases the efficiency of the executive branch of government. It can be argued that a threshold below 2–3 % leads to an atomized party system, where nine or ten political forces may be present in the parliament. At the same time, a threshold of above 5–7 % leads to an excessive narrowing of the ideological field, leaving only three or four parties in it.

2.7 Mixed Electoral Systems

Thus far, we have dealt exclusively with electoral systems that are based on a single principle. However, there are numerous examples of electoral legislation that combine elements of different systems. These are called multi-component systems and are divided into hybrid and mixed.

Hybrid systems that combine different election models within the framework of a single principle (majoritarian or proportional) are rare. In a hybrid majoritarian system, parliamentary elections are simultaneously held using the majority formula in both single- and multi-member districts. This is the case in the Marshall Islands, for example, where elections are held with a plurality and a block system. In a hybrid proportional system, which is used, for example, in elections to the European Parliament, various modifications of proportional systems are used in the EU countries.

In *mixed* systems, majoritarian and proportional elements are combined in a single election. They are in turn divided into *compensatory* (where the components are connected in such a way that the non-representativeness of the majority element is compensated by the proportional element) and *non-compensatory* (where this does not happen). Non-compensatory mixed electoral systems are further divided into *systems with no linkage*, where the majority and proportional elements operate independently of one another, and *systems with seat/vote linkage*, where these elements cannot be separated. Obviously, all compensatory systems by their very definition come with a linkage. Mixed systems are also divided into single-vote and dual-vote, depending on how many ballots the voter places into the ballot box.

The most common type of mixed electoral system is the so-called *parallel* system. This is a non-compensatory dual-vote mixed system with no linkage in which voters are given several ballots in for a single election: one involves voting for a candidate according to the majoritarian system, and the other for a party according to the system of proportional representation. Accordingly, a portion of the seats in the elected body is allocated according to one system, and the remaining seats are allotted according to the other system. For example, in Russia, the State Duma is made up of 450 deputies, who are elected by a parallel system: half are elected in 225 single-member districts using a majoritarian (or relative majority) system, while the other half are elected in a single multi-member district using a proportional system with closed lists. At the same time, it should be noted that elections in Russia involve the creation of not only party lists, but also regional groups (effectively sub-lists), and the allocation of seats according to the proportional part of the system depends on the level of support for a party in a particular group within the single electoral district. What is more, the parties themselves determine the geography of these regional groups and can, for example, include several constituent entities of the Russian Federation as part of a group's territory.

Parallel systems differ above all in their use of the majoritarian component and its combination with the proportional representation component of the party-list system. They may use a version of single-member plurality, a two-round system, a party block system, a block system or a single non-transferable vote system (see Table 2.6).

Table 2.6. The majoritarian component of non-compensatory mixed electoral systems

Majoritarian component	Example
Single-member plurality	Cameroon, Chad, Democratic Republic of the Congo, Guinea, Italy, Japan, Mexico, Nepal, Panama, Philippines, Russia, Seychelles, Tanzania, Thailand, Ukraine, Venezuela, Zimbabwe
Two-round system	Egypt, Georgia, Lithuania, Mauritania, Tajikistan
Party block system	Andorra, Senegal
Block system	Monaco
Single non-transferable vote system	Jordan

The fact that the elements of a parallel mixed system are not formally linked does not mean that the choice of one ballot in an election does not affect the choice of the second. For example, it is a known fact that nominating a strong candidate in a single-member district improves the party's position in the same precincts during list voting. This is called the *contamination* effect.

There are other types of compensatory mixed electoral systems, although these are less common. A unique *non-parallel* system continues to exist in Panama. Its uniqueness lies in the fact that it is a non-compensatory single-vote mixed system with no linkage. Here the components of the electoral system are divided territorially, and voters still have only one vote: voting in single-member districts takes place according to the plurality formula, while voting in multi-member districts takes place using lists. This method of combining electoral models used to be rather popular in Europe, where single-member districts that run on the majoritarian principle were preserved in rural areas, and multi-member districts used proportional voting in densely populated cities.

Italy uses the so-called *Rosatellum* system, which is a non-compensatory yet linked single-vote system. Voters have one ballot only. In single-member districts, votes are automatically cast for the party list – that is, the contamination effect is now a rule here.

Table 2.7. **Types of mixed electoral systems**

		Non-compensatory systems	
		Single-vote	Dual-vote
Unlinked		non-parallel (Panama)	parallel (Russia)
Linked		Rosatellum (Italy)	
		Compensatory systems	
		Single-vote	Dual-vote
Linked majoritarian	Linked by seat distribution	single-vote additional member system (Bolivia, Lesotho)	dual-vote additional member system (Scotland, Wales)
	Linked by vote counting	positive vote transfer system (previously in Romania)	negative vote transfer system (previously in Italy)
Linked proportional		single-vote mixed proportional system (Baden-Württemberg)	dual-vote mixed linked proportional system (Germany, New Zealand)

Compensatory mixed systems combine elements of majoritarian and proportional systems in such a way that this determines the outcome of voting. Depending on which component dominates, compensatory systems can be either *mixed-member majoritarian* (MMM) or *mixed-member proportional* (MMP). Accordingly, the components can be linked when distributing seats (*seat linkage*) or when counting votes (*vote linkage*). Table 2.7 presents a breakdown of mixed electoral systems.

The best-known example of a mixed compensatory system is the *dual-vote mixed linked proportional system*, which Germany and New Zealand use. In this model, voters cast their ballots for both a regional part list and a candidate in a single-member district. However, unlike the parallel system, the faction in parliament is made up first of winners in single-member districts as a share of the votes received for the list as a whole in those districts. The remaining seats are then given to candidates from the party list. An interesting feature of this system is the abundance of so-called overhang seats, which means that the total number of members of parliament is always in flux. Overhang seats are awarded when the regional list receives fewer seats than wins in single-member districts. In this case, the party lists receive the share of seats won, and the winners in single-member districts receive additional seats. A single-vote version of this system operates in the state of Baden-Württemberg in Germany. Here, voters cast their ballots for the candidate only, but the votes also count when distributing seats on the lists.

Another version of compensatory linkage is the *additional member system* (AMS), which also comes in single-vote (Bolivia, Lesotho) and dual-vote (Scotland, Wales) flavours. The majoritarian component dominates here. That is, seats are first distributes in single-member districts, and then additional seats are determined based on the percentage of support for the party as a whole. These additional seats are apportioned in such a way as to ensure the greatest degree of representativeness in parliament as possible. One key difference between this system and the one used in Germany is that the number of seats (basic and additional) is always clearly fixed here. An extreme (and not particularly democratic) example of a mixed linked majoritarian system is the one used in Pakistan, in which 272 seats in parliament are contested through a plurality formula and another 70 are divided among party lists on the basis of the number of majority seats won (rather than on the share of votes won).

Linked majoritarian systems are an even more complex modification of non-compensatory systems, as they are linked not by seat distribution, but by vote counting. They too come in two varieties: dual-vote *negative vote transfer* (NVT), which used to exist in Italy (known as the *scorporo*), and its opposite, single-vote *positive vote transfer* (PVT), which is used in local elections in Hungary and was previously part of the Romanian electoral system. In these systems, after a portion of seats has been allocated on the basis of the majoritarian principle, and additional number of seats is divided between the parties in proportion to the votes they received (on a separate ballot, as with scorporo, or without separate ballots), but minus the votes given to those candidates who won their districts (or, alternatively, who took second place).

So far, we have only talked about mixed systems that are made up of a combination of two elements, although even more complicated systems are, of course, possible. One such example is South Korea's three-component electoral system, whereby 253 of the 300 members of parliament are elected in single-member districts by relative majority, a further 30 are elected using a compensatory additional member system, and the remaining 17 are voted in using a non-compensatory parallel system. The country's multi-component electoral system thus combines elements of the majoritarian and proportional models, as well as elements of linked and unlinked mixed systems.

We can thus conclude that as society becomes more complex in terms of its needs and forms of activity, so too must the models of electoral systems we use. The continued engineering of the latter will produce multi-level, multi-component, asymmetric systems that attempt to strike, for each and every country, a delicate balance between (1) the simplicity and flexibility of the electoral procedure, (2) the personification of the majority principle, and (3) the representativeness of the proportional system.

2.8 Methods of Distributing Seats

Electoral systems face the arithmetic task of determining the best way to divvy up seats in parliament, firstly, between the regions of the country, and secondly, between the winning parties. In both cases, simply dividing the population (or voters) by the number of seats will not give you whole numbers, and so more robust methods are needed to work out the shares. Two methods appeared at around the same time in Europe and the United States, and each aimed to solve a different problem (how to distribute seats among the winning parties in the proportional system in Europe, and how to allocate seats among the states in elections to the House of Representatives in the majoritarian system in America). While these methods ostensibly have two different creators, they nevertheless use the same solutions.

In the following description, we will give the European name first, and the American name in parentheses. The methods used for distributing seats according to the principle used are divided into quota methods (largest remainder) and divisor methods (highest average).

The *Hare–Niemeyer* (*Hamilton*) method uses the quota (largest remainder) method to allocate seats. The problem with dividing the total number of seats by the total number of votes and then multiplying the answer by the number of votes cast for a given party to determine how many seats that party gets is that you might end up with a number that is a fraction. Rounding up or down to a whole number could result in more seats being available in parliament than were actually contested. The following algorithm was proposed to deal with this issue:

1. Calculate the standard quota by dividing the total number of votes by the total number of seats.
2. Allocate seats among the parties by dividing the number of votes received by the standard quota and allocating a number of seats to each party that is equal to the result rounded down to the nearest integer.
3. Determine the remaining number of seats and distribute them among the parties with the largest fractional part obtained by dividing the number of votes cast for each party by the standard quota.

The algorithm for distributing seats between regions will be the same, but here the ratio of the population of the region to the total number of seats will be used to determine the standard quota.

The simplest standard quota (calculated using the method described above) is called the Hare quota, after Thomas Hare, the man who invented to the single transferable voting system (where this method is also used), although other formulas for calculating the standard quota are also used, including the Hagenbach–Bischoff system, the Imperiali quota and, most often, the Droop quota (Table 2.8). Panama has instituted a rather unusual principle: seats are first allocated using the Hare quota, and then using half the quota (the Sartori quota). These quotas are examples of natural (mathematically justified) electoral thresholds (quotas) – the share of votes in proportional elections that need to be obtained in order to win.

Table 2.8. Types of quotas for distributing seats in parliament

Quota	Formula	Usage examples
Hare	$\dfrac{total\ votes}{total\ seats}$	Austria, Cyprus, Honduras
Hagenbach–Bischoff	$\dfrac{total\ votes}{total\ seats + 1}$	Czech Republic, Slovakia
Imperiali	$\dfrac{total\ votes}{total\ seats + 2}$	Ecuador
Droop	$\left(\dfrac{total\ votes}{total\ seats + 1}\right) + 1$	South Africa
Sartori	$\left(\dfrac{total\ votes}{total\ seats}\right) / 2$	Panama

Take the following basic situation: a class of 28 people elects a five-member class council using closed party lists and a proportional system. Three parties take part in the election. The results of the voting were as follows: Green Party – 16 votes, Yellow Party – 7 votes, Red Party – 5 votes. The Hare quota would be equal to 28/5 = 5.6. The "Quotient" column in Table 9 shows the ratio of votes apportioned to each party according to the standard quota. Rounding these numbers up or down to the nearest whole number reveals that, after the first round of counting, the Greens won two mandates, and the Yellows one. There are still two seats left. Looking at the largest remainder, we see that the Greens and Reds should each receive one more mandate. The final result of the voting for the class council is as follows: Green Party – three seats, Yellow Party – one seat, Red Party – one seat.

Table 2.9. Calculation of the distribution of seats using the quota method

Party	Number of votes	Quotient	Seats (1ˢᵗ stage)	Seats (2nd stage)	Number of seats
Greens	16	16/5.6 = 2.86	2	1	3
Yellows	7	7/5.6 = 1.25	1	0	1
Reds	5	5/5.6 = 0.89	0	1	1

As you can see, the number of seats won by each party in quota methods is calculated by dividing the number of votes won (V) by the standard quota, that is, on the ratio of the total number of votes cast to the number of seats up for grabs. The family of divisor methods (highest average), chief among which is the *D'Hondt (Jefferson) method*, is based on another principle of avoiding non-integers when calculating party shares in parliament. Here, a divisor $d(k)$ (a standard quota – a ratio of the total number of votes to the total number of seats) is selected whereby the correct distribution of seats in parliament is obtained after all the quotients have been rounded up or down as necessary.

$$\text{Party share} = \frac{\text{number of votes for the party}}{\text{total votes / total seats}} = \frac{V}{d(k)}$$

Accordingly, divisor methods differ, first of all, in the rounding rule used to obtain the divisors for the calculation. Table 2.10 presents the formulas used for calculating the divisor $d(k)$, where k is the number of seats the party has already won. The divisor for the D'Hondt (Jefferson) method (the most commonly used one) will be $k+1$.

Methods of Distributing Seats

Table 2.10. Distribution of seats using the divisor method

Divisor method	Rule for rounding the divisor	Divisor formula $d(k)$	Divisor range	Usage examples
D'Hondt (Jefferson)	down to the nearest whole number	$k+1$	1, 2, 3, 4, …	Brazil, Belgium, Poland
Koudelka			1.4, 2, 3, 4, …	Some regions in the Czech Republic
Imperiali		$k+2$	2, 3, 4, 5, …	Some regions in Russia
Macao		2^k	1, 2, 4, 8, …	Macao (China)
Adams	to the nearest whole number	k	*, 1, 2, 3, …	
Sainte-Laguë (Webster)	according to the standard rounding rule	$k+1/2$	½, 1½, 2½, 3½, … or 1, 3, 5, 7, …	Bosnia and Herzegovina, Indonesia, Latvia, New Zealand
Modified Sainte-Laguë (Webster)			1.4, 3, 5, 7, …	Nepal, Norway, Sweden
Danish	down to the nearest whole number when the decimal part is less than ⅓	$k+1/3$	⅓, 1⅓, 2⅓, 3⅓, … or 1, 4, 7, 10, …	Denmark
Huntington–Hill	to the geometric mean	$\sqrt{k(k+1)}$	*, 1.41, 2.45, 3.46, …	Distribution of seats among U.S. states
Dean	to the harmonic mean	$k(k+1)/(k+1/2)$	*, 1.33, 2.4, 3.42, …	

* In these methods, the first number in the series of divisors is not possible, so all lists are awarded one ticket each as part of the first step.

The algorithm for calculating seats using the divisor method is as follows:

1. Divide the number of votes received by each party by the divisor (the first number in the list of divisors used in the chosen method

(1 in the D'Hondt [Jefferson] method)) and award the seat to the party with the highest value.
2. Recalculate the score of the party that received the first state by dividing the number of votes it received by the second number in the list of divisors (2 in the D'Hondt [Jefferson] method) and award the second seat to the party with the highest value.
3. Repeat the procedure until all seats have been decided.

Table 2.11. Calculation of the distribution of seats using the divisor method

Stage	Greens	Yellows	Reds
1	16/(0+1) = 16 (1st seat)	7/(0+1) = 7	5/(0+1) = 5
2	16/(1+1) = 8 (2nd seat)	7/(0+1) = 7	5/(0+1) = 5
3	16/(2+1) = 5,33	7/(1+1) = 7 (1st seat)	5/(0+1) = 5
4	16/(2+1) = 5,33 (3rd seat)	7/(1+1) = 3,5	5/(0+1) = 5
5	16/(3+1) = 4	7/(1+1) = 3,5	35/(0+1) = 5 (1st seat)

Let us go back to our example of electing a class council. Table 2.11 lays out the steps for calculating seats using the D'Hondt (Jefferson) method. At the first stage, we divide the number of votes received by each party by the divisor $k+1 = 0+1 = 1$, as per the chosen method. As we can see, the first seat has to be awarded to the Green Party, because $16 > 7 > 5$. At the second stage, the divisor for the Green Party is now $k+1 = 1+1 = 2$, while it will remain $k+1 = 0+1 = 1$ for the Yellows and Reds. Calculating the ratios to the divisors again shows us that the second mandate goes to the Green Party as well, as $8 > 7 > 5$. At the third stage, the divisor for the Green Party is now $k+1 = 2+1 = 3$, while it remains $k+1 = 0+1 = 1$ for the Yellows and Reds. However, the third seat goes to the Yellow Party, because $5.33 \langle 7 \rangle 5$. We continue calculating seats using the same logic and award to fourth seat to the Greens and the fifth to the Red Party. The divisor method has led us to the same distribution of seats as the Hare quota method used above.

According to the Balinski–Young theorem, all of the existing methods are in violation of the quota rule, that is, the number of seats that should be allocated to each region (or party) should be equal to its "ideal quotient," rounded up or down to the nearest whole number, and should not lead to any of the three paradoxes described above. This means that it is mathematically proven that there is no ideal method for allocating seats in parliament, and the choice in each case is determined by context. For example, the Imperiali methods (quotas and divisors) have been shown to slightly favour large parties, while the Adams and Danish methods favour small parties. At the same time, the Hare and Droop quotas and the D'Hondt (Jefferson) and Sainte-Laguë (Webster) divisors appear to be fairer.

2.9 Weighted Electoral Systems

Thus far, we have only discussed electoral systems in which each voter has one vote and, accordingly, all voters and their votes are equal. These are contrasted with *weighted electoral systems*, where votes are not equal. Examples of such systems including voting in the UN Security Council, where five of the 15 members have veto power; on boards of shareholders in the corporate world, where votes are weighted proportionally to the number of shares each voter holds; and in the Electoral College in the United States, where the number of votes in each state is proportionate to the number of citizens living in them. In all these cases, voters have different weights – that is, they have different (and measurable) degrees of influence and power. As we will demonstrate below, there are situations where, formally, all votes are equal, but, in reality, they have greater weight in certain circumstances.

Imagine a group of voters P (also often called a committee, which can consist not only of people, but also of corporations, regions, states, etc.) made up of n members P_1, P_2, ..., P_n, that is $P = \{1,...,n\}$. Each voter has a set number of votes that determines their specific weight. We will use w_1, w_2, ..., w_n to denote the weights of players P_1, P_2, ..., P_n, respectively. We will use V to denote the total number of votes in the system: $V = w_1 + w_2 ... w_n$, and q to indicate the quota – the minimum number of votes required for a decision to be passed. In short, a weighted electoral system will have the following mathematical expression:

$$[q:w_1,w_2,\ldots,w_n],$$

where $w_1 \geq w_2 \geq \ldots \geq w_n$, and $\dfrac{V}{2} < q \leq V$.

Let us look at a few examples of weighted systems with four parties in parliament that need to form a coalition in order to elect a government (the analysis of weighted systems is extremely useful when studying coalition building):

1. [14 : 8,7,3,2] If either of the first two parties join forces with either of the second two, they will not produce the quota. Accordingly, decisions can only be made if the two largest parties reach a consensus, which means that these parties have greater influence than the smaller two parties.
2. [19 : 8,7,3,2] Decisions can only be taken if all four parties reach a consensus. That is, despite the fact that the parties have different voting weights, their influence within the electoral system is equal, which means that weight and influence are not the same here.
3. [11 :12,5,4] Decisions can be taken by the first party no matter what. It does not thus need to enter into a coalition with other parties.
4. [12 : 9,5,4,2] Decisions can only be made if they are supported by the first party, which mathematically has the right to veto the decisions of the remaining parties. Note how this situation differs from the previous one: here, decisions cannot be made unless the player with the right of veto supports it, although this in itself does not mean that the right of veto is enough to guarantee that this player will get what they want. The right of veto is thus only effective when $w < q$ and $V - q < q$.
5. [30 :10,10,10,9] Any decisions taken will not take the opinion of the last party into account, even though it is only slightly behind the others in terms of the number of votes it has.

We use influence indicators to evaluate the level of influence enjoyed by a particular player in the electoral system (including, say, parties in coalition building). These indicators are based on the idea that there is a key player, i.e. voter P, in a coalition $\{P_1, P_2, \ldots, P_n\}$, whose weight is needed for the coalition to win (i.e. without it, the total weight of the coalition will be below the quota). In other words, P is a key player if, and only if, $W - w < q$, where W is the weight of the coalition, and w is the weight of voter P.

The influence index proposed by Banzhaf [205] reflects the influence of voters (parties) in an electoral system (parliament) – that is, their ability to influence the outcome of voting – and is calculated for voter (party) i as the ratio of the share of winning coalitions b_i in which the voter (party) is the key to the sum of coalitions in which there are key players $\sum_j b_j$.

$$\text{Banzhaf influence index}(i) = \frac{b_i}{\sum_j b_j}$$

The distribution of influence according to the Banzhaf index among voters (parties) will reflect the value of each player in the system (which, as we showed above, is not always equal to its weight). Let us consider a weighted system as an example $[4:3,2,1]$. First, we will define all the possible winning coalitions (there will be three), and then we will define the key players for each coalition (Table 2.12).

Table 2.12. Calculation of the distribution of influence according to the Banzhaf index

Winning coalition	Coalition weight	Key players
$\{P_1, P_2\}$	5	P_1, P_2
$\{P_1, P_3\}$	4	P_1, P_3
$\{P_1, P_2, P_3\}$	6	P_1

P_1 will be a key player in three coalitions $(b_1 = 3)$, P_2 will be key in one coalition $(b_2 = 1)$, as will P_3 $(b_3 = 1)$. The sum of the coalitions will equal 5: $\sum_j b_j = 3+1+1 = 5$. The Banzhaf index will be ⅗ for P_1, ⅕ for P_2, and ⅕ for P_3. A quota of $q = 4$ and a weight distribution of $[3,2,1]$ between players will produce an influence distribution of $[60\%, 20\%, 20\%]$ under the Banzhaf index. That is, while P_2 has a weight that is twice that of P_3, it has the same influence in the electoral system.

Another method for calculating influence is the Shapley–Shubik power index [456]. A feature of this index is that it takes the order in which players join a given coalition (i.e. a sequential coalition) into account, and is expressed as $\langle P_1, P_2, P_3 \rangle$, meaning that P_1 was joined in the coalition first by P_2, and then by P_3. The key player is the one that enabled the coalition to win by joining it, which means that, unlike the Banzhaf index, there will be a single key player in each case. Let us look at the same example of a weighted system $[4:3,2,1]$. First, we will determine all the winning sequence coalitions (there will be six of them), and for each coalition, we will establish the point at which the coalition wins – that is, which player joined to make it victorious. That player will be the key (or pivotal) player (Table 2.13).

Table 2.13. Calculation of the distribution of influence according to the Shapley–Shubik power index

Winning sequential coalition	Coalition weight	Key player
$\langle P_1, P_2, P_3 \rangle$	5	P_2
$\langle P_2, P_1, P_3 \rangle$	5	P_1
$\langle P_3, P_2, P_1 \rangle$	6	P_1
$\langle P_1, P_3, P_2 \rangle$	4	P_3
$\langle P_2, P_3, P_1 \rangle$	6	P_1
$\langle P_3, P_1, P_2 \rangle$	4	P_1

The Shapley–Shubik power index will be calculated for voter (party) i as the ratio of the share of winning sequential coalitions s_i in which the voter (party) is key to the sum of all winning sequential coalitions, which is effectively the factorial of all players $n!$ (the product of all natural numbers from 1 to n inclusive).

$$Shapley - Shubik\ power\ index(i) = \frac{s_i}{n!}$$

Accordingly, the Shapley–Shubik power index for the system $[4:3,2,1]$ will be equal to ⅔ for P_1, ⅙ for P_2, and ⅙ for P_3. It turns out that a quota of $q = 4$ and a weight distribution of $[3,2,1]$ between players will produce an influence distribution of $[66,7\%, 16,7\%, 16,7\%]$ under the Shapley–Shubik power index – that is, with account of the order in which coalitions were formed. This means that the influence of P_1 will grow as coalitions are sequentially formed, while that of P_2 and P_3 will go down.

Without providing full calculations here, we will note that each of the five permanent members in the UN Security Council (Russia, the United States, China, the United Kingdom and France) has a Banzhaf influence index of 16.7 % and a Shapley–Shubik power index of 19.6 %, while each of the remaining non-permanent (elected by rotation from all regions) has a Banzhaf influence index of 1.65 % and a Shapley–Shubik power index of 0.19 %. In other words, the power of veto gives the permanent members of the UN Security Council ten times more influence on decision-making in the organization according to the Banzhaf influence index, and one hundred times more influence according to the Shapley–Shubik power index when we take the order in which winning coalitions are formed into account.

2.10 Geographic Favouritism in Electoral Systems

The plurality system is often accused of being unrepresentative when seats are not distributed in a way that corresponds to real public support for various political forces. In other words, this kind of system generates a significant number of *wasted votes* (either lost or redundant). For example, in the 1998 elections in Lesotho, the Democratic Congress party received 61 % of the votes, and 89 % of the seats in parliament, while the Basotho National Party was awarded just 1 % of the seats despite gaining 24 % of the votes. As we can see, parties that enjoy significant support in society may not receive an appropriate share of seats in parliament in a relative majority system, and votes cast for them in this case are called *lost*. This often leads to a *fabricated majority*, that is, a situation where a party that won less than half of the votes receives more than half of the seats in parliament. For example, in the 2003 State Duma elections in Russia, the United Russia party won 37.6 % of the votes yet received 53.3 % of the mandates. Cases of *electoral inversion* – when the party that wins the most votes ends up losing the election – are not uncommon. This is precisely what happened in Belize in 1993, when the People's United Party won 51.2 % of the votes but was only awarded 13 of the 29 seats in parliament, whereas the United Democratic Party, which won 48.7 % of the votes, received 16 mandates. This was because the votes for the United Democratic Party were spread more evenly throughout the country, while the People's United Party won by massive margins in a limited number of districts. In this sense, the over-support for the People's United Party meant that a significant number of votes was effectively wasted – they were *redundant*.

It turns out that the plurality system suffers from *geographic favouritism*, which makes it less sensitive to changes in public opinion and more dependent on the country's geographical division into districts. Geographic favouritism reduces the transparency of the political process because it encourages parties to split the country into safe districts in which victory is preordained thanks to geography. Such a system panders to two types of political party that do not benefit from wasted (in the first case) or redundant (in the second case) votes. On the one hand, these are national-level parties whose supporters are evenly distributed across the country (i.e. there are no excess votes), rather than parties that have broad support in specific areas. On the other hand, they are territorially concentrated small regional or ethnic parties that rely on the support of voters in their ancestral lands (*regional fiefdoms*) and thus do not lose votes to other regions. The plurality system thus limits the representation of geographically dispersed minorities. Moreover, the need for the party to choose the most likely winner as its candidate in the district further reduces the chances of minority representatives winning the vote (there is ample evidence to suggest, for example, that fewer women are elected to parliament in such a system).

Semi-proportional systems frequently lead to disproportionality too, reinforcing the geographic favouritism of the plurality principle. The authorities in Mauritius have attempted to rectify this situation and support minority parties that cannot pass through the filter of block voting by introducing legislation (dubbed the "Best Loser System") that ensured multiple seats for minority candidates who received the highest percentage of votes among those who did not pass the threshold in the multi-member district. Preferential systems differ in that there are no lost votes. However, the geographic favouritism that is typical of this model benefits less polarizing centrist forces in regions where there is fierce competition between opponents at opposite ends of the ideological spectrum.

The proportional principle, as a rule, levels out the geographic favouritism of majoritarianism, as, unlike majority systems, it does not favour large parties with even territorial distribution of the electorate and small parties with stable regional fiefdoms. What is more, while majoritarian systems only benefit territorially isolated minorities, proportional systems allow small parties with territorially scattered support to enter parliament. At the same time, the proportional formula is worse suited to reflect territorially marked demands, such as those of the rural population compared to the urban population, and, accordingly, aggregates regional splits in society worse.

Table 2.14. Degree of geographic favouritism of electoral systems

Degree of geographic favouritism	Type of electoral system	Examples
Extremely high	Majority single-round	– Single-member plurality – Party-block voting
High	Majority multiple-round	– Two-round system
Above average	Semi-proportional	– Block voting – Single transferable vote – Fixed ratio vote
Average	Majority-preferential	– Instant-runoff voting
Below average	Proportional-preferential	– Single transferable vote
Low	Multiple-district proportional	– Party-list proportional representation with open or free lists – Mixed linked (compensatory)
Extremely low	Single-district proportional	– Party-list proportional representation with closed lists

It can be concluded that proportional elements reduce the possibility of geographic favouritism in electoral systems, while majoritarian elements do the opposite. Table 2.14 ranks electoral systems according to level of potential geographic favouritism: it is highest in majoritarian systems, lowest in proportional systems, and average in semi-proportional and preferential systems. The effect of geographic favouritism in mixed linked (compensatory) systems is close to that in multiple-district proportional systems. In non-compensatory systems, the effect is determined by how the majority component works.

Table 2.15. **Influence of elements and properties of electoral systems on geographic favouritism**

Strengthens geographic favouritism	Weakens geographic favouritism
The existence of regional, ethnic and religious parties	Bans on regional, ethnic and religious parties
Single-member plurality or mixed majority	Absolute or qualified majority
Single-round elections	Second ballots
Limited voting	Unlimited voting
Categorical, approval and disapproval voting	Cumulative and preferential voting
Uninominality	Plurinominality
Small multi-member districts	Single and large multi-member districts
Open, free and regional lists	Closed lists
Fixed ratio voting in majoritarian systems	Fixed electoral thresholds in proportional systems
Low electoral threshold	Medium and high electoral threshold
Single-vote	Multiple-vote
Non-parallel mixed systems	Parallel mixed systems
Linked (in terms of seat distribution or vote counting) mixed systems	Unlinked mixed systems
Compensatory mixed majority systems	Compensatory proportional systems
Hybrid majority systems	Hybrid proportional systems

Table 2.15 divides the elements or properties of electoral systems into two categories: those that strengthen geographic favouritism, and those that weaken it. Having evaluated the number of specific elements or properties present in a given country's electoral legislation, we can determine the degree of geographic favouritism of that country's electoral procedure.

There are various ways to assess the degree to which electoral systems distort public preferences. These are called *disproportionality indexes* and are based on the difference between the level (percentage) of support v_i for party i (from the total number of parties n that won seats in parliament) and the share of seats it received s_i.

The most basic of these – the Lijphart and D'Hondt indexes – simply reflect the maximum difference and the minimum ratio.

$$Lijphart\ index = max_{i=1,n} |v_i - s_i|$$

$$D'Hondt\ index = min_{i=1,n} \frac{v_i}{s_i}$$

Some more complicated versions are based on the average ratio of these parameters.

$$Rae\ index = \frac{1}{n} \sum_{i=1}^{n} |v_i - s_i|$$

$$Grofman\ index = \frac{\sum_{i=1}^{n} |v_i - s_i|}{1/\sum_{i=1}^{n} s_i^2}$$

$$Loosemore - Hanby\ index = \frac{1}{2} \sum_{i=1}^{n} |v_i - s_i|$$

$$Rose\ index = 100 - \frac{1}{2} \sum_{i=1}^{n} |v_i - s_i|$$

$$Sainte - Laguë\ index = \sum_{i=1}^{n} \frac{1}{v_i} (v_i - s_i)^2$$

$$Gallagher\ index = \sqrt{\left(\frac{1}{2} \sum_{i=1}^{n} |v_i - s_i|^2\right)}$$

A drawback of the Rae index is that it depends on the number of parties that made it into parliament, in which case the presence of small parties in parliament can distort the overall assessment [432]. The Grofman index deals with this issue by using the effective number of parties as a divisor. The remaining indexes do not depend on the total number of parties. The most accurate is arguably the Gallagher index, which is based on the standard deviation.

It is important to note that disproportionality indexes can be used to analyse more than proportional systems. In mixed systems, for instance, they can be used to assess the results of the proportional component of the system, while in majoritarian set-ups, they can be used based on the total number of votes cast for candidates from a single party.

Disproportionality indexes also allow us to calculate the average level of geographic favouritism. To do this, we need to divide the sum of the disproportionality indexes by the number of regions (or multi-member districts) being analysed. Two possible formulas for the *Geographic Favouritism Index* based on the Lijphart [369] and Gallagher [296] indexes are presented below, although other indexes can be used depending on the specific research objective [119].

$$GFI\,1 = \frac{1}{m}\sum_{n=1}^{m} max_{i=1,n} |v_i - s_i|$$

$$GFI\,2 = \frac{1}{m}\sum_{n=1}^{m} \sqrt{\left(\frac{1}{2}\sum_{i=1}^{n} |v_i - s_i|^2\right)},$$

where m is the number of regions (or multi-member districts) being analysed, n is the number of parties that won seats in a given region (multi-member district), v_i is the share of support for party i in that region (multi-member district), and s_i is the share of seats won by party i in that region (multi-member district).

The Geographic Favouritism Index is necessary for comparative electoral-geographical research, as it allows us to compare election results without having to make adjustments for distortions caused by the regional distribution of votes. The first index is better suited for analysing similar electoral processes, say, several election cycles in the same country, while the second is better for comparing fundamentally different electoral systems.

Key Terms

- relative, mixed, simple, absolute, qualified, and constitutional majority, consensus, veto
- transitivity
- electoral district (single-member, multi-member, single), voting district (local, territorial, regional)
- electoral system: plural, majoritarian, semi-proportional, proportional, uninominal, plurinominal, balanced
- multi-component electoral system: hybrid and mixed, compensatory and non-compensatory, linked and non-linked, single-vote and dual-vote, parallel and non-parallel, linked majoritarian and linked proportional, linked by seat distribution and linked by vote counting
- voting method: categorical, approval, disapproval, cumulative, preferential
- strategic (tactical) voting, electoral inversion, fabricated majority effect, spoiler effect
- wasted, lost, and redundant votes, second ballots
- limited, unlimited, cumulative and conditional voting
- electoral threshold (natural and artificial), majority bonus
- proportional voting lists (closed, open, free, regional), panachage, contamination
- geographic favouritism, regional fiefdoms
- seat distribution method: quota and divisor

Questions and Exercises

1. Determine which type of majority is used to determine the outcome of voting in the following situations:
    1.1. The decision was made at a family meeting involving a father, mother and three children that the choice of holiday destination would be agreed on if no more than one member of the family is against it.
    1.2. At the election of the head of a trade union, the decision was made that the candidate whom the owner of the company specifically votes against cannot be declared the winner.

1.3. A group of friends could not decide which movie to go and see. To resolve the issue, each member of the group named the film they wanted to see and the one that came up most often was declared the winner.

1.4. The four musketeers agreed to make decisions based on the motto "All for one and one for all."

1.5. A casino announces that the winner of the daily super prize would be the one who is ahead of the remaining players by at least 10 % in terms of total winnings at the end of the day.

1.6. A boxer is declared the winner based on the results of voting by a majority of a panel of five judges.

2. Analyse the voting results in the table and answer the questions that follow:

Place	Number of votes			
	69	55	39	33
1	Z	Y	X	X
2	Y	Z	Z	Y
3	X	X	Y	Z

2.1. What would the outcome of the election be if the single-member plurality system is used? What about if a simple majority system is used? Or a mixed majority system?

2.2. Which candidate won the most votes? Which candidate was named last on the preference list most often?

2.3. What does the final preference list look like?

2.4. Who would win in a run-off between X and Y, Y and Z, and X and Z? What does the final preference list look like in head-to-head voting?

2.5. Is the resulting situation transitive?

3. What are the advantages and disadvantages of the various electoral systems in the context of the following elections:

3.1. Secretary-General of the United Nations

3.2. The Pope

3.3. The winner of the Eurovision Song Contest

3.4. The Lord Mayor of London

3.5. The President of Sri Lanka

3.6. The Legislative Council of the Hong Kong Special Administrative Region

3.7. The Parliament of Nauru

3.8. The Council of State in Ancient Greece

3.9. The Doge of Venice

3.10. Decisions at the UN Security Council

4. Look at the map of a pirate state in Figure 4. It is home to ten pirates: nine of them live either by a lake, hills or palm trees, and one lives on a boat. Suggest how to divide the island into:

   4.1. Four voting districts, bearing in mind that the ship must be assigned to one of the districts;

   4.2. Electoral districts for presidential elections (the number of voting districts will be four; this also applies to the items below);

   4.3. Electoral districts for parliamentary elections where three deputies are to be voted in using single-member districts;

   4.4. Electoral districts for parliamentary elections where three deputies are to be voted in using the majoritarian system, with both single-member and multi-member districts;

   4.5. Electoral districts for parliamentary elections where four deputies are to be voted in using several multi-member districts;

   4.6. Electoral districts for parliamentary elections where four deputies are to be voted in using the proportional system;

   4.7. Electoral districts for parliamentary elections where four deputies are to be voted in using a mixed system, where two deputies are voted in using the majoritarian system, and two are voting in using the proportional system;

   4.8. Electoral districts for parliamentary elections where four deputies are to be voted in using a hybrid system.

Source: https://mentamaschocolate.blogspot.com/2015/12/dibujos-de-piratas-y-sus-elementos.html

5. Match the following systems to the countries in which they are used:

(1) Two-round system
(2) Block voting
(3) Hybrid system
(4) Single non-transferable vote
(5) Single transferable vote
(6) Instant-runoff voting
(7) Single-member plurality
(8) Party-list proportional representation
(9) Mixed non-linked system
(10) Mixed linked system

(a) Australia
(b) United Kingdom
(c) Germany
(d) European Parliament
(e) Ireland
(f) Spain
(g) Russia
(h) France
(i) Switzerland
(j) Japan

6. Compare the two most common electoral systems: plurality and closed-list party list. What are the advantages and disadvantages of each? How do the mechanisms for offsetting the costs work in each case?
7. Which electoral systems are most likely to produce the following types of party systems: majority party system, two-party system, moderate pluralism, atomized system?
8. Trace the evolution of the electoral system for electing deputies to the State Duma in the Russian Federation. What prompted changes to the system? What consequences did this have?
9. Assess the impact of the electoral threshold (or majority bonus) on the outcome of the following elections:

    9.1. the 2021 elections in Germany

    9.2. the 2020 elections in Slovakia

    9.3. the 2015 elections in Greece

    9.4. the 2006 elections in Italy

    9.5. the 2002 elections in Turkey

    9.6. the 1995 elections in Russia

    9.7. the 1988 elections in Israel.

10. Is there a geographical pattern to the use of the mixed majority electoral system in the world? What might explain this?
11. Analyse the arguments of those for and against electoral reform in the following cases:

    11.1. The 2011 referendum in the United Kingdom on the transition to an instant-runoff voting system.

    11.2. The 1992, 1993 and 2011 referendums in New Zealand on the transition to a mixed linked proportional system.

    11.3. The 1958 and 1968 referendums in Ireland on the transition to a single-member plurality system.

    How would the electoral landscape in each of these countries have changed if the outcomes of these referendums had been different?

12. Why, and to whom, are votes not transferred in the single non-transferable vote system? Why, and to whom, are votes

transferred in the single transferable vote system? What are the similarities and difference between these systems?

13. Assess the impact of contamination on elections in one of the parallel mixed electoral systems. How can it be calculated?
14. Model a three-component, multi-tiered system that includes majoritarian, proportional, linked and unlinked elements. What tasks can such a complex electoral system solve?
15. Allocate the seats from the assignments in Tables 2.9 and 2.11 using the Droop and Sartori quotas and the Imperiali, Sainte-Laguë and Huntington–Hill divisor methods. What changes?
16. Allocate 450 seats in the State Duma among the constituent entities of the Russian Federation in accordance with the Imperiali quota and the D'Hondt and Adams divisors. Describe the advantages and disadvantages of each method.
17. Research how the distribution of seats among EU countries in elections to the European Parliament has changed over the years. Did any redistribution paradoxes appear?
18. Analyse the impact of geographic favouritism on the results of the following parliamentary election campaigns:

    18.1. the 2019 elections in Canada

    18.2. the 2015 elections in the United Kingdom

    18.3. the 2010 elections in Mauritius

    18.4. the 2010 elections in Japan

    18.5. the 2009 elections in Germany

    18.6. the 2003 elections in Djibouti

    18.7. the 1990 elections in Australia

    18.8. the 1958 elections in France.

19. Select two electoral cycles in the same country that used the proportional system and calculate the Rae and Gallagher disproportionality indexes for them. What do the results indicate?
20. Calculate the Geographic Favouritism Index for two countries (or two electoral cycles in one country). Compare the dynamics of change of the disproportionality indexes for each region within the country. What do the changes indicate?

Chapter 3
Territorial Differentiation of Party Systems

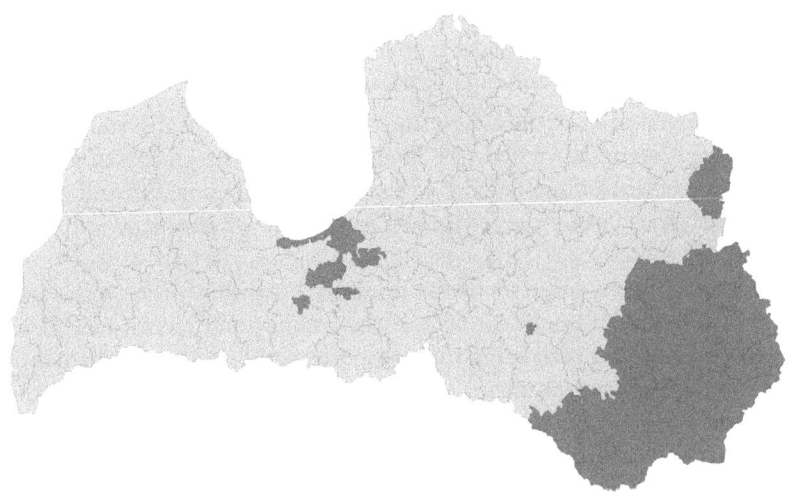

Source: Okunev I. (ed.) *Electoral Geography of Russia's Neighbourhood*, MGIMO University Press, 2024.

- What types of party are more susceptible to territorial differentiation?
- Which variables structure ideologies in the electoral field? Is the "left" always on the left and the "right" always on the right?
- How do centrist, moderate and radical parties differ in their electoral strategies?
- How can we qualitatively and quantitatively assess the magnitude of ideological and political splits in society?
- What historical territorial divisions are reflected in the ideological and political splits in society? What new types of division are characteristic of the modern political process?

3.1 Parties and Their Territorial Differentiation

The key organizing element of the electoral landscape is the *political parties* for which the people vote and which are the conductors of their will in the political process. Here, a party is understood as a sociopolitical association of people striving for power, primarily through participation in elections. It is assumed that this group is united not only by the desire of the people who make it up for power, but also by civic values, a programme of action and organizational discipline.

Political parties perform key functions in the political system. First, they aggregate and articulate the interests of various segments of society, while at the same time helping to shape public opinion. Second, when in power, they ensure the implementation of the government course on behalf of the majority, and when in opposition, they strive to correct this course in the interests of the minority. And third, they transform the ideological and political field of the country and recruit new elites by mobilizing and socializing their supporters and integrating them into politics and state or municipal government. It should be noted that the existence of political parties is not a prerequisite for the political process – many non-democratic countries either do not have political parties at all (they are generally prohibited by law in the monarchies of the Persian Gulf), or non-party candidates receive the majority of seats in parliament (in Belarus, for example). At the same time, it is not uncommon to see *non-party* democratic systems. This often happens in small states or territories where there is no significant fragmentation of society (Nauru, Palau, and some Canadian territories).

Political parties can be classed in different ways based on their structure, ideological platform and electoral tactics [67, 128]. For the purposes of this textbook, it is more important to separate them according to whether or not they are prone to territorial differentiation, by which we mean the uneven spatial distribution of the electorate and thus the party structure and its activity during election campaigns (Table 3.1).

The genesis of formal public associations vying for power can be traced back to the so-called *elite (or cadre) parties*. These parties lacked organizational and territorial structure, and had no members or supporters as such. Such parties grew out of political clubs, where a closed group of likeminded *notables* (members of the aristocracy or, more rarely, industrialists, financiers, landowners, high-ranking military officials, members of the clergy, etc.) formalized the political structure that was already in power in the country and, in the course of democratic procedures, legitimized it as representative of the will of the people. But this kind of elitist democracy was nothing more than sham democracy – power remained in the hands of the same elite, but the appearance had been created that the people had delegated power to them.

Table 3.1. Territorial differentiation of types of political party

Type of party	Prone to territorial differentiation	Not prone to territorial differentiation
Elite (cadre)	Clientelistic parties	Committee-type parties (clubs)
Mass	Pluralist branch-based parties	Proto-hegemonic cell-based and militia-based parties
Identity	Regional and ethnic parties	Parties of non-localized minorities
Electoralist	Programme parties	Catch-all parties
Post-modern	Movement parties	Personalist parties

Elite parties can be divided into two categories according to how prone they are to spatial differentiation. Classical cadre parties that are organized like committees (clubs), act as the governing core for the duration of an election campaign only and convince the people that those who are already in power truly represent them are, of course, not susceptible to territorial differentiation, as they are not mediated by public sentiments in any way. Examples of such parties are the Hats and Caps in Sweden in the 18th century, or the Whigs and Tories in Great Britain in the 19th century. If we reach a little, we can probably include the present-day People's Action Party in Singapore in this list. With that said, the modern *clientelist parties* that have emerged from these clubs are a different beast entirely. These parties are based on quasi-feudal patron–client relations between the elite and its electorate. For example, factory workers will be inclined to vote for the factory owner, and peasants will tend to vote for their landowner – not because of any kind of ideological affinity, but rather because their livelihood depends on the success of their patron. Such parties are extremely susceptible to territorial differentiation, because the area of support is directly tied to the geography of the patrons' economic activity, and the place where the clientele of the given elite group lives. The political machines of the 19th and early 20th centuries in the United States are a classic example of clientelism, as they created a system in which poorer social groups were dependent on their exploiters being in power for their survival.

Mass-based parties were a new kind of political entity that appeared in the late 20th century as a direct challenge to (and polar opposite of) cadre-based political clubs. They came about as a result of the increased political engagement of new social classes – primarily labourers and peasants. The activities of these parties extend beyond the pre-election period, as the tasks of rallying support and representing the interests of those supporters mean that they have to operate even in the periods between elections. Consequently, they require institutionalized mass membership and an extensive organizational and territorial structure. Mass-based parties have a hierarchy of party administrations, grassroots cells, trade unions, religious organizations, youth movements, and other units.

Mass-based parties are also divided into two subgroups depending on how susceptible they are to territorial differentiation. Mass *pluralist parties*, which have an extensive hierarchical structure of grassroots cells based on territoriality, are prone to such differentiation. The first mass-based parties can be described as pluralistic: their scale and heterogenous constituency encouraged these parties to try to embrace as much of the public political discourse as possible, moving it to the party itself and transforming it into a debate between its sections and organizational units. In this sense, mass pluralist parties are extremely susceptible to the influence of the territorial heterogeneity of political views in society, because such differentiation is reflected in their internal party debates, and this ultimately affects the electoral platform of the movement. The Christian democrat parties in Western Europe are a textbook example here.

On the flip side, we have mass *proto-hegemonic parties*, which are not prone to territorial differentiation. Here, the relevant ideology is declared; however, the mass nature and extensive reach of the network are not used to represent the interests of society, rotate staff and develop ideas, but rather to mobilize a broad segment of supporters and build a rigid hierarchy of control over the implementation of the party's ideals. In such parties, sections formed on the basis of territoriality are replaced by *cells* formed at enterprises and institutions that exert greater control over various aspects of voter behaviour. In extreme cases, such parties may become overgrown with a network of *militias* – grassroots paramilitary groupings that resort to coercion tactics. Examples of such parties are the fascist and Nazi movements in Italy and Germany, and it goes without saying that the lack of diversity of opinions meant that significant territorial differentiation in the level of support was impossible.

There are also parties that base their programmes not on an ideological principle, but on the identity of their electorate. These are called *identity parties*, and they appear as a response to the needs of various minorities, for whom ideological differences are secondary to the need to come together to protect their rights. Obviously, regional parties that represent the interest of localized minorities (ethnic, racial, religious, cultural) or an isolated region (for example, the Northern League in Italy) will be extremely susceptible to territorial differentiation. Identity parties do not always represent minorities. For example, *congress parties*, which are also based on common ethnicity and/or faith, often form around a majority group in a country (for example, the Indian National Congress, the Kenya African Union, and the National Front in Malaysia). And, conversely, parties representing non-localized minorities (evangelicals in Costa Rica, emigrant Muhajirs in Pakistan) are less susceptible to the influence of the territorial factor.

The types of political party described above are increasingly being replaced today by new types of political associations, namely, electoralist and post-modernist parties. *Electoralist parties* are less concerned with enticing the electorate over to their side as they are with trying to predict the general mood of voters and tailoring their programmes accordingly. They are interested in electoral success, rather than promoting values and consolidating voters around these values, and thus attempt to offer an ideological compromise that the majority of voters would agree with. What this means in practice is that the ideological core in electoralist parties is constantly being blurred (left to right, and right to left), eventually making them ideologically centrist. Such parties can essentially be divided into *programme parties*, which try to offer the broadest possible compromise in their political programmes (such as the Republican Party in the United States), and *catch-all parties*, which build their campaigns from a position of opposition, by offering the exact opposite to what the programme party is offering, with slogans such as "we are not the left," "we are not the right," etc. (for example, the Democratic Party in the United States). Electoralist programme parties are slightly more susceptible to territorial differentiation, as they more clearly reflect the social divisions in society.

Finally, we have the most recent innovation – post-modernist parties. These are primarily *movement parties*: environmental (green), anti-migrant, libertarian, right- or left-populist, etc. Such groupings do not compete with classical parties on ideological grounds. Rather, they organize their political activity around key social problems, rallying support based on public sentiment regarding these issues. They are also extremely susceptible to territorial differentiation, since the level of engagement of voters with these issues varies greatly from region to region. Less prone to territorial differentiation are so-called *personalist parties*, which, as the name suggests, try not to associate themselves with a specific ideology and instead rally support based on their leader (as the Liberal Democratic Party of Russia did, for example, with Vladimir Zhirinovsky).

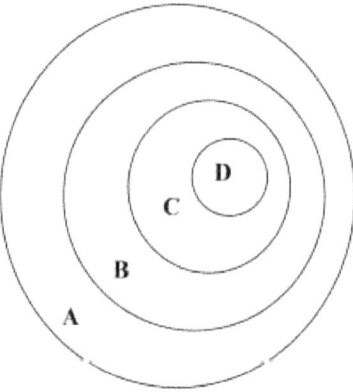

Fig. 3.1. Groups of influence within political parties
Key: A – the electorate; B – supporters; C – members; D – functionaries.

Maurice Duverger proposed a successful model for differentiating political associations according to how involved various groups of influence are in their activities, and this gives us some insight into the reasons behind the evolution of modern political parties (Fig. 3.1) [53]. Cadre parties were typically made up of functionaries and a few members only $(D \to C)$. While this made them maximally flexible politically, it created the risk of being cut off from the fluctuations in public sentiment that were not represented within the party, and opportunities for recruiting new candidates was significantly reduced. Conversely, mass-based parties maximized the number of members and supporters in an attempt to attract the entire electorate $(B \to A)$, which made them both extremely sensitive to public sentiment and extraordinarily inflexible, since any outward action required a broad discussion within the huge party organism. Both models ultimately failed, and this is what led to the development of the types of political party we see today, which are made up of a main body of members supplemented by a broad range of active supporters, not attempting, however, to cover the entire electoral field $(C \to B)$. This makes it possible to recruit new leaders and quickly change the political agenda in response to changes in society. The same logic applies to the analysis of the territorial network of parties, where both extreme strategies – having a party headquarters in the centre with individual representatives in the territories, and running a broad and highly bureaucratized network of branches and cells – turn out to be ineffective.

The electoral process pushes political parties to seek coalitions. Parties may come together in one of two cases: (1) to pass the electoral threshold; and (2) to create a common ruling platform. There are several types of party association, and they should be distinguished from party systems:

- *Bloc* – an asymmetrical association dominated by one party. These can be territorially differentiated, where parties from different regions of a country join forces (for example, the CDU/CSU faction in Germany).
- *Coalition* – a symmetrical association of parties where neither dominates (such as the Civic Coalition in Poland).
- *Alliance* – an informal union of parties that have agreed to cooperate in an election or political process (such as the centre–right alliance in Italy).

Once in parliament, parties or associations form factions, and the party with an absolute majority becomes the *ruling* party. Party coalitions are vital in proportional systems where no party holds a simple majority and a *majority government* is thus formed. The factions that have joined forces to form a government draw up a coalition agreement, and this document is used as the programme of the new *minority government*.

3.2 Regional and Regionalist Parties

Regional parties are a subset of identity political associations in which the emphasis is not so much on ideological tenets as it is on appealing to the community of voters. At the same time, in regional identity parties, this community means belonging to a certain territory and to the general identity of the electorate formed within it. What this means is that these associations are by their very nature susceptible to territorial differentiation – it is one of their defining characteristics. Focusing political activity on a specific territory is typical of ethnic parties that are united by their belonging to a specific geographic area (such as the Party of the Hungarian Community in Slovakia), but can also be characteristic of certain religious (the Catholic Sinn Féin party in Northern Ireland), linguistic (such as the Valdostan Union party in Italy, which represents the interests of the Arpitan-speaking minority), emigrant (parties that focus their efforts on representing the interest of immigrants from the Soviet Union in Israel), and even cultural (for example, the Community of the Lipovan Russians in Romania) regional parties. However, there are also regional parties that cultivate a local identity which is not burdened by ethno-cultural identity (such as the Free Party Salzburg in Austria). There are also ethnic parties with a weak regional component: among territorially dispersed minorities (ethnic minorities, such as the Party of the Roma in Romania, or religious minorities such as the Africa Muslim Party in South Africa); or among ethnic majority parties (the Indian National Congress).

It is worth noting that regional identity parties are effectively an operational mode of a political alliance. For example, the Scottish National Party, acting as a regional party in UK (national) elections, is an electoral programme party in local elections in Scotland. Moreover, regional parties can form a local party subsystem in elections within their region (in the Flemish or Walloon parts of Belgium, for example, where each ethnic group has regional parties with different ideological bases). Also, ordinary parties of EU countries act as regional parties in their political activities in the European Parliament, representing the interests of one country.

Obviously, regional parties build their political platforms around protecting the interests of the given region or minority [123]. These interests typically fall into the following areas:

1. Recognition of a separate local community and the need to preserve its identity.
2. Development of measures to protect and promote the cultural identity of the local community.
3. Legalization of the social activity of the local community in a manner and form in which the community will understand and appreciate, for example, in the native language.
4. Ensuring that the local community is engaged in the political process as a separate force.
5. Expanding the economic independence of the local community.
6. Achieving the transfer of political powers to the local community, which means giving it the right to legislative, executive and judicial activities within the framework of the competencies that have been transferred.
7. Endowing the local community with the right of self-determination.

Political associations that build their platforms around the ideas of self-determination and separatism are considered a separate subgroup of *regionalist parties*. The desire for independence is why regional and ethnic parties are banned in many countries (for example, in Russia) as dangerous to their unity.

3.3 Spatial Typology of Party Systems

The term *party system* refers to a set of political parties operating in a single electoral field. A country with a stable electoral tradition, transparent elections and low disproportionality of the electoral formula will ideally have a party system that reflects all the significant ideological and political divisions in society. In other words, the party system reflects the structure of society and all of its stable ideological and political clusters.

The typology of party systems is based on identifying the number and determining the weight of key nuclei in the electoral field under consideration. Based on this, the ideal types of party system are one-, two- and multi-party. These types are, in fact, continuous rather than discrete. This was best demonstrated, in our opinion, by the Russian political scientist Grigory Golosov, who created a spatial continuum typology of party systems [41]. In this typology, the position of each party system is determined by the coordinates x and y in two-dimensional space. The coordinates of the party system in the continuum are calculated using the formula:

$$x = \frac{s_2 + s_r}{s_1 + s_r}, y = \frac{s_3 + s_r}{s_1 + s_r}$$

where s_1, s_2 and s_3 are the shares of seats won by the top three parties, and s_r is the total share of seats won by all the other parties. Since $s_3 \leq s_2$, then $y \leq x$, which means that the continuum of party systems will look like a triangle in which the points represent the ideal types of one-party (A, $x = 0, y = 0$ – all the seats belong to one party), two-party (C, $x = 1, y = 0$ – there are only two parties and they have an equal number of seats), and multi-party (B, $x = 1, y = 1$ – there are more than two equal parties) systems. The sides of the continuum are, respectively, the hypotenuse $y = x (AB)$ and the two legs: the horizontal line (AC) and the line $x = 1 (BC)$.

The CE bisector divides the continuum in half in such a way that only party systems with a party that holds an absolute majority will be below it, and systems without such a party will be above it. Accordingly, this bisector divides two-party systems that tend to $\angle ACB$ into two-party systems that have a propensity for one party to dominate (*two-party monovalent*) (in $\angle ACE$), and two-party systems that have a propensity for political multiplicity (*two-party polyvalent*) (in $\angle ECB$).

The median line AF also divides one-party systems that tend to $\angle BAC$ into one-party systems that have a propensity for bipartism (*one-party bivalent*) (in $\angle FAC$), and one-party systems that have a propensity for political multiplicity (*one-party polyvalent*) (in $\angle BAF$). Finally, the median line BD splits multi-party systems that tend to $\angle ABC$ into multi-party systems that have a propensity towards one-party domination (*multi-party monovalent*) (in $\angle ABD$), and multi-party systems that have a propensity for bipartism (*multi-party polyvalent*) (in $\angle DBC$). As a result, the centroid G, which is the intersection of the bisector CE and the median lines AF and BD, divides the continuum into six equal segments, each of which describes one of the six types of party system.

The following is a classification of party systems, including their ranges on the continuum, with examples of the distribution of seats among four parties:

1. One-party (*AEFG*):
   a. mono-party (point A, 100 – 0 – 0 – 0),
   b. one-party bivalent (ΔAGD, 80 – 15 – 3 – 2),
   c. one-party polyvalent (ΔAEG, 65 – 15 – 15 – 5);
2. Two-party (*CDGF*):
   a. two-party monovalent (ΔCDG, 60 – 35 – 3 – 2),
   b. two-party polyvalent (ΔCGF, 45 – 40 – 10 – 5);
3. Multi-party (*BFGE*):
   a. multi-party monovalent (ΔBGE, 45 – 25 – 25 – 5),
   b. multi-party bivalent (ΔBFG, 35 – 30 – 25 – 10).

The main characteristic of one-party systems is that a single party consistently plays the leading role in the political process. At the same time, such systems differ significantly from one another. *Mono-party* systems (sometimes simply called one-party systems) consist of a single party only, any other parties are officially or unofficially banned. Today, such systems exist in China, North Korea, Cuba, Vietnam, Laos and Eritrea, where the electoral procedure does not offer an alternative. One-party systems with alternative choices are divided into hegemonic and dominant. *Hegemonic systems* allow alternative parties, although they do not really oppose the parties in power and can never claim victory in an election (such a system existed in Mexico in the second half of the 20th century). The purpose of such systems is to create the impression that discussions are being held with small but active segments of society. Unlike mono-party and hegemonic systems, which can only exist under non-democratic regimes, *dominant systems* may indeed involve competitive elections. Such systems develop over a long period of rule by the dominant party, which proves capable of responding to the broadest range of public sentiment (examples include the Swedish Social Democratic Party in 1932–2006 and the Liberal Democratic Party in Japan in 1955–2009). One-party systems with alternatives are, more often than not, polyvalent, meaning that they have several weak parties (for example, India in 1952–1984), although bivalent systems – where the ruling party has one main (but weak) opponent – are not uncommon (such was the situation is South Africa in 1948–1989).

Two-party systems, as the name suggests, involve two leading political parties which spend time both in power and in opposition. These systems typically use the majoritarian principle in elections, especially when a plurality formula is utilized. Two-party systems are common in English-speaking countries – the United Kingdom, the United States and Canada, for example, all have a two-party system. Importantly, a two-party system does not mean that other parties are banned from carrying out political activities. What usually happens is that minor parties are forced to play the role of spoilers. That is, the system is effectively a two-party monovalent one, although there are cases where a third party with strong support emerges. In these cases, the system is referred to as a two-and-a-half-party system, and on the continuum, it corresponds to two-party polyvalent systems (such a system exists, for example, in Australia).

Multi-party systems typically function in democratic regimes with proportional electoral systems. Here, there are a number of leading parties with roughly the same level of support, none of which can win an absolute majority on their own. Accordingly, they are forced to seek coalitions with other parties in order to form a government. In multi-party monovalent systems, one party finishes ahead of the rest, but does not have an absolute majority (this is the situation in France). Multi-party bivalent systems differ in that there is no clear leader (as happens in Brazil). Giovanni Sartori separated multi-party systems into those with limited pluralism (four to five parties), extreme pluralism (six to eight parties), and atomized (nine or more parties) [447]. It is clear that more parties equals greater fragmentation of society, and in such a situation the ability of the party system to effectively run the country diminishes.

3.4 Effective Number of Parties

Up until now, we have been talking about the qualitative assessment of the configuration of party systems. Calculating the effective number of parties in a given system is the result of a quantitative assessment of the degree of fragmentation of the political system. Doing so allows us to identify how many significant parties are involved in the political process or, in other words, how many stable ideological segments have developed in society under a functioning electoral system. The original formula for calculating the effective number of parties index was proposed by Markku Laakso and Rein Taagepera [360]:

$$N_{LT} = \frac{1}{\sum_{i=1}^{n} s_i^2},$$

where n is the nominal number of parties, and a s_i is the share of the i-th party in elections or in the legislature. When $N \to n$, the parties in the elections or in parliament are equally significant. When $N < 1.8$, the party system is considered to be one-party; $1.8 \leq N \leq 2.4$ means a two-party system, and $N > 2.4$ denotes a multi-party system. Obviously, the number of effective parties can vary at the national, regional or district levels, meaning that territorial differentiation can also change from territory to territory.

Thus, the effective number of parties can be calculated both for the electoral process (when it is taken from the number of parties participating in the elections, called the *effective number of electoral parties, ENEP, N_E*), and for the legislative process (when it is calculated from the number of factions in parliament, called the *effective number of parliamentary parties, ENPP, N_P*). $N_E = N_P$ when the party system does not distort the will of society. However, we know that such a situation is unlikely, and that $N_E > N_P$ is more typical of majoritarian systems. Rein Taagepera and Mathew Shugart came up with the following criteria for assessing the disproportionality of electoral systems:

– criterion of the absence of absolute reduction $N_E - N_P = 0$

– criterion of the absence of relative reduction $\dfrac{N_E - N_P}{N_E} = 0$

There are other ways to calculate the degree of fragmentation of the party system that yield more accurate results. The index proposed by Juan Molinar, for example, is more accurate when it comes to distinguishing between one- and two-party systems and how dominant the leading party is [392]. Using the index, the winning party is immediately assigned a unit, while only the degree of fragmentation of the remaining field is assessed.

$$N_M = N \left(1 - \frac{s_i^2}{\sum_{i=1}^{n} s_i^2} \right) + 1$$

This problem is overcome using an additional index proposed by Taagepera as a supplement to the main effective number of parties calculation. For parties with $s_i > 0.5$, the level of dominance of the leading party is calculated as $N_T = 1/s_i$. In this case, the level of dominance will be within the range $1 \leq N_T < 2$.

Another option for calculating the effective number of parties that is gaining recognition for its accuracy is the Golosov index [42].

$$N_G = \sum_{i=1}^{n} \frac{s_i}{s_i + s_1^2 - s_i^2}$$

For more mature party systems where votes are distributed relatively evenly, the figures that this index produces will not differ greatly from those of the Laakso–Taagepera model. However, it is more accurate for analysing new democracies and electoral autocracies, where votes are distributed far less evenly.

3.5 Nationalization and Regionalization of Party Systems

In a normal situation, electoral choice tends towards territorial differentiation, for the simple reason that people in different regions are interested in different issues. However, this aspect of the regionalization of the electoral field is violated by the centripetal forces of nation- and state-building. Centralization and the creation of nation-states support the unification of political preferences. Territorial differences fade into the background when there are disputes across the country about the correct development path. This aspect of the nationalization of the electoral landscape leads to the weakening of territorial differences within states and the strengthening of such differences between countries. As a result, significant electoral splits can be observed along state borders, where people in one region that is divided between several countries (for example, Alsace) who should be concerned about similar issues actually belong to significantly different electoral and political systems.

Daniele Caramani used two criteria (the level of electoral support for national parties and the presence of significant regional political associations) to identify four types of party system in terms of the degree of their nationalization (Table 3.2) [234]:

- *nationalized* party systems with homogenous support for all political parties, and the absence or extreme weakness of regional parties (Greece, Denmark, Iceland, Norway);
- *territorialized* party systems, in which there are also no strong regional parties, but, due to social divisions, there is uneven territorial support for the leading parties (the United Kingdom, France, Switzerland);
- *segmented* party systems, which are characterized by the presence of at least one significant regional party with homogenous support from other political forces (Italy, Germany);
- *regionalized* party systems, which are characterized both by territorially uneven support for national parties and by the presence of significant regional associations (Belgium, Finland).

Table 3.2. **Types of party systems by level of nationalization**

		Support for national parties	
		heterogenous	homogenous
Significant regional parties	Yes	regionalized	segmented
	No	territorialized	nationalized

We should note here that the process of nationalizing the electoral field is not necessarily a continuous process that takes place first in the centre and then concentrically outwards in stages. It can take place in fits and starts, similar to the process described in Torsten Hägerstrand's theory of spatial diffusion of innovations. According to this theory, the spatial hierarchy has both a territorial structure (the farther from the centre, the stronger the periphery) and a functional structure (the lower the territory's ranking, the stronger its periphery). Patterns that exist in the centre can thus be distributed according to territorial structure (*continuous diffusion* – from neighbour to neighbour), or according to functional hierarchy (*hierarchical diffusion* – from the capital to the main agglomerations, then to major cities all the way to small villages).

Finally, these two types can be combined into what we call *wave* (or *cascade*) *diffusion*. An example of this would be an innovation spreading from Moscow along the first wave both territorially to Moscow Region and functionally to St. Petersburg and then, along the second wave, territorially across the Central Federal District and functionally to Yekaterinburg and Novosibirsk.

There are a variety of methods for assessing the degree of territorial differentiation of a given party system [160]. The simplest are the so-called competition indices, which determine the share of constituencies in which a party participates. The territorial coverage index proposed by Caramani does this through the ratio of the number of constituencies in which a party has put a candidate forward to the total number of constituencies in the country [234]. The Rose–Urwin competition index measures the degree of nationalization of a party in terms of the number of seats it does not contest by not putting any candidates forward, while the Cornford Index measures the degree of competition in terms of the number of "safe" seats, that is, uncontested districts.

Another option is variation indices, which evaluate how a party's results in regional elections deviate from those in national elections. The Rose–Urwin index, for example, is calculated using mean absolute deviation – that is, the sum of deviations from the national percentage for all regions of the country (regardless of whether these deviations are positive or negative) and is determined for each party separately [443]. A variation of this is the Lee index, in which the denominator is always 2. The Russian researcher Alexei Zimokha offered his own variation, which he called the regional coefficient of heterogeneity [63]. The Caramani variability index would be more accurate here, as it is not sensitive to party size and the number of regions. A simple index that uses standard deviation would also do the trick.

$$Rose-Urwin\,variation\,index = \sum |s_i - s|/n$$

$$Lee\,variation\,index = \sum |s_i - s|/2$$

$$Zimokha\,variation\,index = \frac{\sum s_i/s - 1}{n}$$

$$\text{Caramani variability index} = \sqrt{n \frac{\sum |s_i - s|}{2(n-1)\sum s}}$$

$$\text{Standard deviation variation index} = \sum (s_i - s)^2 / n$$

where $s_i - s$ is the difference between the share of votes for a party in region i and nationally, and n is the number of regions.

Somewhat similar to the concept of a variation index is the territorial distribution index proposed by Panciano, which returns to the idea of estimating cumulative regional inequality developed by Rose and Urwin.

$$\text{Territorial distribution index} = \sum |s_i / s - v_i / v| / 2$$

where v is the total number of votes cast in the country, and v_i is the number of votes cast in region i.

Pradeep Chhibber and Ken Kollman developed a rather unique way to assess the level of party aggregation that involves calculating the average effective number of parties at the regional level (ENP_i in region i, divided by the number of regions n) from the effective number of parties at the national level (ENP).

$$\text{Party aggregation index} = ENP - \frac{1}{n}\sum_i ENP_i$$

Finally, the opposite approach – identifying the degree of regionalization of the party system and the presence of typical and deviant regions within it – is also possible. This can be done by calculating the Euclidean distance between the region being analysed and the typical statistical average.

$$d(x, y) = \sqrt{\sum_{k=1}^{n}(x_k - y_k)^2}$$

where x is the share of votes for party k in region n, and y is the arithmetic mean of the regional percentage of the party in the country.

The Euclidean distance can be corrected by dividing the resulting value by the number of parties involved.

3.6 The Ideological Spectrum of the Electoral Field

One of the key elements of political choice for voters are their ideas about how and in what direction the community should develop. A systematically organized set of such ideas is called an *ideology*. And it is ideology that is the key parameter for most political parties, and it is what determines their place in the electoral field.

Up until the end of the Enlightenment, ideologies were predominantly the subject of philosophical debate, the belief being that the ideal path for the development of society could be thrashed out in an academic setting. However, at the end of the 18th century, the two most developed political and philosophical ideological currents of the time arrived at a question that humankind cannot answer. We are creating an ideal society of people, but are people themselves born ideal? If people are born ideal, a notion that serves as the basis of *liberalism* (more precisely, the liberal humanism that came before it), then the programme for building the perfect society should, within limits, aim to ensure that these wonderful people have complete freedom of action (let us call this position *individualism*). The proponents of the other leading trend – *conservatism* – believed that we are born imperfect, and it is the public institutions of morality, tradition and culture that are capable of unlocking the beauty within us. And the only way to build a better society is to protect and develop these public institutions (this position can be called collectivism, or, more precisely, *communitarianism*). The fact that there is no answer to this question led these ideologies to drift away from philosophy and towards politics, and the task of creating the future no longer falls to academics and thinkers, but to voters and the elites.

The division into "left" liberalism and "right" conservatism defined the leading differentiating axis of the *ideological spectrum*, which in turn configures the electoral field. However, a second axis appeared in this spectrum (or even a second dimension, in which case we can refer to it as an ideological *compass*). The idea of a left-right divide in politics came about during the French Revolution, when supporters of liberal reforms sat in parliament on the left, and followers of conservative principles sat on the right. This, unfortunately, led to confusion with the rather unscientific use of the concept of "liberal" to describe those who support progress and are thus reformist, and "conservative" to describe those who defend tradition and are thus reactionary. Politically speaking, these designations are nonsensical: if, for example, the liberals are in power for a long time, then the conservatives in opposition are the ones who will be calling for reform and progress.

In the 19th century, the political (and not only philosophical) debate between liberals and conservatives about how best to achieve the ideal society has moved, first of all, from questions of individual and collective rights and freedoms to economics, which has only confused people even more. Initially, liberals believed that an optimally functioning economy could only be achieved if the government had as little involvement in it as possible (classical economic liberalism), while conservatives were convinced that state regulation would ensure the preservation of the necessary social order and tradition (*mercantilism*). However, a different alignment came to dominate in the 20th century: left-wing ideologists now believed that the state was crucial to building the optimal society, providing, through the redistribution of resources, equal opportunities to all citizens (the notion of equality, central to *egalitarianism*), while right-wing politicians started to criticize state intervention in the economy as it undermines the main driver of evolution that is necessary to motivate people to evolve – competition between people and segments of society (competition, as opposed to equality, is at the heart of the concept of *elitism*). As a result, a thoroughly confusing situation developed during the last century whereby classical economic liberalism became the core of the ideological platform of modern conservatives.

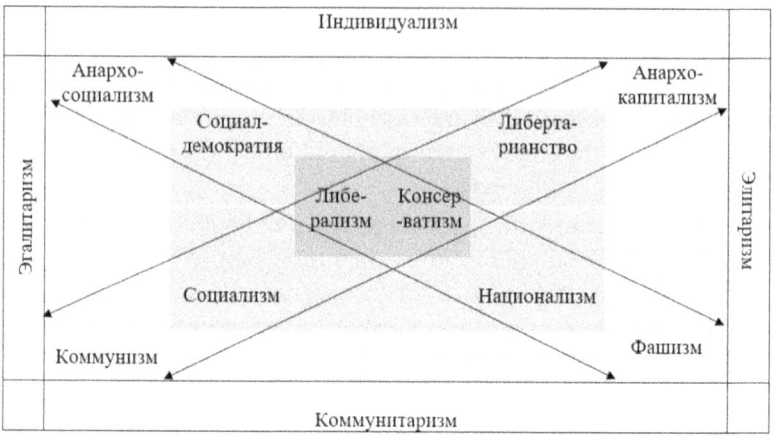

Fig. 3.2. The ideological spectrum of the electoral field

Dark grey indicates centrist ideologies; grey indicates moderate ideologies; and light grey indicates radical ideologies

Эгалитаризм	Egalitarianism
Индивидуализм	Individualism
Элитаризм	Elitism
Коммунитаризм	Communitarianism
Анархо-социализм	Social anarchism
Коммунизм	Communism
Анархо-капитализм	Anarcho-capitalism
Фашизм	Fascism
Социал-демократия	Social democracy
Социализм	Socialism
Либертарианство	Libertarianism
Национализм	Nationalism
Либерализм	Liberalism
Консерватизм	Conservatism

Another factor in the transformation of the ideological attitudes of liberalism and conservatism in the 20th century was essence of the political process. As the debate over the future of society gradually moved from the offices of academics and philosophy clubs to the polls, it turned out that the desire for electoral victory led to significant ideological changes.

In 1948, Duncan Black proved the median voter theorem, which proves that the electoral process is set up in such a way that the optimal strategy for winning elections is to offer a programme that is closest to the position of the median voter [213]. The median voter denotes the *centre of the ideological spectrum*, and it is their vote that the liberal and conservative parties fought to win, as this was a guaranteed path to victory. This is why they came to be called *"centrist"* parties. But it turned out that in order to win over the median voter, the left had to make a small compromise by adopting certain conservative ideas (that is, they needed to "move to the right"), while the right had to allow certain liberal ideas in (to "move to the left"). After many electoral cycles of liberals "moving to the right" and conservatives "moving to the left," these centrist parties became almost indistinguishable from one another. They formed the core of the ideological spectrum and took turns leading the country, professing extremely close principles and arguing on certain pressing issues only. This encapsulates the key feature of the democratic political process – in a normal situation, the electoral process prompts political forces to abandon their original ideologies in favour of a compromise, which comes about through an open, broad and ongoing discussion.

The formation of this liberal-conservative core of the ideological spectrum has given rise to new types of ideology that we call *moderate*. Figure 3.2 clearly shows how the spectrum expands from the core in two dimensions: on the left, the combination of egalitarianism and communitarianism gave rise to *socialism* (a term that in this case refers to a party ideology, rather than a command economic system), and egalitarianism and individualism created *social democracy*. Meanwhile, on the right, the combination of elitism and communitarianism led to the development of *corporatism*, in which tradition and social order are maintained through the support of stable social groups. Since the most stable groups are nations, this led to *nationalism* becoming the most viable form of corporatist ideology. Finally, the combination of elitism and individualism created the ideology of *libertarianism*, which is closest to classical economic liberalism, that is, support for capitalism as an economic system and its main quality – the market economy. The electoral strategy of moderate parties is to take votes away from the centrist parties that they have lost in their desire to win over the median voter. For example, the core electorate of the liberal party, unimpressed with how quickly the party is "moving right," may decide to vote for the social democrats, while conservative voters may see the movement of their party to the left as a betrayal and thus vote for the nationalists. While this type of party is farther from the centre, it only confirms the tendency of the political process to find a compromise in society, so these ideologies are partly centrist too.

However, some ideologies still exist that do not allow for compromise, as this would prevent the goal of building an ideal society from being achieved. These currents moved in opposite directions, away from the centre of the political spectrum, pushing the original ideologies to the far right and far left, which is why they are called *radical* (or fringe). The combination of egalitarianism and communitarianism gave rise to an extreme form of socialism (namely *communism*), while elitism and communitarianism created extreme forms of corporatism and nationalism (*fascism, Nazism* and *religious fundamentalism*), and extreme individualism degenerated into *anarchism*, which exists in both left (*social anarchism*) and right forms (*anarcho-capitalism*). Some would argue that the median voter theorem dooms such ideologies to inevitable failure, although this is not necessarily the case. Extreme ideologies shine when there is a crisis in society, when people are disappointed in the incessant search for compromise. It is at these moments that representatives of extreme trends come to the fore and say that the problem with the system lies precisely in the half-measures taken by the centrist parties and their departure from "pure" ideologies, declaring that the only way to build the ideal society is to apply the ideology in its radical form. Having won over voters in a moment crisis and coming to power, extreme movements always proclaim a monopoly on ideology and ban other points of view, because they would not be able to hold onto power otherwise, that is, in the normal course of the political process. Radical ideologies thus represent the *boundaries of the electoral spectrum*, although they are typically represented by marginal movements today.

Our short history of political ideologies would not be complete without mentioning movements that focus on a key issue, say feminism or environmentalism, which are extremely heterogenous in terms of their content and can occupy quite different positions on the ideological spectrum.

3.7 Cleavages in the Ideological and Political Space

Parties and party systems are fairly stable institutions, while ideological disputes and electoral programmes are extremely dependent on market conditions. This is because the latter reflect deeper and more entrenched divisions in society that have developed over a long period of time. The political scientists Seymour M. Lipset and Stein Rokkan researched information on European political systems and identified the most characteristic demarcations (*cleavages*), arguing that the territorial differentiation of society played a key role in their formation [7]. Susanne Pickel identified four signs of ideological and political cleavage:

1. the presence of a socio-cultural basis in the form of a conflict (or split) between population groups with different views;
2. a long tradition of confrontation;
3. the transition of the conflict into a political process, where parties or their factions (wings) turn into institutions that represent the interests of conflicting groups;
4. stable voting of conflicting groups for the political forces that represent them, which consolidates the coalition between them [425].

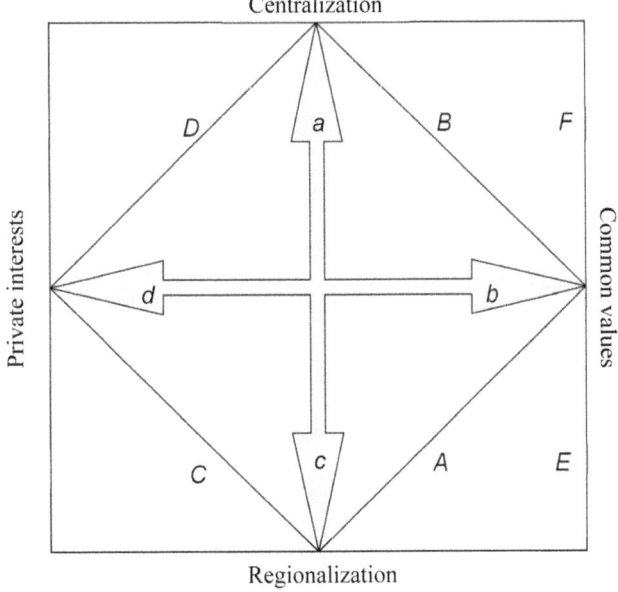

Fig. 3.3. Spatial model of ideological and political cleavages

There are two types of ideological and political cleavage: territorial (for example, between city and rural dwellers) and functional (for example, between supporters and opponents of secularization). The spatial continuum of these cleavages thus develops in two mutually intersecting dimensions (Fig. 3.3): territorial (the ac axis) and functional (the bd axis). The outer edges of the territorial dimension of cleavages are: (1) aimed at centralization and standardization ("centralization"); and (2) aimed at preserving and protecting the identity of local communities ("regionalization"). In other words, the territorial dimension oscillates between forces that are aimed at reducing or increasing the degree of spatial differentiation, and this is how it should be. The poles of the functional dimension of cleavages are: (1) ideas aimed at satisfying group interests – for example, classes or other segments of society ("private interests"); and (2) ideas that support development based on common values ("common values"). The functional dimension, therefore, is also a choice between unification and diversity, but in society rather than in space.

The crisis that came in the form of ideological and political cleavage in different political periods left their scars on society, and this is reflected in the configuration of the party system (Table 3.3). These cleavages appear according to the continuum described below:

- regional, national or supranational cleavages appear along the territorial axis (when states start to act as an integral unit in supranational bodies);
- value-based (national, civilizational) cleavages, material cleavages that focus on the private interests of individual social groups, as well as post-material cleavages that are also based on non-material values but have acquired a global character, appear along the functional axis.

Table 3.3. Cleavages in the ideological and political space

No.	Split	Type of split	Key event
I		Centralization of nation states	
1. (A)	Centre–periphery	regional, value-based	The Reformation and Counter-Reformation of the 16th–17th centuries
2. (B)	State–church	national, value-based	The French Revolution of 1789
II		The Industrial Revolution	
3. (C)	Town–village	regional, material	The Industrial Revolution of the 19th century
4. (D)	Owner–worker	national, material	The Russian Revolution of 1917
III		The Post-Industrial Revolution	
5. (E)	Globalization–sovereignization	national, value-based	The Cold War and decolonization of the second half of the 20th century
6. (F)	Survival–self-expression	national, post-material	The May 1968 events in France

Lipset and Rokkan identified four key splits that were characteristic of Western European countries and which developed until the first quarter of the 20th century (A–D in the table). Other researchers (primarily Lijphart) would go on to add splits that occurred later, some of which are effectively carbon copies of the "original" splits (E and F in the table). It would be a mistake to simply superimpose the Lipset–Rokkan splits onto other societies and eras, although allusions to them are often made.

The *centre–periphery* cleavage – the oldest of those that influenced party splits at the time – dominated during the emergence of representative politics in Europe and was associated with the spread of Protestantism and religious wars of the 16th and 17th centuries. This is a cleavage between the forces of centralization and standardization and is associated with the processes of national of imperial construction, on the one hand, and the desire to preserve ethno-linguistic and religious uniqueness in the provinces, on the other. At the same time, this process had two dimensions: (1) the emerging nation states acted as provinces of the Catholic Church; and (2) absolutism had already pushed them to suppress their internal regional identity (for example, Gallicization in France). The unification of the national language thus became a tool to combat the Roman Curia (the transition from Latin to local languages in church services), as well as a way to assimilate ethnic minorities. This cleavage occupies a position in the southeastern sector of the spatial model (point A), as it is regional (or supranational in the case of an imperial space in which states act as regions) and ideological (that is, built around common values).

The institution of parliamentarianism that took shape during the split in the early periods had the advantage of a territorial function, as it was a body of representatives from distant regions. This ensured that they would be part of a single political process, and it also contributed to the centralization of nation states. The map at the beginning of this chapter showing the shares of votes for the Republican People's Party in Turkey in 1950–1957 is an example of how such a cleavage works today: the areas shaded in dark grey are the 15 regions located on the eastern periphery of the party in which the party achieved the best results; the light grey areas denote the 15 regions – mostly near Istanbul – where it received the fewest average votes.

However, parties are more than simply "agents" of the split between the centre and the periphery, as they can also act as a tool for integrating remote regions into a single electoral space. This phenomenon is called *"mobilization of the periphery."* Including the population living in the border regions serves to legitimize the central government, and parties – even those that protect local interests but operate at the national level – contribute to the formation of a common ideological discourse for the country and thus set the framework for seeking a consensus in society. The recruitment of local politicians in the political process is particularly important here, as it provides an opportunity for regional voices to be heard when developing a national strategy.

The state–church cleavage was associated with secularization and anti-clericalism, which were a big part of public discussions during the French Revolution. This is a national, ideological cleavage that occupies the northeastern section in the model (point B). Key here is the rivalry between the secular and church authorities for a monopoly on the establishment of social norms, and nowhere did this manifest itself more intensely than in the dispute over control of the institution through which these norms should be projected – the education system. And the fact that the liberal position was linked to secularization had little to do with this, rather, it was that the conservative position was associated with the traditionalism of the church. For example, support for religious parties could have indicated the desire of parents to have a bigger say in the content of their children's education and to be less dependent on secular authorities. This split was not as pronounced in the countries of Northern Europe, where national Protestant churches were subordinate to the state. It did, however, make a mark on the political landscape of Catholic countries and countries where more than one religion dominated. Second (as was the case in the Netherlands, Switzerland and Germany, for example), this split had both a functional and a territorial dimension, the latter being due to the fact that different religious communities often lived in specific geographic areas. The Christian democratic movements effectively became the prototype for mobilizing mass parties of the proletariat, and wherever they existed, they partly channelled the social stratification of society.

As we can see, the first two cleavages were the result of the centralization of nation states, the so-called "national revolution." A common feature among them – their axiological nature – was a result of the fact that national values are the leading mechanism of state building. At the same time, while the centre–periphery cleavage was unambiguously territorial, the state–church cleavage was mostly functional. The second generation of ideological and political cleavages consisted of oppositions brought about by the Industrial Revolution and the subsequent emergence of the classes of city dwellers (the "bourgeois") and workers. A common feature of these cleavages was social particularism – the desire to isolate and protect the interests of individual groups in society. We can thus say that the axiological component was replaced by material, common values, that is, by private interests. The second generation of cleavages saw the development of both a territorial split (town–village) and a more functional one (owner–worker).

The *town–village* cleavage originated in the confrontation between the rapidly growing wealthy artisan, and then industrial, bourgeoisie and the landed aristocracy that once dominated (as can be seen, for example, in the centuries-old rivalry between the Whigs and Tories in Great Britain). In actual fact, the roots of the conflict ran far deeper than the contradictions between urban and rural production – it was the struggle between a system of social status based on origin and connections and a system of recognition based on personal success and prosperity. This is what shaped the ideological polarization between the liberal views of the urbanized part of society and the more conservative foundations of those who lived in the countryside. The cleavage soon acquired an economic dimension: while urban industry was interested in economic liberalization and open competition, including at the global level, agriculture relied on protectionism and closed markets. The latter formed a stable ideological base for the agrarian and peasant parties, which made up a key part of the electoral landscape in the countries of Northern Europe. This split continues to be relevant for those countries that experienced large-scale urbanization somewhat later, for example, in Argentina, Iran, Egypt, and so on. We can therefore say that this is a split of a regional and material nature, which occupies the southwestern sector in our spatial model (point C). Today, this cleavage is developing along the "city–suburb" line, since in some Western countries, the United States in particular, the city has become a place of residence for the less well-off segments of the population, including those in receipt of state benefits, while the more prosperous segments of the population move to suburban areas. This has effectively flipped the cleavage on its head, forcing the previously agrarian forces to aggregate the interests of city slums, and the former urban movements to build support networks in small settlements.

The *owner–worker* cleavage also has a material basis, but it is implemented at the national, rather than the regional, level. It occupies the northwestern sector in the model (point D). The split led to the rapid rise of the working class that followed industrialization, and its key bifurcation point was the October Revolution in Russia in 1917. The basis of this cleavage was the dispute regarding the optimal economic model for society (socialist or capitalist) and the accompanying political and ideological superstructure. It was this split that shaped the basic axes of the ideological spectrum that defines the electoral field in most countries today. This is the axis between individualism and communitarianism, on the one hand, and between egalitarianism and elitism, which we discussed earlier, on the other. The communitarianism and egalitarianism of the communist movements that relied on workers was the point from which the entire modern ideological spectrum was reconfigured.

The Lipset–Rokkan cleavages cover the period before the formation of the basic structure of the modern ideological spectrum and, accordingly, the party system, although this does not mean that new types of splits will not continue to fragment communities in the future. Interestingly, some cleavages continue the logic of the divisions we have looked at, just in new conditions. We can thus conclude that ideological and political cleavages are cyclical in nature.

The first such cleavage is the clash between the forces of globalization, on the one hand, and sovereignization on the other. This was first caused by the formation of the bipolar geopolitical system after the Second World War, and then by its dissolution after the collapse of the Soviet Union. It is expressed, depending on the context, in the choice of geopolitical orientation to the West or East (say, in Eastern Europe and the post-Soviet space), or between positioning oneself as belonging to the Global North or the Global South (in Latin America, Africa and South Asia). In essence, this geopolitical choice represents either the orientation of societies towards global institutions and models, or the desire to preserve their civilizational and economic identity. The anti-globalization camp is essentially opposed to Westernization, which forms the basis of global values and norms today. In essence, this split is a repeat of the very first centre–periphery cleavage, but on a planet-wide scale. In some communities, this division is marked by the pitting of democratic choice against communist or authoritarian models. This split also sublimated the town–village cleavage, or, to be more precise, the key dispute between a free market and national protectionism, once again taking it to the supranational level. It has a territorial and ideological content, but is implemented at the supranational, inter-country, level, rather than at the regional level, and is also located in the southeastern sector of our model (point E). This cleavage forms the "elitism–egalitarianism" poles of the ideological splits in the modern world. In a way, it serves as a territorial embodiment of this axis, since it projects the struggle between egalitarianism and elitism onto itself through the opposition of territorial unification and spatial differentiation.

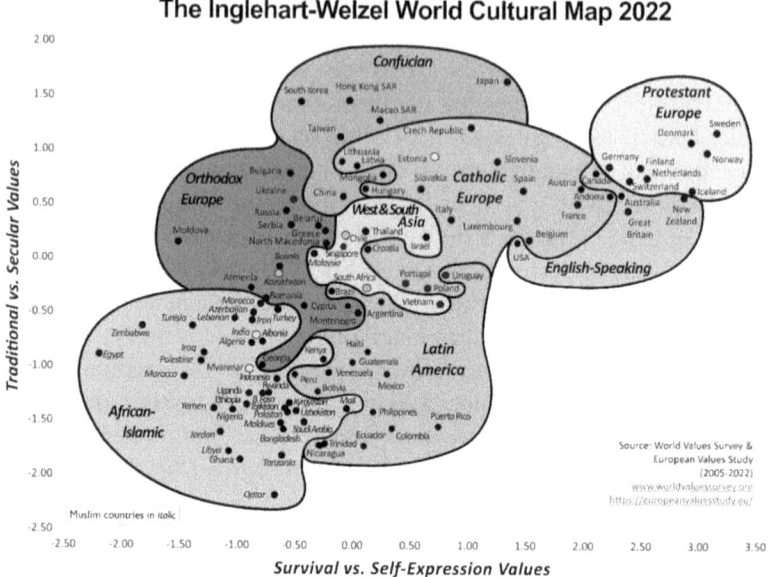

Fig. 3.4. Values map of the world. Source: World Values Survey, 2022

A version of the state–church cleavage is playing out in the world today, this time in connection with the struggle between the personal and the collective in determining social norms. It was best described by Ronald F. Inglehart, who introduced the "survival–self-expression" axis of values into his global study [327, 328]. For Inglehart, the inclusion of this variable in the axiological alignment complements the struggle between traditional and secular values, which fits into the logic that this is an old variable that was absorbed by the state–church cleavage during the confrontation between the two (Fig. 3.4). The May 1968 events in France were both an anti-state (anarchist) and atheist movement. A new answer was given to the question of whether the norms should be set by society or by the church – namely, that norms are needed to the extent to which they arise during the course of the natural interaction between people. Any external frame is considered superfluous, which is expressed in the famous postulate that A person's freedom ends where another man's freedom begins.

Moving away from the issue of who should set the norms in society to the question of whether anyone should set these norms at all, we see that this split has significantly expanded the horizon of the discussion today. The values of "survival" make the need for social institutions as a projection of social order and development universal, while those of "self-expression" ultimately reach towards the imperative of individual rights and freedoms in all aspects of life. Arend Lijphart also positioned this opposition as a dilemma of the values of post-materialism, which fits into the logic that this cleavage had already overcome the stage of materialism during the Industrial Revolution [370]. In the economic sense, this cleavage took shape in a new iteration of the dispute between owners and workers about whether the economy is market- or command-driven, only today we are talking about the limits of the state's involvement in the market economy. This debate has also taken on a new quality, as it is now centred around the choice of regulatory function of the state as either a conductor of universal equality for individual self-realization and complete freedom, or a guardian of collective traditions and norms. As we can see, this kind of cleavage supports the "individualism–collectivism" poles on the ideological spectrum.

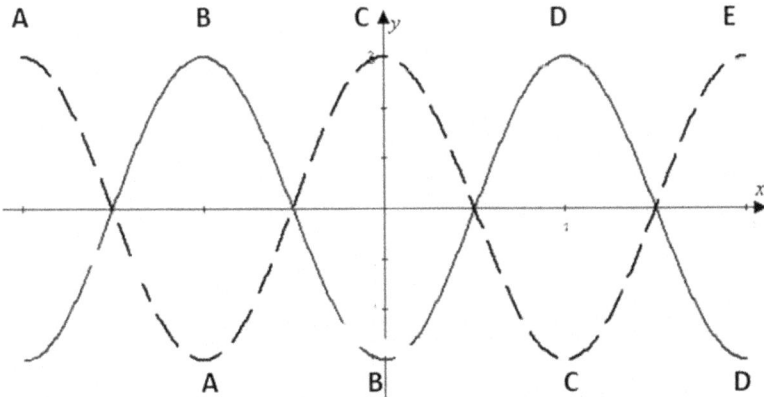

Fig. 3.5. The cyclical nature of ideological and political cleavages

Figure 3.5 shows the cyclical nature of the functioning of ideological and political cleavages: the x-axis denotes temporal dynamics, and the y-axis represents the degree to which a given type of cleavage (marked in Latin letters) can be felt. Cleavage A (centre–periphery) cleavage fades and flows into Cleavage C (town–village), then into Cleavage E (globalization–sovereignization). Similarly, Cleavage B (state–church) fades and flows into Cleavage D (owner–worker), and then into Cleavage F (survival–self-expression). At the same time, two dominant principles of split coexist in each era – territorial (the dotted cosine ACE), which projects the ideological axis egalitarianism–elitism, and functional (the solid cosine BDF), which supports the ideological axis "individualism–communitarianism." We can thus conclude that it is the cyclicality, coexistence and resonance of ideological and political cleavages that ensures the continuity of political and party systems.

Thus, party systems today are dominated by two types of splits that are cyclical repetitions of previous cleavages and maintain two key axes of the ideological spectrum. Exactly how deeply they impact the structure of party systems, of course, varies from country to country. In the United States and Western Europe, For example, the dispute between survival and self-expression prevails, while in Russia and Eastern Europe, the old dispute over geopolitical orientation has not yet disappeared, although they generally complement each other and support the dynamics of the current political process.

Key Terms

- political parties: elite (cadre), clientelist, mass-based, pluralistic, proto-hegemonic, regionalist, electoralist, programmatic, inclusive, personalist, ruling, centrist, moderate, radical
- committee-type parties, branch-based parties, cell-based parties, militia parties, congress parties, movement parties
- bloc, coalition, party alliance, faction
- majority and minority government, non-party system
- party system: one-party (mono-party, hegemonic, dominant, bivalent, polyvalent), two-party (monovalent and polyvalent), multi-party (monovalent and bivalent), nationalized, regionalized, segmented, territorialized
- ideology: anarchism, liberalism, libertarianism, communism, conservatism, nationalism, social democracy, socialism, fascism
- ideological spectrum (compass): core (centre), axis, contour, cleavage, mobilization of the periphery

- diffusion of the electoral field: continuous, hierarchical, wave (cascade)

Questions and Exercises

1. What type of political party are the following parties? Evaluate their propensity for territorial differentiation. The Women's Party in Australia, the All-India Muslim League, the Evangelical People's Party of Switzerland, the Chinese Communist Party, the Byrd machine in the United States, Podemos in Spain, the Party of Regions in Ukraine, the Pirate Party in Sweden, the Justicialist Party in Argentina, and the Jacobins in France. Find one more example for each type of party.
2. Rank political parties that form factions in the following parliaments according to their susceptibility to territorial differentiation:

    2.1. The Bundestag in Germany
    2.2. The House of Commons of the United Kingdom
    2.3. The National Assembly of France.
    2.4. *The Lok Sabha in India.

3. Look at how Belgium ratified the Treaty of Lisbon (or other international agreement). At how many stages were representatives of the leading ethnic parties involved? At which stages did they act as regional parties? And at which stages did they act as programme parties? At what stage were they best able to defend regional interests?
4. How does classical economic liberalism differ from modern liberalism? What modern ideology would Adam Smith support?
5. In what situation is a conservative a reformist and a liberal a reactionary? Why has it typically been the other way round in history?
6. How do the electoral strategies of centrist, moderate and radical political parties differ?
7. What is the difference between a republic, a democracy and a liberal system? Are illiberal republics, illiberal democracies, undemocratic republics, undemocratic liberalism, non-republican democracy and non-republican liberalism possible?
8. How can the emergence of radio be connected with the spread of radical political ideologies?

9. Classify the party systems of the countries of Central America by evaluating the coordinates on the spatial continuum of party systems for each.
10. Choose a country with alternative elections and use data on the last four electoral cycles to calculate the place of the party system in the spatial continuum. Comment on your results.
11. Calculate the effective number of parties (the Laakso–Taagepera, Molinar and Golosov indices) for all Duma electoral cycles in Russia. Compare and analyse the results.
12. Take two neighbouring countries and, using numerical indicators, determine the degree of territorial differentiation of their party systems. What explains the possible difference?
13. How can we conduct cross-temporal electoral research if new political parties regularly appear in the political landscape? Propose some algorithms.
14. Analyse the case of the inclusion of the territory of Burgenland in the electoral field of Austria. What were the potential cleavages and how were they identified by existing political parties in the country? What is this phenomenon called? Find similar examples for comparison.
15. Choose a country and trace how the structure of its party system took shape through the history of the ideological and political divisions of the electoral field. What splits continue to be relevant in the country? What makes its party-ideological evolution unique compared to other countries in the region?
16. Study the dynamics of changes on Ronald F. Inglehart's value map. How does it correlate with the cycles of ideological and political cleavages?

Chapter 4
Spatial Models of Voting

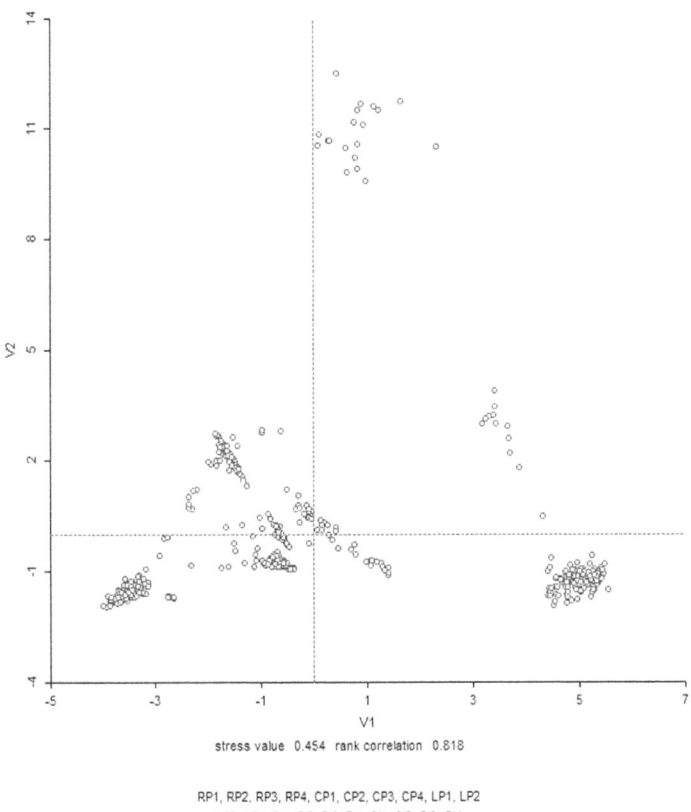

RP1, RP2, RP3, RP4, CP1, CP2, CP3, CP4, LP1, LP2
LP3, LP4, RI1, RI2, RI3, RI4, CI1, CI2, CI3, CI4
LI1, LI2, LI3, LI4, RN1, RN2, RN3, RN4, CN1, CN2
CN3, CN4, LN1, LN2, LN3, LN4

Source: Okunev I. (ed.) *Electoral Geography of Russia's Neighbourhood*, MGIMO University Press, 2024.

- What factors determine the electoral choice of a voter?
- How can we tell which candidate's (or party's) programmes is closest to the voter?

- Is there such a thing as an average voter? What would his or her political position be?
- How beneficial is it for parties to focus on the average voter?
- Are party political platforms stable? How do they change?

4.1 Theories of Electoral Behaviour

An important part of political science is the development of theories that try to explain the logic of people's electoral behaviour. Theories of electoral behaviour fall into three main groups:

1. According to the sociological (Columbian) school (Paul F. Lazarsfeld, Bernard Berelson and Hazel Gaudet [362]), the voter's belonging to a specific social group (primarily class, ethnic group and religion) is the key factor determining electoral behaviour. Weak political literacy and a disinterest in politics push people to focus on their immediate environment when deciding on who to vote for: colleagues, neighbours and relatives, who most often belong to the same social stratum. Basing political choice on the preferences of one's social group reinforces the voter's sense of belonging to that group, which in turn increases their status among the members of that community and boosts their self-esteem. The differentiation of the electoral field is thus manifested primarily in terms of socio-economic and demographic characteristics and turns out to be rather stable. This is where the so-called "*funnel of causality*" in politics comes from: the social structure forms the value orientations of voters, which, in turn, determine their party identification on the basis of social affiliation [66, 104, 107].

2. The psychological (Michigan) school (Angus Campbell, Philip E. Converse, Warren E. Miller, and Donald E. Stokes [232]) identifies early political socialization as the key factor in the development of the norms of electoral behaviour. In this regard, intergenerational familial ties turn out to be stronger: children absorb and repeat the ideological portrait of their parents. Political orientation is seen as part of one's own individual identity, rather than a social identity, and the voter goes to the polling station with a strong psychological (emotional) attachment to one of the political forces. Sticking with the traditional choice provides a sense of spiritual comfort and a belief in the stability of their way of life. At the same time, since the conflict between generations is

a constant, people will oscillate between periods of distance from the party (*dealignment*) and reunification with it (*realignment*).

3. The political economy (Rochester) school (Anthony Downs [266], William H. Riker, Morris Fiorina [279–280]) is based on the theory of rational choice and assumes that the voter's choice is based on an analysis of his or her own interests, needs and risks, while also making it as useful as possible. According to this model, the voter compares the political platforms of the parties and determines the *election differential*, that is, the difference between the various potential outcomes of the electoral process for him or her personally. Since the key criterion of rationality is the correspondence between ends and means, in which the maximum result is achieved at the minimum cost, parties model their programmes in such a way as to meet the aspirations of the majority of voters. It turns out that parties create political courses in order to win elections, rather than winning elections in order to be able to formulate and implement political courses.

Spatial models of electoral behaviour use spatial (or, more simply, geometric) abstraction to describe the patterns of electoral behaviour. In a sense, it can be argued that the spatial model is better than others when it comes to helping us understand the nature of people's behaviour during elections.

4.2 The Hotelling–Downs One-Dimensional Model

We often hear the question: Which candidate is closest to you in terms of their political views? As if the electoral process is like looking for the nearest bus stop. What this metaphor demonstrates is that spatial proximity can act as a metaphor for ideological similarity. This is the idea that underlies the subfield of political economy concerned with building a spatial model of voting.

Any space, whether it be real (physical) or abstract (mathematical), has two characteristics. First, it has a *coordinate system* that makes it possible to determine the location of elements in it and the distance between them. This can be measured in degrees, metres, or, as in the case of electoral space, ideological proximity. Second, the elements in this space form a certain *structure of spatial connections*, which allows us to talk about its topological characteristics: neighbourhood, connectivity, homogeneity, continuity, dimension, etc. Space can be one-dimensional, two-dimensional, three-dimensional, or multi-dimensional, and the question we posed at the beginning of this chapter indicates that the person asking assumes the ideological spectrum in which he or she wants the respondent to place the candidates has one dimension.

Imagine that each voter has their own views on how and in what direction the community should develop. These views will typically focus on a specific issue (raising or lowering taxes, whether or not to legalize abortion, etc.), or a set of issues. In the mathematical spatial model of voting, individual issues that are relevant to voters are taken as a dimension of space. The voter's preferences about how society should develop act as the centre (ideal point) of this model, and they vote for the candidate or party that is closest to this point. Since the voter's views are usually a combination of several key positions, this means that his or her choice is made in a multi-dimensional space. In reality, the voter's choice is projected by the established party system, which, in a sense, is a generalization of the dimensions of political choice that exist in society, expressed in a multi-dimensional model of the ideological spectrum (compass).

It is important to note that the choice a voter makes cannot always be represented as a single point. Instead, there exist *single-peaked* preferences, where there is a single maximum point on the choice scale, and as the distance from that point increases, the support level falls steadily, and *multi-peaked* preferences, where several maximum points or preferences cannot be described linearly C . The very fact that people have multi-peaked preferences creates the so-called Condorcet paradox (discussed earlier), which prevents the perfect electoral system from ever coming into being. Further, we will describe the evolution of spatial voting patterns based on single-peaked preferences, but we must be mindful of the complications in the reality of the political process.

Вариант P_1	Variant P_1
Вариант P_2	Variant P_2
Ранги предпочтений	Preference rank
Предпочтения	Preferences

The first approaches to the spatial model of voting were set out in Harold Hotelling's linear city model [20]. Imagine a motorway along which cars travel from point O to point L and back – that is, a one-dimensional model of space (Fig. 4.1). There are two petrol stations along the road, at point A and point B. The horizontal line reflects our space, while the ordinate axis represents the cost (in terms of money and time), which increases as we travel farther away from the petrol stations (the closer we are to a petrol station, the quicker and cheaper it is to get to it). We also assume that the price of petrol is fixed (say, by the government). That is, the only source of competition between the petrol stations is their location in space. As there are only two petrol stations, sections a and b are monopoly zones for petrol station A and petrol station B, respectively, while the section between points A and B is the zone of competition between them, where point x would be the market division boundary, since it is equidistant from the two petrol stations and thus the point of maximum cost between them.

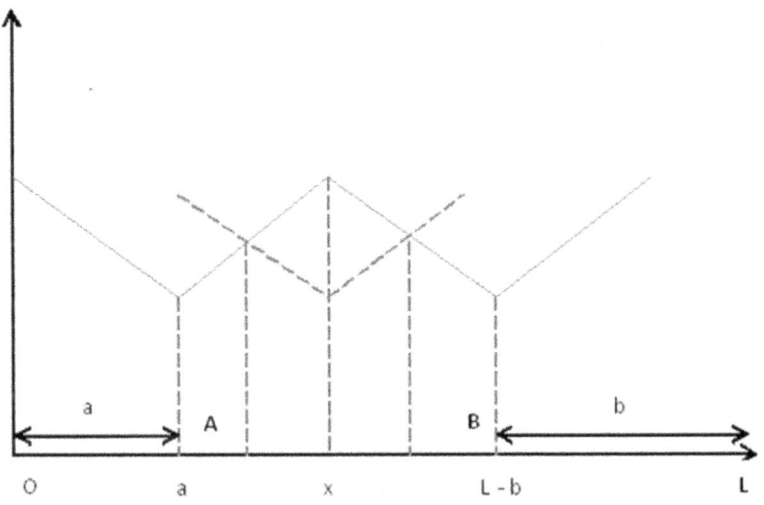

Fig. 4.1. Hotelling's linear city model

Suppose that petrol station A has already been built. Where would it be most profitable for the owner to place petrol station B? The correct answer here is "as close as possible to petrol station A, only a little farther from city O," as this would allow it to maximize its zone of monopoly dominance and minimize the zone of competition. If the owners of the two petrol stations had come to an agreement beforehand on where to build their petrol stations, then they, using the same logic, would place them to the left and right of the equilibrium point x. Such an equilibrium (the Nash equilibrium) would be the optimal solution, since, at any given location, none of the petrol stations can increase its income by moving in either direction, unless the other petrol station also moves location. It is also the optimal solution because demand for petrol increases the further you move away from the cities and reaches its peak at precisely the middle point on the motorway. Note that the decision to place the petrol stations as close to each other as possible flies in the face of conventional logic, which suggests that competing companies should be located as far as possible from one another.

Anthony Downs transferred the Hotelling model to the realm of politics, seeing the ideological spectrum between parties on the right and left as a one-dimensional space [266]. It turns out that the optimal electoral strategy for a political party, particularly in two-party systems, is, oddly enough, to adopt a position that is as close as possible to that of the opponent. And this plays out to a tee in political practice, where, in countries that have a stable electoral tradition, it is almost impossible to tell the difference between the political positions of the dominant centrist forces.

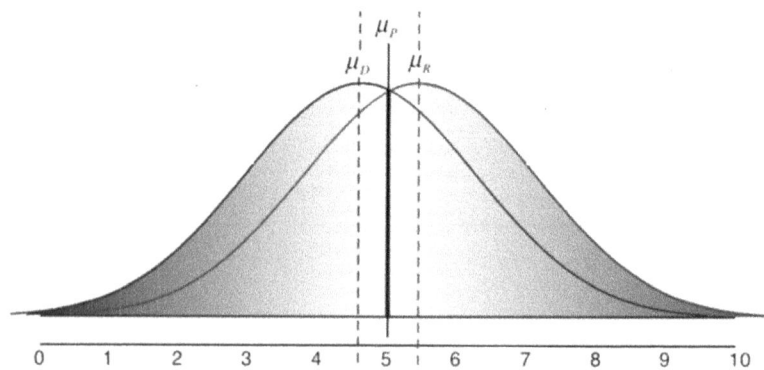

Fig. 4.2. Downs' spatial model of voting

Let us suppose that the ideological preferences of voters and the political platforms of parties can be represented as a one-dimensional space, in which point 0 corresponds to extreme leftist views, and point 10 corresponds to extreme rightist views (Fig. 4.2). The distribution of votes is approximated using a Gaussian bell curve, meaning that the majority of votes are in the centre of the ideological spectrum, and the share of votes decreases as you approach the ends of the spectrum. Parties and their candidates offer the voter a platform, which they evaluate based on how closely they align with their own political views. In other words, parties try to place their programmes on the spectrum in such a way as to ensure that it is close to as many ideal points of voters as possible. The distance between the positions of the party and the voters is calculated as the module of the difference in coordinates $dist(\mu_D, \mu_R) = |\mu_D - \mu_R|$.

Let us first consider the situation where an election is effectively being contended by two centrist parties: the centre-left occupies the position of μ_D, while the centre-right occupies μ_R. In this position, each party has an equal chance of winning, as their electorates are the same size: voters with views ranging from 0 to μ_p will vote for the centre-left, and those with views ranging from μ_p до 10 will vote for the centre-right. How can a party that starts off at the point μ_D win the election? The answer would be to shift its platform towards point μ_R, that is, to become as close as possible ideologically to the centre-right party.

Now let us imagine that we have three parties: left, centre-left and centre-right. In the previous electoral cycle, the centre-left and centre-right parties propose platforms that were close to the point α_p, and the centre-left party emerged victorious. The core electorate believes that the position of their centre-left party has moved too far towards the right, and now the left party needs to offer a platform that corresponds to the point α_p in order to win. No matter how many parties take part in them, elections effectively turn out to be a process that seeks to capture the point of view of the average voter. Kevin Quinn and Andrew Martin summed this up by proposing to use game theory to identify the best strategy for a political party to adopt in the electoral field, which depends on the positions of other parties in it [430]. In other words, the voter's choice is shaped not by the proximity of a given party to his or her own views, but by the entire structure of the mutually determined configuration of parties in the ideological and political space.

Duncan Black came to a similar conclusion about how the electoral field is structured, albeit in a different way, to prove his *median voter* theorem, which states that the optimal strategy for any party or candidate in a one-on-one fight is to choose a platform that aligns with the views of the median voter [213]. Imagine we have an odd number of voters and at least two candidates, and the position of all of them can be represented on a one-dimensional spectrum. Here, the median voter is simply the one who is in a position of ideal equilibrium, that is, where the candidate who shares his or her views most closely would win a one-on-one competition with any other candidate. The median voter in Fig. 4.2, for example, will be μ_P, and the candidate closest to him or her ideologically will win a one-on-one against opponents who occupy the positions μ_D and μ_R. Accordingly, the party or candidate that adopts a position that coincides with the views of the median voter will find itself in a position where it cannot lose to any other opponent in a one-on-one contest. This explains why politics in a democratic society is designed in such a way that elections are effectively the search for the best compromise among the people, which is inevitably located at the centre of the ideological spectrum.

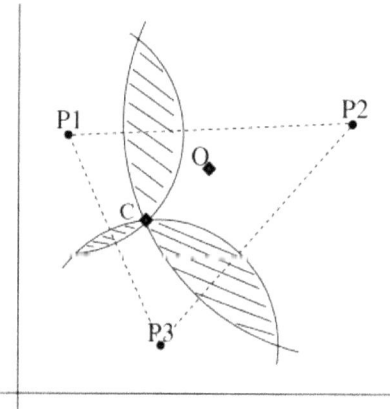

Fig. 4.3. Two-dimensional Downs' spatial model of voting

Similar patterns can be found in multi-dimensional space. For example, Fig. 4.3 shows a two-dimensional ideological space in which the positions of voters $P1$, $P2$ and $P3$ are determined by their views on two ideological axes, say, "individualism–communitarianism" and "egalitarianism–elitism." The Euclidean distance in n-dimensional space will now be calculated using the formula:

$$d(x,y) = \sqrt{\sum_{i=1}^{n}(x_i - y_i)^2}$$

In an election contested by parties O and C, the latter will win, since it will be exactly at the centre of the ideological spectrum (the median voter).

4.3 The Enelow–Hinich Linear Model

If you scrutinize almost any specific election campaign, you will soon see that it is effectively conducted on two planes. On the one hand, politicians and voters have long since developed their ideological platforms and preferences – the coordinates that position them in the multidimensional political spectrum. On the other hand, there are always a few key issues in the campaign around which the main points of contention unfold: taxes, certain rights and freedoms, or even something as seemingly trivial as the candidates' attitude toward a particular event or statement. It is often believed that such agendas are nothing but a projection of ideological divisions, a field for candidates to do battle. However, James Enelow and Melvin Hinich took a new approach to this issue and proposed a spatial model of voting where ideological and opportunistic differences represent two different dimensions of space in which the electorate makes its choice in parallel.

Imagine a two-dimensional space in which the horizontal line corresponds to the position of the party on the ideological spectrum (Fig. 4.4). This space does not change, because, on the one hand, it reflects stable values, while on the other, it is associated with the "brand" of the given political movement, as well as with its history and traditions, and it is difficult to dissociate itself from these things. Meanwhile, the y-axis reflects the party's position on a specific issue, say, the legalization of the death penalty. The party can change its position on the y-axis as the political situation requires, provided that it remains still on the x-axis

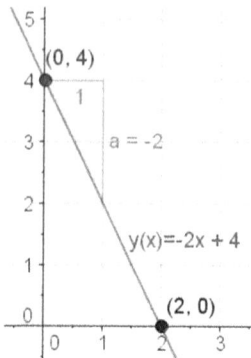

Fig. 4.4. The Enelow–Hinich perceptual voting model

Since election campaigns invariably revolve around a number of controversial issues of the time, parties carry out their search for the median voter to ensure victory by calibrating their positions along the y-axis only. The problem is that there are typically several burning issues, which means that there will be more than one y-axis. As Andrei Akhremenko rightly points out, we are not dealing with a multidimensional space in which agenda items make up the dimension; rather, we are dealing with a multitude of one-dimensional agenda spaces, each of which is projected onto the ideological spectrum of the x-axis. What is more, different issues will be of greater or lesser importance in politics, and their specific combination may differ from voter to voter or from territory to territory.

The methodology developed by Enelow and Hinich helps reduce this set of one-dimensional agenda spaces to the ideological spectrum by projecting them onto the x-axis [271–273]. Enelow and Hinich proposed considering such a projection as an ordinary linear function $y(x)$. Imagine Candidate A with ideological views $x = 2$ and a position on the current agenda of $y = 4$. It follows that the relationship between the ideological platform and the agenda can be described by the linear function $y(x) = -2x + 4$. Now imagine that we know Candidate B has ideological preferences expressed as $x = 1$. Even if the current agenda is not at the top of the voter's list of priorities, the linear function tells us that it will be equal to $y = 2$, and the voter will evaluate how suitable the candidate is for him or her based on this calculation. Conversely, knowing the current agenda of the party regarding a particular issue, the voter has a rough idea of where it belongs on the ideological spectrum, and this is an indicator of where the party stands on issues that are important to the voter. This is why the authors of the model called it probabilistic, since it allows the voter to calculate the probable agenda and ideological orientation of parties and candidates. Moreover, the candidate's position turns out to be not so much an objective reality as a subjective assessment by voters of his or her views, which allows us to look at the model as perceptual in nature.

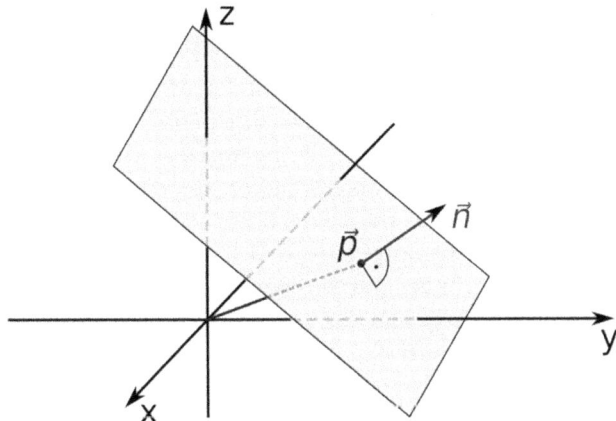

Fig. 4.5. The three-dimensional Enelow–Hinich voting model

The variables of the linear function of the Enelow–Hinich spatial model of voting $y(x) = kx + b$ can be interpreted as follows. Slope k shows how a voter's ideological position correlates with the party's stance on a particular issue (strongly, weakly, positively or negatively). When $k = 0$, the linear function is parallel to the horizontal axis, which can be interpreted as complete unanimity of all parties on this issue. The constant b is an indicator of the ordinate of the point of intersection of the linear function with the y-axis (the point at which $x = 0$) and can be interpreted as the parties being indifferent to this item on the agenda. The voter, of course, does not perform linear function calculations. Kinder and Laskin developed the theory of schemes, which states that voters create schemes that represent their ideological attitudes: liberals must necessarily support one thing, while a conservative must always defend another, etc.

If the ideological spectrum is multidimensional, or at the very least has two dimensions, x and y (which is how it is most often described), then the dimensions of the agenda will be plotted along axis z (Fig. 4.5). Accordingly, knowing the function $z(x, y)$, we can project the position of a given party on a particular agenda issue onto the two-dimensional ideological spectrum, or vice versa.

4.4 The Granberg–Brown Parabolic Model

Imagine voter A has views that correspond to point 0 on the x-axis (Fig. 4.6). In relation to this voter, point 1 represents the conservative party's programme, point 2 the nationalist party's programme, −1 the liberal party's programme, and −2 the socialist party's programme. In the spatial model of this voter's choice, the distance between him or her and the position of the conservative party is equal to the distance between the conservative party and the nationalist party, and the distance between him or her and the liberal party is equal to the distance between the liberal party and the socialist party. But this is not how voters think: the conservative and liberal parties are understandable to them; they are ready to choose between the two, and will never vote for the nationalists or socialists, as the views of these parties seem extremely far from their own. Now imagine voter B, whose views align with the position of the conservative party at point 1. This voter is happy with what the Conservatives are doing, but he or she could, under certain circumstances, vote for the nationalists. In other words, voters A and B perceive the distance (difference in views) between the conservatives and nationalists completely differently. That is, the distance between the views of political parties is different depending on the starting point.

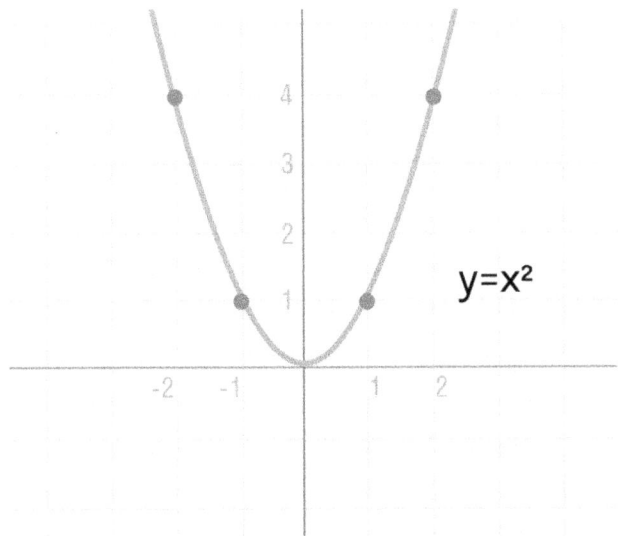

Fig. 4.6. The Granberg–Brown parabolic model of voting

Donald Granberg and Thad A. Brown took this conclusion and turned it into what they called a parabolic spatial model of voting [308]. This model involves plotting the voter's perception of the proximity of the parties' positions to his or her own on the y-axis, rather than the actual proximity of views. This perception can be described by the parabolic function $y(x) = x^2$. That is, the further the party is from the voter, the stronger (in arithmetic sequence) the voter's perception of the distance to it. The voter's perception of how their views differ from those of the candidates also affects their electoral manoeuvrability: the further the voter believes a given political party is from his or her own views, the more difficult it is (again, in arithmetic sequence) for him or her to support them, say, in a second ballot.

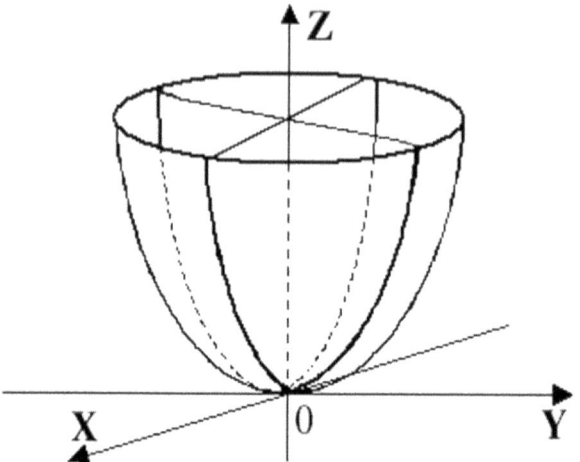

Fig. 4.7. The Granberg–Brown three-dimensional parabolic model of voting

Since we are talking about the two-dimensionality of the ideological spectrum, the perception of the difference between the parties will be plotted along the y-axis and expressed as a paraboloid with the function $z(x, y) = x^2 + y^2$ (Fig. 4.7).

4.5 The Rabinowitz–Macdonald Vector Model

Imagine that the key issue in an upcoming election is taxes. Say the tax rate is 20 % and some parties are in favour of lowering it, while others want to raise it. To assess the proximity of their positions to the electoral preferences of a given voter, we first need to understand what tax rate that voter advocates. In reality, the voter's decision tree looks something like this: (1) either change the tax rate or keep it as it is; (2) either raise or lower the rate; and (3) raise/lower the rate significantly or slightly. In other words, the voter is less interested in where the party is positioned in the electoral field than in the vector of the party's movement. If it aligns with his own sensibilities, then that party is for him. This logic underlies the vector spatial model of voting proposed by George Rabinowitz and Stuart Macdonald [374, 431].

In such a model, the electoral space has a field of both positive and negative values (say, if the voter calls for a reduction in the tax rate).

The Rabinowitz–MacDonald Vector Model

Distance will then be calculated using the formula $dist(a,b) = a_{ik} * b_{jk}$, where a_{ik} is the position of Party i on Issue k, while b_{jk} is the position of voter b on Issue k. If their positions coincide in terms of their vector (in terms of their sign: where a_{ik} and b_{jk} are either positive or negative), then the product will always be positive. If their positions do not align (one of a_{ik} or b_{jk} has a negative value. If a voter is in favour of maintaining the status quo, then the product will be zero. The creators of the model also suggested introducing a certain threshold value that would remove excessively radical parties and their obviously unrealistic programmes (for example, those calling for taxes to be abolished completely) from the equation.

Fig. 4.8. The Rabinowitz–MacDonald three-dimensional parabolic model of voting
Source: A.S. Akhremenko, "Spatial Modelling of Electoral Choice: Development, Current Issues and Prospects (II)", *Polis*, no. 2 (2007), p. 168.

Социальный либерализм	Social liberalism
Левый экономический курс	Left-leaning economic course
Социальный консерватизм	Social conservatism
Порог интенсивности	Intensity threshold
Правый экономический курс	Right-leaning economic course

Let us consider the model in practice using the example proposed by Akhremenko (Fig. 4.8) [15]. Here, we see a two-dimensional ideological spectrum and an intensity threshold, beyond which the electorate will not approve of the position of Party y. All the other parties reflect different vectors of changes and their intensity, and try to guess the vectors of voters' aspirations. An undeniable advantage of this model is that it reflects the dynamics of the political system and its ability to adapt to new agenda issues – something that the models we have looked at thus far appear to lack, as parties in them based their electoral behaviour on hovering around the point of the median voter.

Our analysis of spatial models of voting leads us to the conclusion that none of them taken alone is capable of explaining electoral behaviour in its totality. The problem is that voter preferences appear to be formed in two projections at the same time (Table 4.1).

Table 4.1. **Projections of spatial models of voting**

Type of projection	Key variables	Spatial models	Situation where such models dominate
Static projection	Ideological and political divisions and the ideological spectrum	The one-dimensional Hotelling–Downs model and Black's median voter theorem	Two-party system, stable electoral tradition, "old" parties
Dynamic projection	Current political agenda	The Enelow–Hinich linear model, the Granberg–Brown parabolic model, the Rabinowitz–MacDonald Vector Model	Multi-party system, periods of bifurcation, "new" parties

First of all, there is a *static projection* that reflects deep ideological and political divisions and the distribution of the electorate and parties along the ideological spectrum. The Hotelling–Downs model is best suited for analysing this projection, though it must be kept in mind that the ideological spectrum is at least two-dimensional and cannot be assessed in terms of a single dimension. The general logic of the Hotelling-Downs model and the underlying median voter theorem reflect society's search for a broad compromise, which is necessary for stability and to overcome ideological cleavages. This projection will be dominant in explaining two-party systems, countries with a stable electoral tradition, and the logic of the behaviour of "old" parties.

Second, there is a *dynamic projection* that focuses on the current political agenda. It ensures the dynamics of the political process, raises new questions, allows new leaders and movements to move to the forefront of political life, and keeps the electorate engaged in elections. Multi-dimensional spatial models of voting are better suited for interpreting this kind of projection, and the best model is based not on estimates of proximity, but rather on scalar values, for example, the Rabinowitz–MacDonald vector model. This projection will be dominant in explaining multi-party systems, countries undergoing periods of bifurcation, and the logic of the behaviour of "new" parties.

A comprehensive analysis of the electoral process requires modelling the electoral field in two projections simultaneously: a static projection that ensures its stability and continuity, and a dynamic projection that is responsible for its transformation.

Key Terms

- funnel of causality
- election differential, median voter
- single-peaked and multi-peaked preferences
- spatial model of voting: one-dimensional, linear, parabolic, vector
- static and dynamic projections of spatial models of voting

Questions and Exercises

1. Analyse your own voting strategy. Do you vote the same way as your family members? Are your preferences similar to those of your friends or colleagues? Do you read the election programmes of parties? Which theory of electoral behaviour best describes your own electoral behaviour?
2. Trace the features of the electoral behaviour of African Americans in the United States. Who do they vote for most frequently and why? Give examples for and against believing that their vote is a result of the "funnel of causality."
3. A Moscow family of three (a mother, father and their son) is discussing where to go on vacation: their country home in Moscow

Region, the Black Sea, or the Red Sea. The father wants to save money, so he thinks the closer the better. The mother, on the other hand, wants to get away for a while, so the farther, the better, as far as she is concerned. Meanwhile, their son wants to go to the Red Sea, because he has never been there before. At the very least, he would agree to go to the countryside, because his friends are there. He is categorically against going to the Black Sea, because they went there last year. Draw up a chart to help select a destination for the family holiday, where the x-axis represents distance from home, and the y-axis represents the level of desire to go. Who has single-peaked preferences, and who has multi-peaked preferences? Do all of them have transitive preferences?

4. Imagine that your class is electing a head of the group, with the condition that there can be no more than two candidates. Determine which issue is the most important for the members of the group. Ask all the members of the group about their attitude to this issue, giving it a score of 0 to 10. Work out the position the candidate must take to ensure victory in the election.

5. Draw a two-dimensional Granberg–Brown voting model for the last presidential election. Position your views as point 0 on the x-axis and the views of all the candidates according to how they differ from yours, expressed in terms of distance on the model. Assess how your electoral behaviour will change if different pairs of candidates move into the second round.

6. How do spatial voting patterns differ? In what ways do they conflict with each other? How do they complement one another?

7. Which spatial model of voting best describes the following electoral campaigns:

   a. The most recent elections to the U.S. House of Representatives.

   b. The elections following the Velvet Revolution in Poland.

   c. The elections to the State Duma of the Russian Federation of the First Convocation.

   d. The elections to the UK House of Commons following Brexit.

8. Can the spatial model of voting be applied to international relations? In what ways would countries be spatially proximate in this case? Try to create such a model based on three rounds of voting in the UN Security Council.
9. What is the difference between static and dynamic projection of spatial models of voting? In what case is the explanatory power of each of the projections stronger?
10. Analyse the most recent election campaigns for elections to the lower (or only) house of the national parliament of your country in both the static and dynamic projections of spatial models of voting.

Chapter 5

The Spatial Effects of Voting

Source: "The Gerry-Mander," *Boston Gazette* (March 26, 1812).

- How can the stable geographic differentiation of electoral behaviour be explained?
- Do friends and neighbours tend to vote for the same political party?
- How can equal representation of territories in parliament be ensured?
- Can the boundaries of electoral districts affect the outcome of elections?
- How can the compactness of electoral districts be measured?

5.1 The Scalar Effects of Voting

A number of factors influence the electoral behaviour of voters, some of which have a pronounced geographical nature. These are called the spatial effects of voting, which we will discuss in greater detail. *Scalar* spatial effects of voting include those in which electoral behaviour is determined by the properties of a given place (say, the fact that an election campaign was successful in one region, but unsuccessful in others). *Vector* spatial effects of voting are those where electoral behaviour is determined by the location of a given place (for example, the fact that the region is located close to the area where a given candidate can traditionally rely on support). We can thus say that scalar voting effects are based on the principle of vertical conditionality in geography, where the properties of an object are determined by the properties of the place in which it is located, and vector effects are based on the principle of horizontal conditionality in geography, where the properties of an object are determined by its location relative to other objects – that is, they arise as a result of its relations with the positions of other objects in space (Table 5.1).

Table 5.1. The scalar and vector spatial effects of voting

Effect type	Geographic principle	Explanatory factor	Examples of effects
Scalar	Vertical conditioning	Properties of the place	Friends-and-neighbours effect, campaign effect, issue voting effect
Vector	Horizontal conditioning	Location of the place	Neighbourhood effect

The most obvious scalar geographic factor in electoral behaviour is the *friends-and-neighbours* effect (sometimes called the localism effect), which describes the propensity of voters to vote for people from the same region. There are, in fact, dependencies of two different scales here.

At the local level, the friends-and-neighbours effect can be seen in a narrow sense: the closer voters live to the hometown (district, street, home) of a candidate, the more likely they are to vote for them.

At the regional level, the friends-and-neighbours effect is evident in the broad sense: when voting for a candidate to represent them in the authorities, the electorate is more inclined to support someone local, that is, a candidate who was born in or spent a significant part of his life in the same region (district, city, country) as his or her voters. Accordingly, candidates get more support in territories they have a personal connection with.

There are two key explanations for this dependency [159]. First, candidates are traditionally more recognizable in their native regions and it thus is easier for them to rally support there, because proximity to the candidate (geographical, but also social) equals trust, since it is far more likely that interactions with this candidate will take place in the future than with a candidate who is not from the region. Second, voters will naturally expect a "hometown" boy or girl to pay more attention to local interests, which means that it is easier for these candidates to convince doubters or get indifferent voters to be more politically engaged. The first explanation is more typical of situations where a narrow understanding of the *friends-and-neighbours* effect is evident, while the second is truer of situations where the broader understanding reigns. Moreover, opening up the scale of the analysis makes the first explanation irrelevant – that is, isolated local cases aside, voters tend to cast their vote for local candidates, even when they know just as much about them as they do about their rivals.

The effect was first described by Valdimer Key Jr, in his 1949 book *Southern Politics in State and Nation* [348], but in a negative connotation: it turned out that the localist factor was more important than the political platform of the candidate. Next came the methods proposed by Raymond Tatalovich in 1975 [475] and John Van Wingen and Joseph Parker in 1979 [487] that estimate the friends-and-neighbours effect by calculating the correlation between the level of support for a candidate and the distance between voting districts and that candidate's hometown. These approaches differ in that the first describes the dependence as linear, while the second describes it as logarithmic.

Tatalovich friends-and-neighbours effect $V(D) = aD + b$,

Van Wingen–Parker friends-and-neighbours effect $V(D) = b\left(\dfrac{1}{D}\right)^a$,

where V is the percentage of support for the candidate, D is the distance to the candidate's voting district, b is the level of support in the candidate's voting district, and a is the strength of the friends-and-neighbours effect.

Another obvious scalar spatial effect of voting is the *campaign effect*, according to which a voter is more likely to support a candidate in areas where they have campaigned more actively.

During an election campaign, candidates and parties are forced to choose constituencies and regions into which more resources will be channelled in order to secure victory. This is done based on two main factors. First, attention is paid to areas where the electoral strategy will be most effective, that is, areas with the highest numbers of undecided voters who are more likely to lean towards the chosen candidate. Second, this electoral strategy tries to adjust the inherent geographic favouritism of the electoral system in order to prevent wasted votes. For example, in single-member plurality systems, large parties try to avoid receiving excess votes, as they would prefer a more even distribution of voters across the country to hypertrophied support in certain regions. At the same time, small parties are interested in consolidating their supporters in individual districts, since equal distribution across the territory necessarily leads to a large number of lost votes and, subsequently, seats. These considerations, among others, affect the decision whether a candidate should visit a region on his or her campaign trail for rallies and meet-and-greets with voters, or whether a media campaign will suffice.

Researchers are effectively left guessing as to the size of the resources that candidates direct to various regions during their election campaigns. Indicators they look to typically include the distribution of regional shares in the overall campaign budget, or the volume of campaign materials, although gaining access to such data is far from easy. This is why the main indicator for evaluating the campaign effect is the number of times a candidate visits individual districts, and for how long, as well as the number of pre-election events held there – we could just as well call it the pre-election visit effect. In our opinion, these metrics could be supplemented by an assessment of the nature of the spatial distribution of these visits (in terms of turnout and regularity).

Eric Mintz is considered a pioneer in the analysis of the geography of election campaign tours following his 1985 study into the impact of candidate visits on the 1984 Canadian federal election [390]. One of his followers, Thomas Holbrook, used regression analysis to prove that each of Harry S. Truman's campaign visits during the 1948 U.S. presidential election brought him an average of 0.248 % support [321]. Major election campaigns have effectively turned into a race to see how many regions across the country candidates can visit, because the more meet-and-greets and events the candidate holds in different regions, the greater the spatial effect of his or her campaign will be. For example, during the 2020 U.S. presidential campaign, during the final three weeks before voting, Donald Trump visited 48 cities in 15 states, while Joe Biden visited 23 cities in 10 states. That is, both candidates managed to visit at least two cities per day.

The final scalar effect of voting is the *issue voting effect*, which states that voters tend to support candidates more in areas they pay greater attention to during their election campaigns. For example, a party that is against tearing down residential buildings to make way for new ones in Moscow will find more support in those districts of the capital where this is happening and public opinion is rather negative towards it.

5.2 The Vector Effects of Voting

While a scalar is a value that remains the same when the spatial coordinate system changes, the value of the vector depends on its location relative to other elements of space. Suppose we are studying the relationship between the share of workers and support for the left party in two cities, A_i and B_j, located in places i and j, respectively. The left party received 60 % of the votes in City A_i, where 40 % of the residents are working class, and 40 % of the votes in City B_j, where 20 % of the residents are working class. This territorial differentiation reflects the geography of ideological and political cleavages in society. Imagine we can change the coordinates of these cities, but the dependence remains the same: 20 % of the residents of City A_j are working class, and the left party will receive 40 % of votes; while 40 % of the residents of City B_i are working class, and the left will receive 60 % of the votes. What this means is that the results depend solely on the properties of the pace itself, and that we are dealing with a scalar effect in electoral geography. But what if dependence is determined not by the properties of a given place, but rather by that place's location in space (for example, it belongs to a belt of cities that traditionally supports the left)? Suppose that, now, the left receives 30 % of the votes in City A_j, where 20 % of the population is working class, and just 50 % of the votes in City B_i, where 40 %

of the population is working class. What we have here is an example of the vector voting effect, also known as the *neighbourhood effect*. In this example, the neighbourhood effect will be 10 % in both cities (see Table 5.2). This effect belongs to a large family of contextual voting effects in political science that describe how external factors influence the nature of electoral choice [336].

Table 5.2. The neighbourhood effect

Initial situation		No neighbourhood effect		Neighbourhood effect	
A_i	B_j	A_j	B_i	A_j	B_i
20% → 40%	40% → 60%	40% → 60%	20% → 40%	40% → 50%	20% → 50%

So, let us get this straight: *if you change the location of an object and its properties do not change as a result, then we are dealing with non-spatial (scalar) dependence that is determined by the properties of the object itself. If the properties of the object do change when placed in a different location, then we have spatial (vector) dependence.*

The neighbourhood effect means that support for a candidate (or party) is greater in districts that are adjacent to regions where support for this candidate (or party) is higher than the national average. Accordingly, the inverse neighbourhood effect is when support for a candidate (party) is lower in districts that are adjacent to regions where support for this candidate (or party) is lower than the national average. A more formal expression of the neighbourhood effect is that the share of people who vote for a candidate (party) tends to show a positive correlation with the share of those who vote for the candidate (party) in neighbouring districts. In other words, the neighbourhood effect is observed when there is significant spatial autocorrelation in the level of support for a candidate (party) and, as a result, stable spatial differentiation of the electoral landscape.

The neighbourhood effect was conceptualized in 1969 by David Reynolds in the article "A Spatial Model for Analyzing Voting Behavior" [436], and then further explained that same year by Kevin Cox in his paper "The Voting Decision in a Spatial Context" [253]. Cox based his explanation of this principle on Torsten Hägerstrand's theory of the spatial diffusion of innovations. According to Cox, the regular spatial clustering of specific candidates and parties is somewhat similar to the way in which rumours or diseases spread. The voter's behaviour is determined by the influence of the information that dominates in the area where he or she lives. Each person acts as a node in the network of information flows, acting simultaneously as the addressee, transformer and sender of signals. The effectiveness of connections between nodes, and thus of the dissemination of information in the network, depends on the number of addressees, the distance between them and quality of the political culture. The neighbourhood effect is thus explained by the density and quality of the connections between voters: higher density and quality means that voters tend to repeat the voting decisions of their neighbours. William Miller described the essence of the neighbourhood effect using the simple formula "those who speak together vote together" [389], and this conjecture about the nature of the effect of the communication network on electoral behaviour was confirmed by Robert Huckfeldt and John Sprague's 1995 study on how people in Indianapolis and St. Louis voted in the presidential elections [324].

An important addition to our understanding of the neighbourhood effect was provided by the models of environmental effects identified by Miller based on his observations of the UK electoral landscape in 1977 [389]. Describing the neighbourhood effect, Miller pointed out that the dependence can manifest itself differently for different parts of society (class, strata, ethnic groups, etc.), which partly explains situations when the effect is minimal (Fig. 5.1). Miller identified the following environmental (neighbourhood) effect models:

1. *consensual environmental effect* – the classic neighbourhood effect, where communication between people strengthens the position of the dominant political force in the region;
2. *no environmental effect* – the absence of the classic effect, when communication between people does not affect their electoral behaviour;
3. *reactive environmental effect* – the opposite of the classical type, when communication between people increases their mutual annoyance, anxiety and hostility, and, as a result, this weakens the position of the dominant political force in the region;
4. *Przeworski environmental effect* – when the neighbourhood effect acts differently on individual parts of society: say, one segment of society is influenced by the majority, while another is indifferent to it, and a third votes differently in defiance.

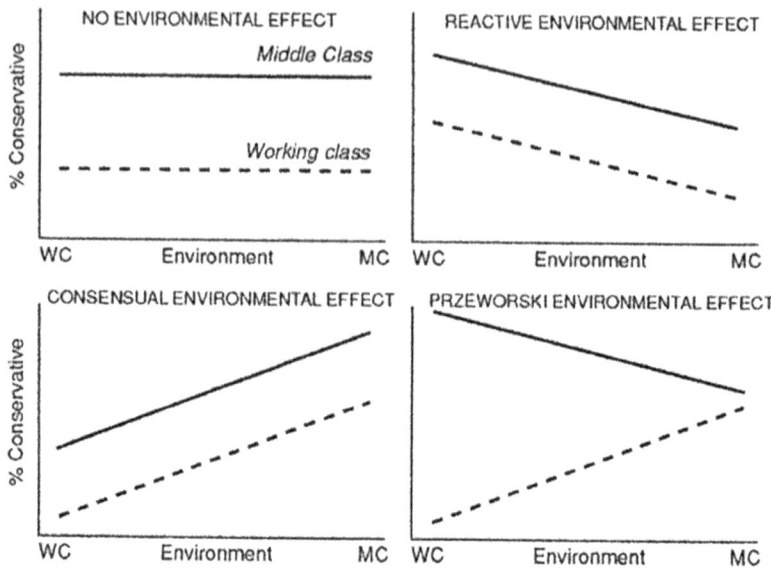

Fig. 5.1. Miller's environmental (neighbourhood) effect
Source: Miller W.L., *Electoral Dynamics in Britain since 1918* (London: Macmillan, 1977), 266.

As we see, the neighbourhood effect is not equivalent to the geography of ideological and political splits in society, including those that are based on territoriality. It can be said that the geography of voting (the electoral and geographic differentiation of society) is a reflection of the geography of ideological and political divisions anchored in space by the neighbourhood effect. The geography of cleavages would create smooth territorial transitions in the spatial continuum of the state, but the neighbourhood effect polarizes society in space by, among other things, creating a territorial dimension for non-territorial (social, ideological and other) splits.

To assess the degree of clustering of the electoral space, either simple indicators of geographic concentration and spatial relationships or spatial statistical analysis methods can be used.

The *localization coefficient* measures the relative distribution (or relative concentration) of supporters of a given party in a specific region in relation to the country as a whole.

$$\text{Localization coefficient} = \frac{X_i / \sum X_i}{N_i / \sum N_i} * 100,$$

where X_i is the number of supporters of the party in region i, $\sum X_i$ is the total number of supporters of that party in the country, N_i is the total number of voters in region i, and $\sum N_i$ is the total number of voters in the country. If the localization coefficient is greater than 1, this means that the party is relatively more concentrated in the region being observed.

While the localization coefficient estimates the relative distribution in space of supporters of one party among all voters, the *geographic disparity index* compares the proportional distribution in space of supporters of two different parties.

$$\text{Geographic disparity index} = \sum_{i=1}^{N} \frac{(X_i / \sum X_i) - (Y_i / \sum Y_i)}{2} * 100,$$

where X_i and Y_i are the number of supporters of parties X and Y in region i, $\sum X_i$ is the sum of supporters of these parties in the country as a whole. The index ranges from 0 to 100, where 0 means absolutely equal spatial distribution of the two parties across the country, and 100 means completely opposite spatial distribution.

$$\text{Geographic segregation index} = \sum_{i=1}^{N} \frac{(X_i / \sum X_i) - (Y_i / \sum Y_i)}{2} * 100$$

The *geographic segmentation index* is similar to the geographic disparity index. The difference is that it compares the performance of one party relative to voters in the country as a whole, rather than the performance of two parties. It is also measured on a scale of 0 to 100 and has the same threshold interpretations.

In order to assess the neighbourhood effect, it is important to both determine the level of space clustering, and to identify the strength of the geographical interaction between objects – i.e. how connected and interdependent they are. This is done using the *gravitational model*, borrowed from Isaac Newton's equation of universal gravitation.

$$I_{ij} = k\left(\frac{P_i * P_j}{D_{ij}^b}\right),$$

where I_{ij} is the force of gravity (spatial interaction) as it is directed from Place i to Place j, k is a constant, D_{ij}^b is the distance between i and j with distance exponent b, and P_i and P_j are the number (or share) of Party P's supporters in regions (electoral districts) i and j. Points i and j are mapped as centroids or administrative centres of regions or electoral districts. For most investigations in absolute two-dimensional space, it is reasonable to use $k=1$ and $b=2$.

5.3 Malapportionment

Unlike the geography of elections, the geography of representation is not interested in the degree of electoral support for candidates and parties. Rather, it is interested in the level of representation in the legislative (or, less often, executive) bodies of power at various levels and the negative experience of unfair distribution. The geography of representation is thus the study of the territorial distribution of seats among the administrative and territorial units of the state, rather than of votes among candidates

The notion of representative democracy is based on the idea that members of parliament express the interests of society, and it is implicitly believed that if the votes of members of parliament are equal, then they should represent equal parts of the community. However, it is practically impossible to divide a country into numerically identical electoral districts. What is more, the population of any given region tends to fluctuate from year to year, which means that even if you start with an optimal distribution of voters, constant calibration of the network is required. Manipulation of the number of electoral districts to ensure that a larger number of voters receives a smaller share of seats is a tool of unscrupulous electoral-geographic engineering called *malapportionment* [274, 394, 424].

Seat distribution is particularly problematic in federative states, as well as in countries with autonomous, remote or sparsely populated territories. In these cases, single-member districts cannot be larger than an administrative division, because if it is merged with another entity, it may lose its representative in parliament. What you get as a result, for example, is Chukotka Autonomous Okrug and Astrakhan Oblast in Russia each electing one deputy to the State Duma, despite a 25-fold difference in population.

Urban and rural single-member districts are also not easily correlated, as balancing them out in terms of representation requires the territory of the latter to be inflated significantly. As a result, the votes of people who live in the countryside typically have greater weight than those of people who live in the city. And this is true the world over.

The simplest way to determine the weight of seats for each specific territory to compare the degree of representation in parliament is through the following formula:

$$Weight\ of\ seat = \frac{total\ population\ /\ total\ seats}{population\ of\ region\ /\ number\ of\ seats\ for\ region}$$

For example, the weight of seats in elections to the European Parliament is below 1 in five countries (Germany, France, Italy, Spain and Poland), meaning that there are more voters per member of parliament than the EU average. At the same time, there is a minimum quota of six deputies per member country, which means that the weight of a single seat for some countries is significantly higher: 7.71 for Malta, 6.19 for Luxembourg, and 4.34 for Cyprus, for example.

Methods for calculating the distribution of seats among territories coincide with the methods for distributing seats among parties, and are also divided into quota methods (largest remainder) and divisor methods (highest average), which we described earlier. The disproportionality indexes covered earlier can be used to assess the degree of territorial distribution, albeit in a somewhat modified form [446]. For example, the variables of the *bad representation index*, based on the Loosemore–Hanby disproportionality formula, will be s – the share of seats won, and v – the share of the population of Territory i, from the total number of seats in parliament and the total population of the country, respectively.

$$Bad\ reprsentation\ index = \frac{1}{2}\sum_{i=1}^{n}|v_i - s_i|$$

Historically, the use of arithmetic methods for allocating seats among regions during parliamentary elections has produced some curious results, which are called paradoxes of redistribution. These paradoxes prompted researchers to develop criteria which the principle of redistribution must meet:

1. *State-population monotonicity.* This criterion is violated when reallocating a fixed number of seats in response to population growth leads to a situation where a region with a relatively large increase loses a seat to a region with a relatively small increase (the population paradox).
2. *House monotonicity.* This criterion is violated when adding a seat in parliament leads to a reduction in the number of seats allocated to the region (the Alabama paradox).
3. *Uniformity/coherence.* This criterion is violated when an increase in the number of seats in parliament by the number of seats allocated to a new region leads to a redistribution of seats among the old regions (the new state paradox).

5.4 Gerrymandering

Gerrymandering is an electoral geographic engineering tool that involves manipulating the boundaries of electoral districts (as opposed to manipulating the population of a district, which is the case with bad representation). The word first appeared in a cartoon published in the *Boston Gazette* in 1812 (we have reproduced it here at the beginning of this chapter) that depicted the boundaries of a Massachusetts electoral district that looked like a salamander [211]. The drawing was a response to the opportunistic redrawing of the electoral district boundaries under governor Elbridge Gerry, which led to the county being dubbed "Gerry's salamander" and the process used to help his party "gerrymandering."

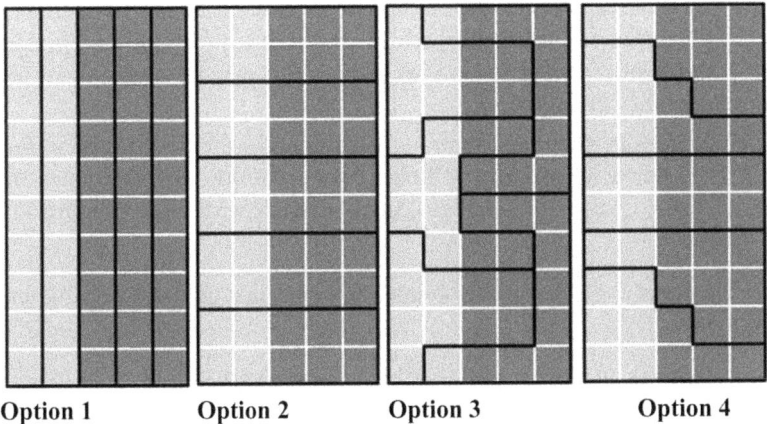

Option 1　　　　Option 2　　　　Option 3　　　　Option 4

Fig. 5.2. Gerrymandering

For example, say we need to distribute 50 voters among five electoral districts of equal size (Fig. 5.2). Twenty voters (40 %) support Party A (the light grey squares), while the remaining 30 (60 %) support Party B (the dark grey squares). The first option for allocating the voters gives Party A two seats and Party B three seats, which corresponds to proportional distribution, but the district boundaries give the impression that some kind of inter-party collusion has taken place in order to avoid competition in any of the districts. The second option awards all the seats to Party B, while the third gives three seats to Party A and just two seats to Party B – both options violate the principle of proportionality. The fourth option is the only one that is consistent with both the principle of proportionality and the desire of society to have competitive elections in all districts.

The following types of gerrymandering can thus be distinguished:

- *Incumbent gerrymandering* – this is done to help the political party that is in power and thus leads to unfair competition.
- *Inter-party gerrymandering* – this is done to help all political parties (as a result of party collusion), which can lead to a proportional distribution of mandates and a reduction in actual political competition.

- *Discriminatory gerrymandering* – this is done in the interests of a certain section of society, which thus reduces the level of representation of the other sections of society (this may be on racial, ethnic, religious, etc., grounds).
- *Legitimate gerrymandering* – this is done in the interests of society as a whole to smooth out inequalities within the electoral space and ensure competitive elections in all districts with a fair proportional result.

There are a number of dishonest ways in which gerrymandering may take place:

- *cracking* or *fracturing* – dividing the electorate into districts in such a way that it is always in the minority (option 2);
- *packing* – consolidating the electorate into several districts where it has an overwhelming majority in order to make it a minority in all other districts (option 3);
- *hijacking* – merging districts that traditionally support two strong candidates;
- *kidnapping* – separating the hometown (district) of the candidate from the district where he or she traditionally enjoys support.
- *status quo* – preserving the traditional division of districts and ignoring changes in the structure of the population and its territorial differentiation.

Gerrymandering is often understood exclusively in the negative sense –changes in the boundaries of electoral districts that lead to an unfair distribution of seats. In this case, legitimate gerrymandering, or changing boundaries in the interests of society as a whole to ensure the fair distribution of seats is called *redistricting* [397]. Redistricting requires, first of all, assessing the level of injustice of the current system of allocating seats. This can be done in two ways – either by evaluating the disproportionality of representation, or by evaluating the geometry of the district, both of which we will look at further on. Second, a strategy to combat unfair gerrymandering needs to be settled on. Independent (political or expert) commissions can be set up at this stage to change the boundaries of electoral districts, attach districts to administrative boundaries, or consolidate the status quo by introducing a ban on altering electoral district boundaries.

Mathematical methods of redistricting can also be used. The simplest of these is the *shortest split-line algorithm* and is used to find the shortest line that divides a territory (polygon) into parts with equal population. If there are several such lines, the northernmost is chosen; if there are several of these, then the westernmost of them is chosen. This procedure is repeated several times, depending on the number of districts into which the country or region (polygon) needs to be divided. If an odd number of districts is needed (say, three), then we look for the shortest line first, dividing the polygon into parts with a population ratio of 1:2, and then we divide the larger district into two parts using the shortest line method, giving us a population ratio of 1:1. At the same time, redistricting is a problem with multiple parameters and cannot be solved using a simple mathematical formula. Factors that need to be taken into account in addition to population size and political orientation include the level of connectivity of the district (including its transport connectivity); the degree to which it resembles other types of administrative, economic, historical, ethno-cultural and physical boundaries; the level of representation of various population groups (ethnic groups, subcultures, social strata, etc.); whether or not the district includes remote or isolated communities (for example, military or mining towns), etc. This is what makes the human factor key in geographical limology.

5.5 The Efficiency Gap

The *efficiency gap* proposed by Nicholas Stephanopoulos and Eric McGhee in 2014 is one of the simplest yet most accurate indicators of the effect of gerrymandering [381, 468]. The idea behind it is to take two types of *wasted* vote into account: *losing votes*, i.e. the number of votes received by the losing candidate(s); and *excess votes*, or the number of votes beyond the number needed for victory.

Table 5.3. Wasted votes and the efficiency gap

District	Party A	Party B	Winner	A wasted votes	B wasted votes
1	6	4	A	0	4
2	1	9	B	1	3
3	6	4	A	0	4
4	1	9	B	1	3
5	6	4	A	0	4
Total	20	30	A – 3, B – 2	2	18

Take a look at Table 5.3, which describes option 3 for the distribution of voters in Figure 5.2. Party A won the first district with six votes. To win, it needed to receive 50 % + 1 of the votes, that is, it needed six votes. Accordingly, none of the votes it received was wasted here. At the same time, the four votes for Party B in this district were losing, and thus wasted votes. Party B won the second district with nine votes. But it only needed six to win, meaning that three votes were excess votes. Meanwhile, Party A lost just one vote in this district. The efficiency gap is calculated as the difference between the number of wasted votes divided by the total number of votes. An efficiency gap of greater than 7 % is seen as sufficient grounds to claim that the electoral district boundaries are unfair.

$$Efficiency\ gap = \frac{18-2}{50} * 100\% = 32\%\ \text{in favour of Party A}.$$

Let us translate all this into the language of mathematics. Wasted votes W for Party P in won districts i and lost districts j are calculated from the total number of votes received by the party in each district V^P using the formula:

$$W^P = \sum E_i^P + \sum L_j^P\ \text{(wasted votes)},$$

where $E_i^P = V_i^P - \frac{V_i}{2}$ for District i, which Party P won (excess votes); and $L_j^P = V_j^P$ for District j, which Party P lost (losing votes).

Or in a generalized form for all electoral districts d, where Γ^P is the set of districts which Party P won:

$$W_d^P = \left\{ V_d^P \text{ for } d \notin \Gamma^P, \left(V_d^P - \frac{V_d}{2} \right) \text{ for } d \in \Gamma^P \right\}$$

The efficiency gap for parties A and B in this case would be equal to:

$$\text{Efficiency gap}(EG) = \frac{W^A - W^B}{V}$$

Note that if the votes in a given district are split 60/40, then, according to the Stephanopoulos–McGhee formula, the first party will have $60 - 50 = 10\%$ excess votes, while the second party will have 40% losing votes, and 10% of the votes will be unaccounted for by either party. In reality, however, it turns out that the first party only needed 40% + 1 of the votes (i.e. not 50 % + 1), which means that 20%, rather than 10%, are considered excess votes. A solution to this problem that would make it possible to calculate the efficiency gap for systems with more than two dominant parties was offered by John Nagle, who proposed calculating a weighted efficiency gap based on the difference (λ) between the number of votes won by each party, as well as a relative efficiency gap based on the proportion between the percentages of support for parties [250, 402].

$$\text{Weighted efficiency gap}(EG_\lambda) = \frac{W^A(\lambda) - W^B(\lambda)}{V}$$

$$\text{Relative efficiency gap}(REG_\lambda) = \frac{W^A(\lambda)}{V^A} - \frac{W^B(\lambda)}{V^B}, \text{where}$$

$$W_d^P(\lambda) = \left\{ V_d^P \text{ for } d \notin \Gamma^P, \lambda \left(V_d^P - \frac{V_d}{2} \right) \text{ for } d \in \Gamma^P \right\}$$

Let us return to our example with the distribution of voters according to option 3 (Table 5.4). Party A won the first district with six votes. Five votes were needed to win, because the second party only won four votes, meaning that one vote was wasted. At the same time, the four votes for Party B in this district were also losing votes. Party B won the second district with nine votes, although it only needed two votes to win, meaning that seven votes were excess votes. Meanwhile, Party A lost just one vote in this district. The weighted and relative efficiency gaps in this case will be 42 % and 62 %, respectively, in favour of Party A.

$$EG_\lambda = \frac{26-5}{50} * 100\% = 42\%$$

$$REG_\lambda = \left(\frac{26}{30} - \frac{5}{20}\right) * 100\% = 62\%$$

Table 5.4. Wasted votes and the weighted efficiency gap

District	Party A	Party B	Winner	A wasted votes	B wasted votes
1	6	4	A	1	4
2	1	9	B	1	7
3	6	4	A	1	4
4	1	9	B	1	7
5	6	4	A	1	4
Total	20	30	A – 3, B – 2	5	26

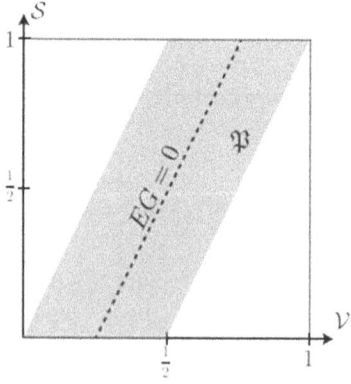

Fig. 5.3. Efficiency gap
Source: Tapp, K., "Measuring Political Gerrymandering," *The American Mathematical Monthly*, Vol. 126, No. 7 (2019), p. 602.

The efficiency gap can be visualized on the graph (Fig. 5.3), on which axis V denotes the share of votes received by the party, and axis S is the number of seats it won [474]. Parallelogram P represents the area of possible variations of the relationship between V and S. The straight line denoting the efficiency gap passes directly through the centre of the parallelogram.

5.6 Compactness of Electoral Districts

It is generally believed that electoral districts should be compact, that is, they should have a closely spaced area without gaps. A low compactness index score indicates the likelihood of unfair gerrymandering. There are various approaches to assessing the degree of compactness of a district's shape: isoperimetric, dispersed, inertial, and others. Most of them are based on comparing the geometry of a given district with basic shapes. The abundance of methods for assessing the compactness of the shape of a territory is due to the fact that different tasks and different cases require different approaches to measurement. The simplest methods would be to compare the absolute deviation from the country's centroid (the distance to the farthest point from it), or the sum of the distances between the country's centroid and all the voting districts (possibly weighted by the population in each district). However, such methods do not actually give an idea about the shape of the district.

Compactness is most often assessed using isoperimetric methods – that is, methods based on the size of the district's perimeter. The simplest approach is to calculate the total length of the country's borders: the more complex and non-compact its configuration, the greater the length of the border. The downside of this approach is that the length of the boundary of the geographical feature is calculated as the sum of the perimeters of its constituent parts, which means that it depends on the scale of the analysis and the available data.

The most common method for assessing the degree of compactness of an electoral district is the so-called Polsby–Popper test, named after the lawyers who developed it (Daniel D. Polsby and Robert Popper) [426]. A similar test was developed by Eric Cox back in 1927 to tackle problems in paleontology. The test is based on the isoperimetric coefficient: the ratio of Area A of Electoral District d to the area of a circle whose circumference is equal to the perimeter of the shape of Electoral District P_d.

$$The\ Polsby-Popper\ test = 4\pi A_d\ /\ P_d^2$$

The Polsby–Popper test is measured in the interval $[0,1]$, with 0 representing no compactness at all, and 1 representing absolute compactness (where the district is shaped like a circle – the most compact of all the geometric shapes). A drawback of the test is its sensitivity to the resolution of cartographic data, which distorts the sinuosity of the country's boundaries (especially coastlines). Non-continuous electoral districts – those consisting of several parts – present special challenges. In this case, the test must be performed separately for each of the parts of District i.

$$Non-continuous\ Polsby-Popper\ test = 4\pi \frac{\sum_i^n A_i}{\sum_i^n P_i^2}$$

Similar to the Polsby–Popper principle is the test developed by Joseph E. Schwartzberg, which is based on the ratio of the perimeter of the district to the length of the region bordering the circle with an area equal to that of the electoral district [451]. Norman Pounds put forward a similar idea whereby the percentage that a district's perimeter is from the perimeter that makes up the circumference of the circle with an area equal to that of the electoral district is calculated. The Schwartzberg test, as the Soviet geographer Yuri Frolov points out, is an application to electoral geography of an old approach known at least since Nagel's formula, which was proposed in 1835 [166]. For comparison: the Schwartzberg test will give us a score of 1.13 for a square (if we use the range $[0,1]$, that is, assign it a value between 0 and 1, it will be 0.88), 1.29 (0.77) for a triangle, and 1.95 (0.51) for a five-pointed star.

$$Schwartzberg\ test = P_d \big/ \left(2\pi \sqrt{A_d/\pi}\right)$$

$$Attneave-Arnoult\ test = 1 - \frac{2\sqrt{\pi A_d}}{P_d}$$

Another measurement of compactness based on the perimeter-to-area ratio is the Attneave–Arnoult test proposed by psychologists. It is difficult not to notice that all three isoperimetric tests – the Polsby–Popper (PP), Schwartzberg (S) and Attneave–Arnoult (AA) are variations of each other, since $S = (1/PP)^{1/2}$, and $AA = 1 - PP^{1/2}$. The original method of calculating the isometric index of compactness was discovered by Mark Flaherty and William Crumplin, who proposed using a set ratio of the area to the perimeter of the district in calculations that is equal to 0.282: $A_d / 0,282 P_d$ или $A_d / (0,282 P_d)^2$.

The second group of methods for assessing compactness is dispersion methods, which, in turn, are divided into two subgroups: the first is based on the area of the territory, and the second on its length and width. Territory-based dispersion methods estimate the ratio of the area of a given electoral district to the area of the smallest geometric shape that would completely enclose it: a circle (Reock score [434]), rectangle (rectangular Reock test), hexagon (Geysler test), or a polygon – a convex hull (Niemi test [404]). Other options include identifying the ratio of the area of the maximum inscribed circle to the area of the maximum escribed circle (Skew test), or to the area of the electoral district itself (Ehrenberg test); calculating the ratio of the area of intersection of the shape of the electoral district and a circle of equal area to the sum of their areas (Lee–Sallee test); or, finally, estimating the ratio of the area of the electoral district in the denominator to the overlapping area after superimposing it on top and its reflection along the horizontal or vertical axes in the numerator (X- and Y- symmetry tests).

$$Gibbs\,test = A_d \Big/ \Big[\pi (L/2)^2 \Big]$$

These methods are all rather complicated. The Gibbs test is the only one that provides a more or less simple approach, as it measures the ratio of the area of the electoral district to the area of a circle with a diameter equal to the length of the largest axis of District L, which can be carried out if you only know the area of the district and the diameter between its limiting points. Figure 5.4 compares the basic principles for measuring the compactness of electoral districts based on area.

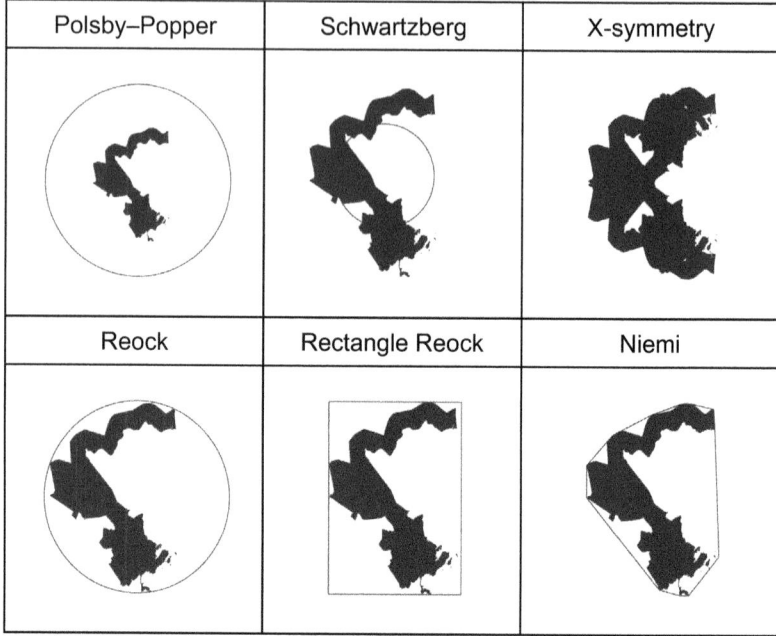

Fig. 5.4. Main compactness tests for districts
Pictures taken from: https://fisherzachary.github.io/

The next subgroup of dispersive methods bases measurements on the length and width of the electoral district. The following methods are used for this [206, 284]:

- the ratio between the maximum diameter of the district as the denominator and the longest perpendicular to that diameter as the numerator;
- the ratio between the width and length of the minimum bounding rectangle surrounding the district;
- the maximum ratio between the width and length of the bounding rectangle surrounding the district and touching its four extreme points;
- the ratio between the width and length where the width is the longest axis of the district and the length is the longest axis of the bounding rectangle surrounding the district and touching its four extreme points;

- the ratio between the diameter of the maximum circle inscribed in the district and the diameter of the smallest circle escribed outside the district;
- the ratio between the radius of a circle that is equal to the area of the district and the radius of the circle escribed outside the district;
- the ratio between the minimum and maximum distance between the district's boundaries;
- the difference between the length of the maximum diameter of the district and its longest perpendicular;
- the difference in the distance between the northernmost and southernmost points and the westernmost and easternmost points of the district.

One final and rather unique approach is to find the moment of inertia in the electoral district. In general terms, the moment of inertia is equal to the sum of the products of the point masses and the square of their distances to the base set (point, line or axis). For an electoral district, the moment of inertia is the normalized variance from all points in the district to its centre of gravity.

$$\text{Moment of inertia of an electoral district} = A_d \Big/ \sqrt{2 \iint_D r^2 dD},$$

where r is the distance from the district's centroid, and D is the set of points in the shape of the district that represent a given variable. Here, both the distance to the centre of the district boundaries and, say, the population in individual voting districts can be used as variables.

Key Terms

- scalar and vector spatial effects of voting: the friends-and-neighbours effect, the campaign effect, the issue voting effect, the neighbourhood effect
- the neighbourhood effect: consensual, no effect, reactive, Przeworski
- electoral-geographic engineering tools: unfair distribution, gerrymandering (incumbent, inter-party, discriminatory, legitimate), redistricting

- gerrymandering techniques: cracking, packing, hijacking, kidnapping, status quo
- shortest split-line algorithm, the efficiency gap, compactness test

Questions and Exercises

1. Analyse the map of votes for Vladimir Putin in the 2000 Russian presidential election. It shows the deviations in the level of support in the regions from the national average. What spatial effects of voting can be inferred from the map?

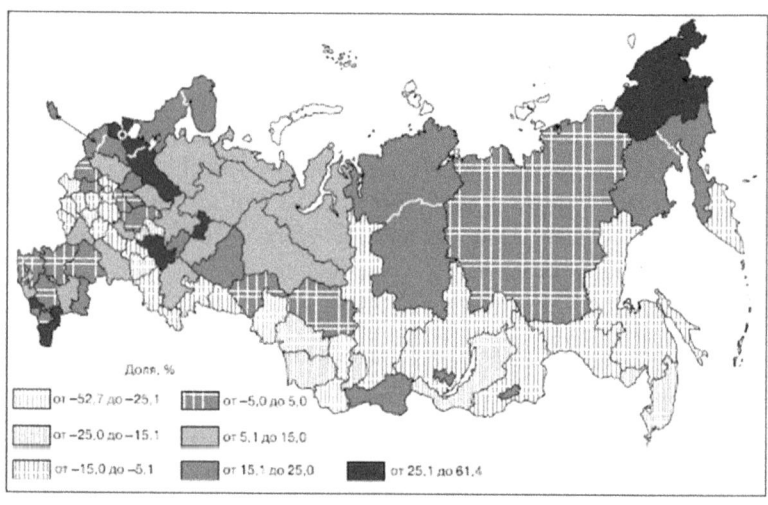

Source: V.P. Maksakovsky, *Geographic Map of the World* (Moscow: Drofa, 2004), Vol. 1, p. 23.

Доля, %	Share, %
от ... до ...	from ... to ...

2. Analyse the effects of electoral geography in the following feature films: *Brexit: The Uncivil War, Election Day, Wag the Dog*.
3. Besides the theory of spatial diffusion of innovations, what other explanations are there for the neighbourhood effect?

4. Analyse the most recent parliamentary elections in the United Kingdom. What types of neighbourhood effects can be seen for each of the parties?
5. Calculate the weight of seats for all regions of Russia for single-seat elections to the State Duma. How does it change when new regions join Russia?
6. Look at the following map of a pirate state. It is home to ten pirates: nine of them live either by a lake, hills or palm trees, and one lives on a boat and comes to the island to vote. We know that the pirates living along the island's largest river have similar electoral preferences. Suggest how to divide the island into two single-member districts for parliamentary elections according to the majoritarian system so as to:

   a. ensure that electoral competition is as high as possible;
   b. ensure that electoral competition is as low as possible.

Source: https://mentamaschocolate.blogspot.com/2015/12/dibujos-de-piratas-y-sus-elementos.html

7. Divide the pirate state from the previous task into two or three electoral districts using the shortest split-line algorithm.
8. Study the following U.S. Supreme Court cases of gerrymandering, the arguments of the parties, and the methods used to assess disproportionate results:

    a. Gaffney v. Cummings (1973);

    b. Miller v. Johnson (1995);

    c. League of United Latin American Citizens v. Perry (2006);
    d. Gill v. Whitford (2016).

9. Plot tables similar to Table 1 for options 1, 2 and 4 for allocating seats in Fig. 1. Calculate the base, weighted and relative efficiency gaps for them.
10. Find the least compact electoral district in Moscow.

Chapter 6

Spatial Analysis in Electoral Geography

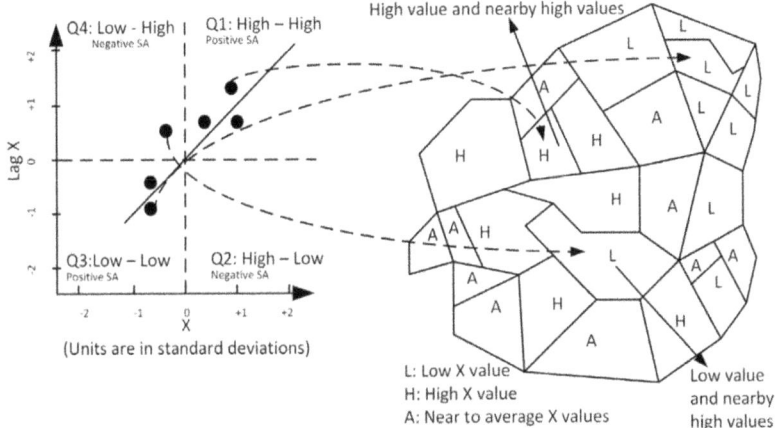

Source: Grekousis, G. (2020). Spatial Analysis Methods and Practice: Describe – Explore – Explain through GIS. Cambridge: Cambridge University Press, p. 230.

- How can electoral statistics be visualized on a map?
- What is the normal (standard) distribution of votes for a candidate by territory?
- How can the electoral space be clustered?
- How can regions that differ from their neighbours in their electoral behaviour be identified?
- In statistics, can a dependent variable explain itself?

6.1 Electoral-Geographic Maps

Electoral geography works with data (primarily the percentage of support for candidates or parties, but also, for example, with voter turnout and certain other parameters) that is differentiated in space, meaning data that is distributed in space according to certain patterns, rather than randomly. Electoral geography tries to identify the patterns in the distribution of electoral behaviour in space, and thus establish the nature of the influence of space on the electoral process. Various methods are used to do this, the simplest and most revealing of which is electoral-geographic mapping – the plotting of data on a map to help visualize patterns in the distribution of electoral behaviour in space and make assumptions about their causes.

A geographic *map* is a generalized image of the plane of the Earth's surface It is based on the mathematical law of construction, which transforms the spherical shape of the Earth into a plane. However, in electoral geography, data is not displayed on geographic maps; rather, it is displayed on chorochromatic maps (or *choropleth maps*) – simplified geographic models that reflect the respective spatial patterns. Chorochromatic maps do not include many of the elements you would expect to see on a map: the environment of the object of study, a coordinate grid, scale bar, etc.

The most commonly used level of analysis in electoral geography is a *polygon*, that is, an areal object (electoral district, administrative unit, country), or, far less frequently, a point (voting district, city) or linear (border, highway) object. This is why we often work with data in vector format, the most common type of which is a shapefile (*.shp*).

One of the main problems for analysis is how to determine the boundaries of electoral district polygons for the purposes of using statistics for spatial analysis. There is an international project dedicated to this very work, the shapefile repository of the Center for Political Studies at the University of Michigan.[2] If you are researching countries where districts coincide with administrative divisions, then you can use the DIVA-GIS repository.[3] Data on Russia and neighbouring countries can be found in the catalogue of the Center for Spatial Analysis in International Relations at the Institute for International Studies of MGIMO University.[4]

---

[2] K. Kollman, A. Hicken, D. Caramani, D. Backer, D. Lublin, J. Selway, and F Vasselai. *GeoReferenced Electoral Districts Datasets*. Ann Arbor: Center for Political Studies, University of Michigan, 2019. http://www.electiondataarchive.org.

[3] R. Hijmans, E. Rojas, M. Cruz, R. O'Brien, and I. Barrantes. DIVA-GIS. Berkeley: University of California, 2011. https://www.diva-gis.org/.

[4] *Spatial Data*. Moscow: Center for Spatial Analysis in International Relations, Institute for International Studies, MGIMO University under the Ministry of Foreign Affairs of the Russian Federation (2022). https://mgimo.ru/about/structure/ucheb-nauch/imi/geo/docs/spatial-data/

Most countries with a strong electoral tradition and open data policies publish shapefiles with distinct boundaries on the websites of electoral commissions or other election authorities.

The examples in this chapter use a shapefile with the results of the 2012 and 2016 U.S. presidential elections produced by the Center for Spatial Data Science at the University of Chicago.[5] The shapefile contains data on 74 variables, including both electoral results and socioeconomic statistics for 3,108 observations (state counties) for 2007–2016. Most of the mapping and analysis procedures will be carried out using the GeoDa geoinformation software.

The main (and most common) task of an electoral map is to represent the spatial differentiation of voting results for a specific candidate or party, as well as voter turnout. In other words, the map must plot the change in a relative parameter, measured as a percentage from 0 to 100. This is usually accomplished using an interval chorochromatic map with a uniform colour scale, where the colour saturation changes smoothly to denote the quantitative change in the parameter. As the name suggests, interval chorochromatic maps divide the data set into intervals – this is their key characteristic. A number of methods are used to determine intervals:

- The *equal classes (quantile) method* produces chorochromatic maps with a uniform scale, in which the entire dataset is divided into classes with an equal number of observations, each of which corresponds to a separate shade of colour. These types of chorochromatic maps typically use quartiles,[6] in which the data is divided into four equal classes, or quintiles, where the data is divided into five classes. Chorochromatic maps with equal classes with an odd number of classes show, among other things, the median class of the parameter's distribution.

- The *equal interval method* produces chorochromatic maps with a uniform scale; this time it's the scale itself, not the observations, that is divided into equal parts. Each part is assigned a separate

---

[5] 2012 and 2016 Presidential Elections. Chicago: Center for Spatial Data Science, University of Chicago, 2017. https://geodacenter.github.io/data-and-lab//county _election_2012_2016-variables/.
[6] Broadly speaking, any equal-quantity interval can be called a quantile.

shade of colour (in this sense, an equal interval chorochromatic map is consistent in its logic with a histogram).

- The *natural interval method* is used to generate chorochromatic maps in which observations are divided into groups based on the optimization method of the American cartographer George Jenks [330]. With this method, the boundaries of intervals are set where the greatest differences between groups of similar values of the variables are found. That is, this approach allows you to define classes that have the smallest in-class variance, but at the same time differ as much as possible from one another. To do this, the number of required classes is set first, and all possible combinations of dividing the data by the given number of classes are determined. Then, the standard deviations between all the average values of the classes are calculated, and the combination of intervals with the smallest standard deviation is selected.

- The *geometric interval method* allows us to create chorochromatic maps that have both a relatively average number of observations in classes and relatively average value ranges of the variable. They are created by minimizing the sum of the squares of the elements in each class.

- The *exponential interval method* is needed when you want to build chorochromatic maps with such intervals that the number of observations increases (or decreases) exponentially with each subsequent interval.

- The *head–tail* method is used for data with a large number of observations, low variable values, and a high degree of variance. The data set is divided into two classes according to the arithmetic mean, then the class containing the maximum value is further divided into two classes according to the arithmetic mean. In other words, the "tail" is cut off from the "head" until the required number of classes is reached.

- The *specified interval method* is used for chorochromatic maps in which the values of class boundaries are determined by the cartographer or researcher. This approach can lead to data manipulation, but it is sometimes required in order to solve specific research issues.

- The *percentile method* produces chorochromatic maps with hard interval boundaries, but an unequal scale, for example: 0–1 %, 1–10 %, 10–50 %, 50–90 %, 90–99 %, 99–100 %, which can be

Electoral-Geographic Maps

used to identify observations with extremely high or extremely low values (*outliers*).

- The *range method* is based on the idea of a box and whisker plot. It involves dividing the data into six classes: two classes of high and low outliers and four classes of equal quartiles.
- The *standard deviation method* is similar to the idea of the range method, but it uses a different principle for determining outliers. The resulting chorochromatic map also shows the extent to which the observations deviate from the mean, but class breaks will be set in increments of 0.25σ (0.25 standard deviation), 0.5σ and σ (sometimes even 3σ) on either side of the mean. The result of the classification shows how the attribute values of objects differ from the average.

It is clear that the first two methods – equal classes and equal intervals – produce chorochromatic maps with a uniform scale, and the others do not. The last three methods of determining intervals, which focus more on identifying outliers, are best used with a two-colour scale, since this will clearly show data located in classes above and below the average value.

To illustrate some of the types of interval chorochromatic maps, let us look at the map of support for Republican candidate Donald Trump in the 2016 U.S. presidential election. We divided the quintile chorochromatic map of support for Trump (Fig. 6.1) in the 3,108 counties that kept electoral statistics into five equal classes, giving us 621 or 622 counties in each class. The median class was thus made up of counties with levels of support ranging from 62 % to 70 %, which includes an arithmetic mean of support across all counties of 64 % and a median of 67 %. However, the problem lies in the fact that this approach gives us a huge range of values in the first class (4–51 %), and a very limited range of views in the fourth class (70–77 %).

Fig. 6.1. Chorochromatic map of five equal classes (quintiles)

Note that, when mapping the equal intervals, it was the scale of the parameter, rather than the observations (counties), that were divided into five equal parts. In our case, this was the share of votes for Trump, which ranged from 4 % to 95 % (Fig. 6.2). This chorochromatic map divides the range into almost equal intervals of 18.2–18.3 %: 4.1–22.4 %, 22.4–40.6 %, 40.6–58.8 %, 58.8–77 % and 77–95.3 %. Now, however, the counties are unevenly distributed: 51 in the first quintile; 233 in the second; 716 in the third; 1,502 in the fourth; and 606 in the fifth. The distribution of counties by class also gives us a histogram of five equal intervals (Fig. 6.3). The intervals here are more representative, but the median interval has now shifted to the range of 41–59 %, which is far from the arithmetic mean and median. The fourth interval can also be considered problematic here, as it includes almost half of the observations.

Electoral-Geographic Maps   183

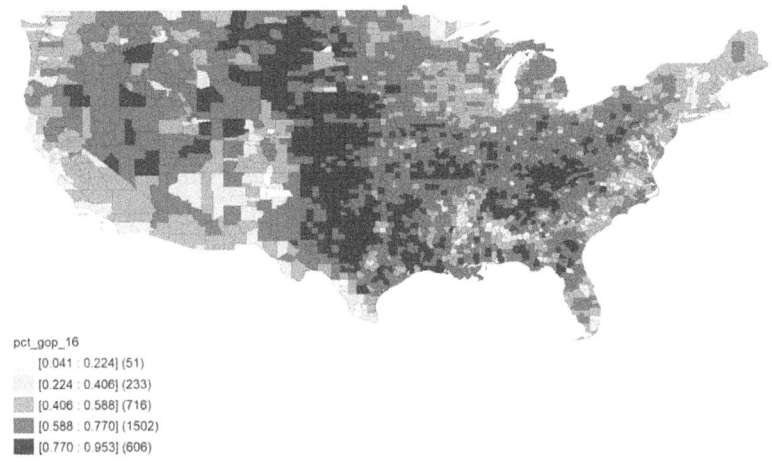

Fig. 6.2. Chorochromatic map of five equal intervals

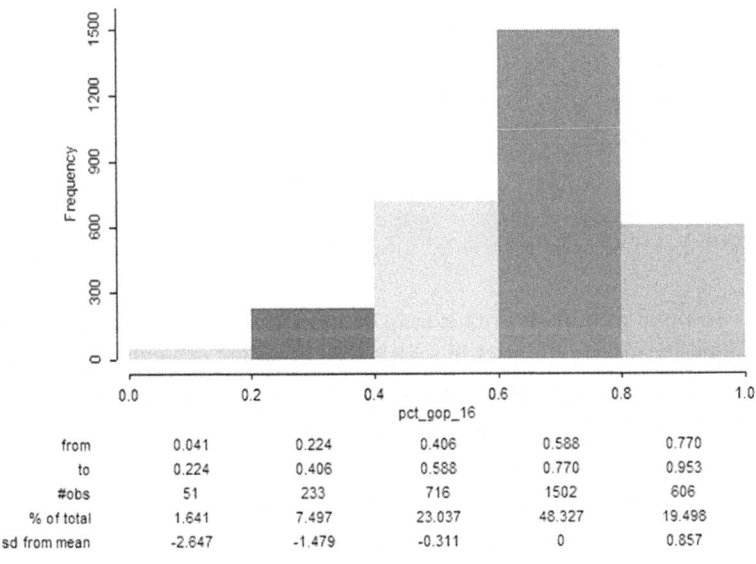

Fig. 6.3. Histogram of five equal intervals

Now let us look at a chorochromatic map of natural intervals, where Jenks optimization was used to determine classes with the following ranges: less than 35 % (187 counties); 35–53 % (500 counties), 53–67 % (855 counties), 67–77 % (939 counties), and more than 77 % (627 counties) (Fig. 6.4). This method turns out to be the most successful, because, first of all, both the observations and the parameter are relatively evenly distributed, and, second, the median class coincides with the mean value of the variable.

Fig. 6.4. Chorochromatic map of five natural intervals

To build a chorochromatic map of specified intervals, we set the task of identifying the counties in which the candidate overcame psychologically significant barriers: 1/3 of the votes, 1/2 of the votes (that is, an absolute majority and guaranteed victory), 2/3 of the votes, and 3/4 of the votes. It turned out that Trump received sufficient support to win in 2,527 of the 3,108 counties in the country (the third, fourth and fifth classes).

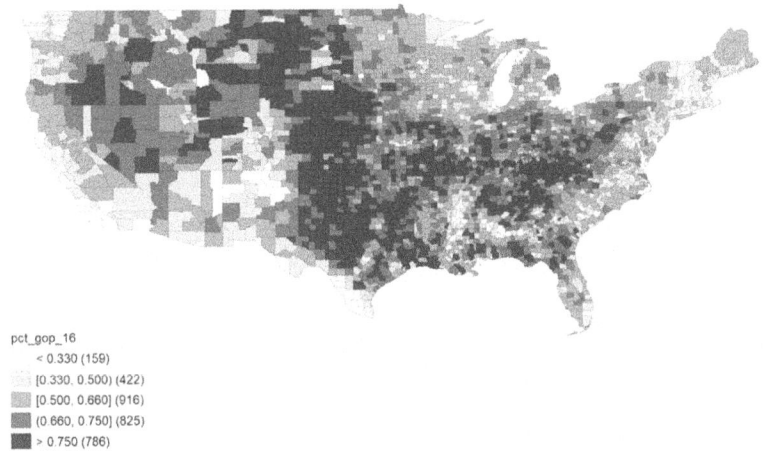

pct_gop_16
< 0.330 (159)
[0.330, 0.500] (422)
[0.500, 0.660] (916)
(0.660, 0.750] (825)
> 0.750 (786)

Fig. 6.5. Chorochromatic map of specified intervals

6.2 Exploratory Data Analysis

Mathematically speaking, two sets take part in the electoral process: voters and the people (things) they are voting for (candidates, parties, or, less often, decisions – for/against, as in referendums). In this sense, elections are the process of mapping a set of voters to a set of people (things) they are voting for, dividing voters into non-overlapping classes of electoral preferences. Imagine that voters A in region i voted for party P_1, which can be represented as set A_i intersecting with set $P_1 : A_i \cap P_1$. The sum of the elements within the subsets resulting from this partition equals the original set, which means that the sum of votes cast for individual candidates is equal to the total number of those who took part in the elections.

The starting point for electoral-geographic statistical analysis is *exploratory data analysis* (*EDA*), which involves assessing the distribution of variables (typically the share of votes), identifying deviations and anomalies, and verifying dependencies between individual variables.

The first basic statistical operation is to determine the average value of the distribution of a variable, say, to clarify the average distribution of votes for a party by region. This can be achieved in several ways, but the most common techniques are to find the arithmetic mean and median.

The *arithmetic mean* ($\underline{x}$ or μ) is calculated as the sum of a collection of numbers divided by the count of numbers in the collection.

$$\underline{x} = \frac{\sum x_i}{N}$$

The *median* is the middle number in a set (series) of numbers. For example, say we have a series of numbers 1, 2, 3. The arithmetic mean will be equal to (1+2+3)/3=2, and the median will also be equal to two. But say we add a fourth number to this series, giving us 1, 2, 3, 6. Now the arithmetic mean will increase to 3. Meanwhile, the median is calculated for an even number of observations by taking the arithmetic mean of the two middle observations (2 and 3), giving us 2.5. Thus, the arithmetic mean, unlike the median, can be distorted if there are significant outliers (large deviations) in the sample. This is why the mean is not considered *robust*, as it is not resistant to various kinds of noise.

Another valuable tool for electoral-geographic analysis is *weighted average*, which takes into account the degrees of importance of the values being averaged. Average values in weighted voting systems can be determined in this manner. However, weighted averages are even more important for spatial analysis, when the geographical location of an observation (say, its proximity to or distance from a point under observation) can be described by a weighting coefficient. The weighted arithmetic mean is calculated using the following formula:

$$\underline{x} = \frac{\sum_{i=1}^{n} w_i x_i}{\sum_{i=1}^{n} w_i} = \frac{w_1 x_1 + w_2 x_2 + \ldots + w_n x_n}{w_1 + w_2 + \ldots + w_n}$$

where w is the weighting coefficient, assuming $w_i \neq 0$.

The second basic statistical operation is an estimation of the *variation* (scatter) of a variable. This is done by using the range of a variable, that is, knowledge of its maximum and minimum values and the difference between them, or its *interquartile range* (*IQR*) – the difference between the 75th and 25th percentiles (or the third and first quartiles). More important than this, however, is to understand the degree to which the average variable deviates from its centre in the scatter. This involves calculating the *variance* (σ^2), which is defined as the ratio of the sum

Exploratory Data Analysis

of the squared deviations from the arithmetic mean to the total number of observations.

$$\sigma^2 = \frac{\sum (x - \underline{x})^2}{N}$$

Variance is an intermediate value. In order to bring it to the same values in which the variable is measured, we need to calculate the standard deviation of the variable (σ), which gives us the square root of the variance.

$$\sigma = \sqrt{\frac{\sum (x - \underline{x})^2}{N}}$$

When calculating variance for a sample (rather than the entire data set), the value $N-1$ is taken as the denominator. This is what is usually known as the sample standard deviation.

$$sd = \sqrt{\frac{\sum (x - \underline{x})^2}{N-1}}$$

In probability theory, Gauss and Laplace proved that with a *normal distribution* of a single-peaked (unimodal) phenomenon, the probability of a data point deviating from the arithmetic mean by 3σ is extremely unlikely. Figure 6.6 shows the probability of deviation from the mean according to the Gauss–Laplace normal distribution. This is why statistical outliers are considered values $> 3\sigma$ (less often $2\sigma \approx 1,5 IQR$).

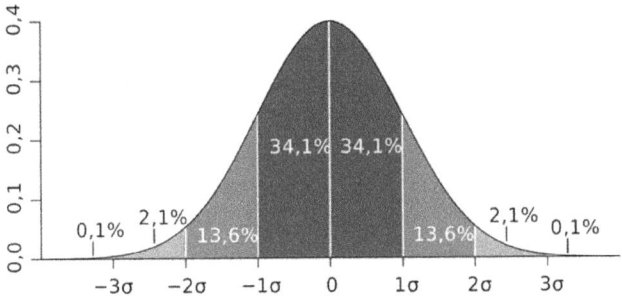

Fig. 6.6. Gauss–Laplace normal distribution

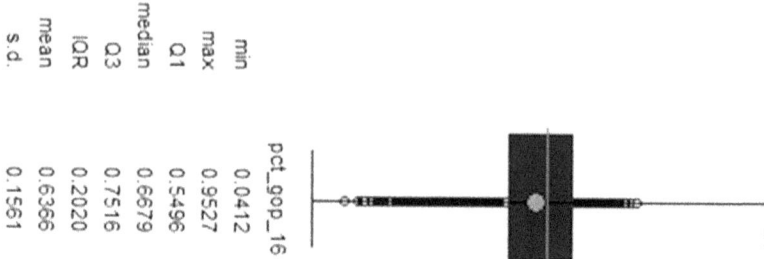

Fig. 6.7. Box-and-whiskers plot with a standard deviation of 3σ

Let us analyse, for example, the distribution of votes for Trump in the 2016 presidential elections (Fig. 6.7). The range is 91 % (a minimum of 4 % and maximum of 95 %), while the interquartile range is just 20 %. The mean (64 %) is close to the median (67 %), and the standard deviation is 16 %. However, the distribution is asymmetric, as there are significantly more extremely small values than there are extremely high values. However, none of the indicators deviate by more than three standard deviations, which means that we can say we have a normal distribution with no outliers, and further statistical analysis can be carried out with it.

Science seeks to establish laws, that is, relationships between variables, and assumptions about those relationships ("the deeper into the forest, the more firewood you will find") are hypotheses. The law in its generalized form can be represented as a formula $y_i = f(x_i)$, where y is the dependent variable, x is the independent variable, and f is the link between them. The link is considered *correlational* when the distribution of the independent variable functionally coincides with the dependent variable (it is *covariant* to it) but we cannot establish a cause-and-effect relationship between them, and *causal* when the independent variable is the cause of the distribution of the dependent variable. In our example hypothesis, the link is correlational. However, in the social sciences, where it is impossible to conduct a true experiment, we do not deal with laws – rather, we deal with patterns and regularities, conclusions obtained with statistical probability. They are described using the formula $y_i = f(x_i) + \varepsilon$, where ε are uncontrolled or unaccounted factors. The lower the weight of ε in the model, the more *valid* the regularity.

Exploratory Data Analysis

The classic method for detecting linear correlation is to calculate the *Pearson coefficient* (r). It is calculated as the difference between the value of the variable and its arithmetic mean, divided by the standard deviation of the variable. The sum of the products of all such calculations for both variables is further divided by the number of observations. The correlational relationship ranges from –1 to 1, where 1 is an absolutely positive relationship, –1 is an absolutely negative relationship (say, the more x, the less y), and 0 is no relationship at all.

$$r_{X,Y} = \sum_{i=1}^{n}\left(\frac{x_i - \underline{x}}{\sigma_x}\right)\left(\frac{y_i - \underline{y}}{\sigma_y}\right) / N = \frac{\sum_{i=1}^{n}(x_i - \underline{x})(y_i - \underline{y})}{\sqrt{\sum_{i=1}^{n}(x_i - \underline{x})^2}\sqrt{\sum_{i=1}^{n}(y_i - \underline{y})^2}}$$

The square of the Pearson coefficient is the simplest example of the coefficient of determination (R^2). Table 6.1 presents the ranges of the Pearson coefficient (in absolute value and squared) that are generally considered to be significant.

Table 6.1. **Levels of determination of correlational links**

Level	$|r|$	R^2
Strong	> 0.7	> 0.5
Average	0.7⟨...⟩0.5	0.5⟨...⟩0.3
Weak	< 0.5	< 0.3

Scatterplots and *regression lines* are used to represent the correlation. In the diagram, the y-axis denotes the independent variable, the x-axis the dependent variable, and the line represents the mathematical expectation of the absolute relationship between them. The regression line (S) is determined using the least squares method, in which its position is calculated as the shortest distance from all points to the line.

$$S = \sum_{i=1}^{n}(y_i - \breve{y}(x))^2 \rightarrow min$$

Figure 6.8 is a scatterplot that tests the hypothesis that Republicans (the dependent variable) are more likely to vote in municipalities with a larger proportion of white population (the independent variable). The hypothesis was not confirmed. Moreover, the regression line actually shows a weak negative correlation $(R^2 = 0.307)$.

Let us also note the additional statistical test values, namely, the standard error, t-statistic and significance level (p-*value*). These are given for two parameters: the constant α and the regression line β. The constant α (or y-intercept) is a coefficient that shows what y will be if the dependent variable is equal to zero; in other words, it is an estimate of those residuals (uncontrolled and unaccounted factors) denoted by ε in the basic formula $y_i = f(x_i) + \varepsilon$. Then the regression line β describes the functional relationship between the dependent and independent variables, as if there were no residuals – i.e. the formula would be $y_i = f(x_i)$.

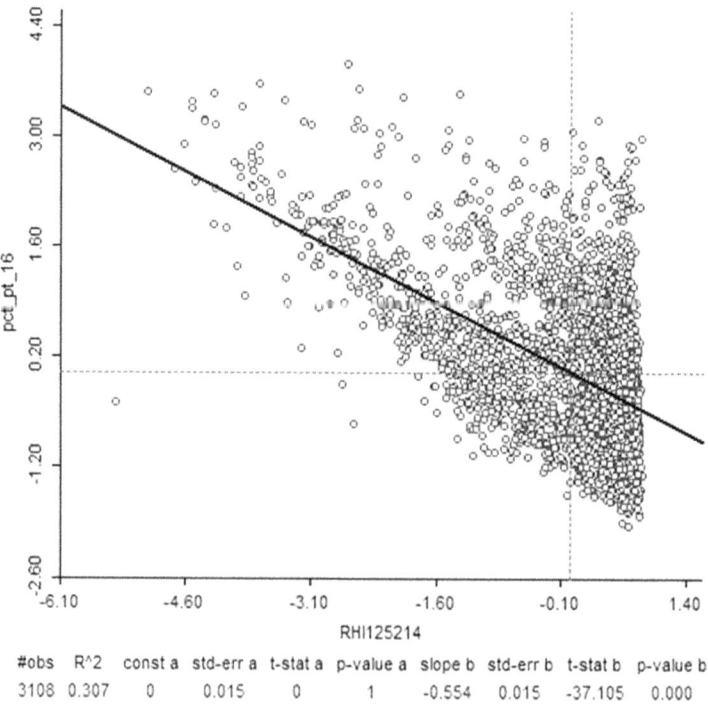

#obs	R^2	const a	std-err a	t-stat a	p-value a	slope b	std-err b	t-stat b	p-value b
3108	0.307	0	0.015	0	1	-0.554	0.015	-37.105	0.000

Fig. 6.8. Scatterplot and regression line

Standard error (*SE*) is the ratio of the standard deviation to the square root of the number of observations and is a measure of the uncertainty of a test (in this case, constants and regression lines), that is, it describes how the value of the test statistic changes from sample to sample. The smaller the value, the more valid the dependence. When working with small samples, it is better to use $\sqrt{n-1}$ as the denominator.

$$SE = \frac{\sigma}{\sqrt{n}}$$

The t-statistic reflects the Student's t-test, which describes the similarity of the means in two variables. A t-statistic value of $>|2|$ allows us to state that the differences between the values being compared are statistically significant, no matter the number of observations (unless there are a restrictively small number of observations). The t-test is applied to variables with a normal distribution and equal variance. Welch's t-test is used for unequal variances, while the Mann–Whitney U test is used when there is no normal distribution.

The most important parameter for correlation model verification is the *significance level* (*p-value*). This criterion evaluates the likelihood of obtaining the same distribution of data assuming that our hypothesis about the dependence of the two indicators is absolutely wrong (that is, the *null hypothesis*, which states that these variables are not related in any way, is true). The lower the significance level, the more valid the model. However, a significance level of $p < 0.05$ is acceptable for the social sciences.

Spearman's rank correlation coefficient (p) is used if we need to recheck the Pearson coefficient, and is measured in the same range. Here, each value of variable x is assigned a rank R_{xi} in accordance with the natural order $i = 1, 2, ..., n$. The rank of each indicator y is defined as the rank of R_{yi}, which corresponds to the rank of the pair (x, y), for which rank x is equal to i. The following formula uses the rank difference $d_i = R_{xi} - R_{yi}$.

$$p = 1 - \frac{6 \sum d_i^2}{n(n^2 - 1)}$$

In electoral research, correlation analysis is most often used to analyse dependencies: 1) between the results of parties and candidates in the same elections; 2) between the results of parties and candidates in different electoral cycles in order to analyse the continuity of electoral preferences; and 3) between the election results and socioeconomic indicators.

6.3 Exploratory Spatial Data Analysis

The next step in electoral-geographic statistical analysis – one that is specific to the geographical approach – is *Exploratory Spatial Data Analysis, (ESDA)*, which consists of three similar stages:

1. *Centrography* identifies the average values of the spatial distribution of the phenomenon and standard deviations from them;
2. *Spatial relative analysis* determines the nature of the distribution of variables relative to space (including their spatial lag and relative risk);
3. *Spatial autocorrelation analysis* tests spatial dependencies between individual variables (that is, hypotheses about the presence of spatial autocorrelation at the global and local levels) [281–282].

Common sense tells us that the closer you get to an ill person, the more likely you are to catch whatever they have. This knowledge manifests an intuitive understanding of the fundamental law of existence – its spatio-temporal organization. You cannot catch a cold from someone before they have fallen ill, because the world operates according to a temporal (historical) paradigm. Similarly, you cannot catch a cold from a person who is miles away, because the world also operates according to spatial (geographic) laws. And, as physics tells us, time is just one of the dimensions of space, we can say that our existence is determined by the coordinate grid of the space-time continuum in which it is located.

Many political and socioeconomic phenomena spread in accordance with the very same laws as a virus. For example, it has been proven that people vote more actively for candidates in areas neighbouring those in which he or she campaigned, or that the appearance of an upscale neighbourhood increases the wealth of neighbouring districts. In geography, it is the spatial connections between objects that explain phenomena. For example, it would be wrong to say that democracy is only possible in small states, while the hypothesis that democracy is more likely in a

Exploratory Spatial Data Analysis 193

country surrounded by democracies is quite valid. The basic hypothesis of geography is that the spatial organization of Earth predetermines the territorial arrangement of objects on its surface. But this is often not quite the case, as the nature of the territorial alignment of political forces (at both the domestic and international levels), for example, is influenced by a multitude of other factors in addition to the spatial organization of the planet, which is shaped by nature, society and humans. Understanding the extent to which the basic hypothesis of geography is correct and capable of explaining natural and social processes is the goal of geographic research.

In the natural sciences, this law governing the spatial organization of the existence of matter is called the principle of locality (short-range interaction), which states that an object is influenced only by its immediate surroundings. Despite the fact that quantum mechanics has proved that this postulate is not valid in all cases (quantum entangled particles affect one another, even though they are located far apart), even Albert Einstein insisted that if this axiom were not true, the world around us would be magical, unknowable in any rational sense. As a branch of geography, spatial statistical analysis (sometimes called spatial econometrics or geostatistics) applies methods of mathematical statistics to confirm or refute such patterns. This is done via data modelling in geographic information systems, primarily QGIS, ArcGIS AND GeoDa, and the statistical packages Stata and R.

There are two typical problems in spatial statistical analysis that can lead to significant distortions in research results and which thus require special attention during the initial examination of the data.

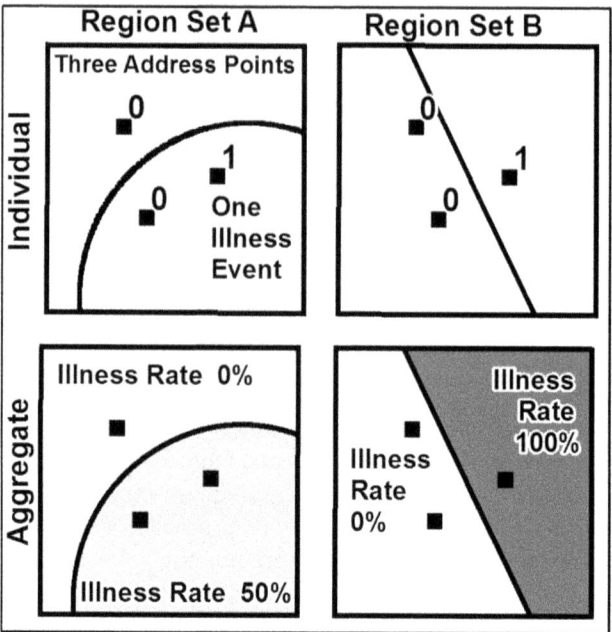

Fig. 6.9. The modifiable areal unit problem

1. The *modifiable areal unit problem (MAUP)* arises as a consequence of the fact that aggregating data into territorial units may yield significantly different results at different scales of analysis (see Fig. 6.9) [417]. For example, if an electoral district includes both urban and rural settlements, then the total figures for it smooth out the internal differences between them, which may be important for assessing functional cleavages in society. This problem compelled the Italian statistician Giuseppe Arbia to reformulate Tobler's first law of geography as: "Everything is related to everything else, but things observed at a coarse spatial resolution are more related than things observed at a finer resolution," or, in other words, the larger the scale of analysis, the clearer the heterogeneity of space and the more reliable the analysis results. A separate problem for electoral geography is the change in the boundaries of electoral and voting districts or the division of the country into administrative units, which makes it difficult to compare election results cross-temporally in a single country.

Exploratory Spatial Data Analysis

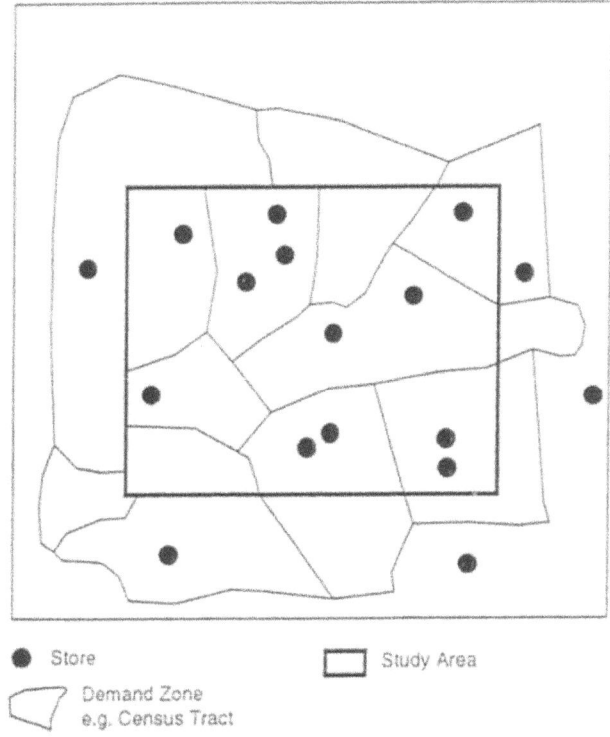

Fig. 6.10. The boundary problem
Source: S.A. Fotheringham S.A., and P.A. Rogerson, "GIS and Spatial Analytical Problems," *International Journal of Geographical Information Systems*, Vol. 7, No. 1 (1993), p. 8.

2. The *boundary problem* arises when geographical limits are imposed on the area of analysis [289–291]. To deal with this, the researcher ignores the influence that observations outside the area of analysis have on the analysis and which make data on the nature of the neighbourhood of cells asymmetric: those in the centre of the sample have a full-fledged neighbourhood, while those on the boundaries only have a selective neighbourhood in certain directions (see Fig. 6.10).

3. The *multiple comparisons problem* states that the more spatial objects we study, the higher the probability of error in estimating spatial dependencies in relation to a given observation [230]. What makes this problem worse is the fact that neighbouring

observations in space have a similar environment, which increases the likelihood of a statistical error in relation to a single observation based on comparing a large number of neighbouring objects.

When analysing data in space, we should also be mindful of the *distance decay effect*. Spatial influence does not spread evenly (or exponentially), rather, it has a decay cutoff – the point at which it ceases to be significant. For example, expanding the study to cover more orders of neighbourhood can distort the results. What is more, spatial influence can be asymmetric in terms of its strength in different vectors, which can lead to the phenomenon of *spatial inversion*, when distant objects are more connected than closer ones.

Other areas of spatial statistical analysis in addition to exploratory spatial data analysis include spatial regression analysis, spatial cluster analysis and geostatistics (spatial interpolation and kriging). That said, the latter is almost never used in the social sciences these days, as it deals with the analysis of continuous data.

6.4 Centrography

Our journey into the exploratory spatial analysis of data begins with *centrography*[7] – the identification and visualization of basic statistical indicators on a chorochromatic map: the average values in the spatial distribution of a phenomenon and the standard deviation from these values. They will be symmetrical to non-spatial statistics (Table 6.2).

Table 6.2. Basic indicators in spatial statistics

Non-spatial statistics	Spatial statistics
Mean	Geographical average (geographic midpoint)
Weighted mean	Geographical weighted average
Median	Median centre, central object
Standard deviation	Standard distance, standard deviation ellipse
Outlier	Spatial outlier
Gauss–Laplace normal distribution	Bivariate Rayleigh normal distribution

---

[7] The centrography method was first described by Dmitri Mendeleev in his 1901 work "To the Knowledge of Russia."

The *geographical average* (or midpoint), which is analogous to the arithmetic mean in non-spatial statistics, is the average of the x and y coordinates of all objects in the area we are looking at. It is used to compare the spatial distribution of various observational parameters, namely their central trend (Fig. 6.11). The coordinates $(\underline{X},\underline{Y})$ of the geographical average are calculated as follows:

$$\underline{X} = \frac{\sum_{i=1}^{n} x_i}{n}, \underline{Y} = \frac{\sum_{i=1}^{n} y_i}{n},$$

where x_i and y_i are the coordinates for object i, and n is the total number of objects. If polygons (say, electoral districts) or lines are used in the analysis, they are first converted to centroids, and then the geographical average is calculated based on the coordinates of those centroids. For multi-part polygons, the centroid is calculated using the weighted average of the centre of all parts of the object.

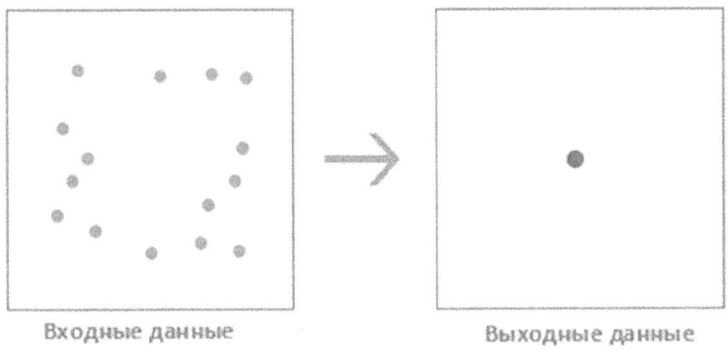

Входные данные   Выходные данные

Fig. 6.11. Algorithm for calculating the geographical average
Source: https://pro.arcgis.com/ru/pro-app/2.7/tool-reference/spatial-statistics/mean-center.htm

The geographical average does not necessarily need to be calculated on the basis of geographic coordinates, as it can be done for a given statistical indicator too. Say, for example, we want to compare the geographic midpoint of a country with the geographical average for each of the political parties that operate in it and thereby estimate the vector of their displacement relative to the common centroid. To do this, we would use the geographical weighted average $(\underline{X}_w, \underline{Y}_w)$. In this case, the weight w_i is the actual value of the parameter being analysed.

$$\underline{X}_w = \frac{\sum_{i=1}^{n} w_i x_i}{\sum_{i=1}^{n} w_i}, \underline{Y}_w = \frac{\sum_{i=1}^{n} w_i y_i}{\sum_{i=1}^{n} w_i}$$

The *median centre* is the point of the shortest distance to all other objects in the dataset. The algorithm for finding the median centre is based on selecting coordinates (X^t, Y^t) so that the sum of their distances to the median centre $(\underline{X}, \underline{Y})$ will be minimal.

$$(\underline{X}, \underline{Y}) = \sum_{i=1}^{n} d_i^t \rightarrow min, \text{ where } d_i^t = \sqrt{(X_i - X^t)^2 + (Y_i - Y^t)^2}$$

The median centre is just a point in space which may not correspond to any given object, but sometimes you want to define a specific object in a sample (say, a voting district) that has the shortest distance to all other objects in the dataset. This is the so-called central object, and it is calculated using a formula identical to the one that gives us the median centre, only d_i^t will not be the distance to the random point, but to one of the objects in the sample.

The median centre and central object can also be weighted if we add the parameter weight to the algorithm: $\sum_{i=1}^{n} w_i d_i^t$.

Standard distance (SD) is the variance parameter of a variable that expresses the spatial compactness of the dataset. It is a circle whose centre is the geographical average and whose radius is the standard distance of the variable (Fig. 6.12). It can also be weighted by a non-spatial dimension tied to territorial units.

$$SD = \sqrt{\frac{\sum_{i=1}^{n}(x_i - \underline{X})^2}{n} + \frac{\sum_{i=1}^{n}(y_i - \underline{Y})^2}{n}}$$

$$SD_w = \sqrt{\frac{\sum_{i=1}^{n} w_i (x_i - \underline{X}_w)^2}{\sum_{i=1}^{n} w_i} + \frac{\sum_{i=1}^{n} w_i (y_i - \underline{Y}_w)^2}{\sum_{i=1}^{n} w_i}}$$

Centrography

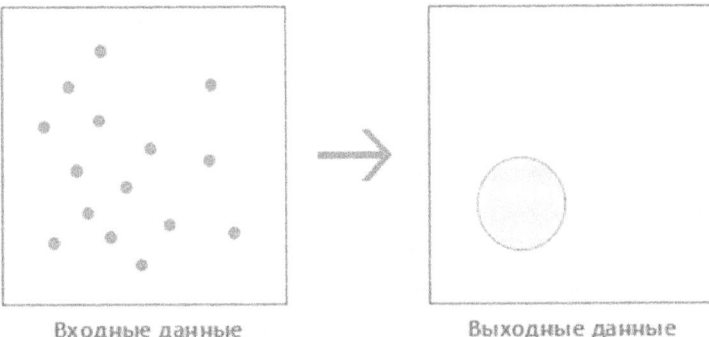

Fig. 6.12. Algorithm for calculating standard distance
Source: https://pro.arcgis.com/ru/pro-app/2.7/tool-reference/spatial-statistics/standard-distance.htm

A circle is not the most accurate method for estimating the variance of data in space. A standard deviation ellipse is better suited to the task, as the standard distances are calculated separately for the X and Y axes.

$$SD_X = \sqrt{\frac{\sum_{i=1}^{n}(x_i - \underline{X})^2}{n}}, SD_Y = \sqrt{\frac{\sum_{i=1}^{n}(y_i - \underline{Y})^2}{n}}$$

$$SD_{wX} = \sqrt{\frac{\sum_{i=1}^{n} w_i (x_i - \underline{X}_w)^2}{\sum_{i=1}^{n} w_i}}, SD_{wY} = \sqrt{\frac{\sum_{i=1}^{n} w_i (y_i - \underline{Y}_w)^2}{\sum_{i=1}^{n} w_i}}$$

The angle of rotation of ellipse θ (the angle between the north along the X axis and the major axis of the ellipse, assuming that the major axis is rotated clockwise if the tangent is positive, and anti-clockwise if it is negative) to deviate points from the midpoint $x'_i = x_i - \underline{X}$ and $y'_i = y_i - \underline{Y}$ is calculated using the formula:

$$tan\theta =$$

$$\frac{\left(\sum_{i=1}^{n} x'^{2}_{i} - \sum_{i=1}^{n} x'^{2}_{i}\right) + \sqrt{\left(\sum_{i=1}^{n} x'^{2}_{i} - \sum_{i=1}^{n} x'^{2}_{i}\right)^{2} + 4\left(\sum_{i=1}^{n} x'_{i} \sum_{i=1}^{n} x'_{i}\right)^{2}}}{2\sum_{i=1}^{n} x'_{i} \sum_{i=1}^{n} x'_{i}}$$

In its generalized version, the ellipse formula can also be represented as:

$$SD = \begin{pmatrix} var(x) & cov(x,y) \\ cov(y,x) & var(y) \end{pmatrix} = \frac{1}{n}\begin{pmatrix} \sum_{i=1}^{n} \underline{x}_{i}^{2} & \sum_{i=1}^{n} \underline{x}_{i}\underline{y}_{i} \\ \sum_{i=1}^{n} \underline{x}_{i}\underline{y}_{i} & \sum_{i=1}^{n} \underline{y}_{i}^{2} \end{pmatrix}, \text{where}$$

$$var(x) = \frac{1}{n}\sum_{i=1}^{n}(x_{i} - \underline{x})^{2} = \frac{1}{n}\sum_{i=1}^{n}\underline{x}_{i}^{2}$$

$$var(y) = \frac{1}{n}\sum_{i=1}^{n}(y_{i} - \underline{y})^{2} = \frac{1}{n}\sum_{i=1}^{n}\underline{y}_{i}^{2}$$

$$cov(y,x) = \frac{1}{n}\sum_{i=1}^{n}(x_{i} - \underline{x})(y_{i} - \underline{y}) = \sum_{i=1}^{n}\underline{x}_{i}\underline{y}_{i}$$

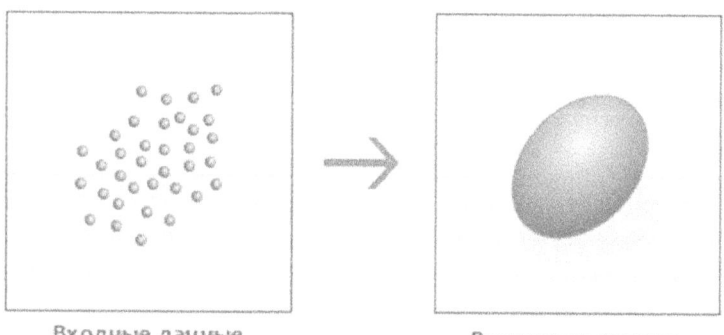

Входные данные → Выходные данные

Fig. 6.13. Algorithm for calculating the ellipse of the geographical average
Source: https://pro.arcgis.com/ru/pro-app/2.7/tool-reference/spatial-statistics/directional-distribution.htm

This kind of ellipse describes a *bivariate Rayleigh normal distribution*, which is a two-dimensional analogue of the Gauss–Laplace normal distribution. Here, too, approximately 63 % of cases are described within one standard distance, 98 % are described within two standard distances, and almost all cases are described within three standard distances. Accordingly, observations that fall outside the three standard distances will be considered *spatial outliers*.

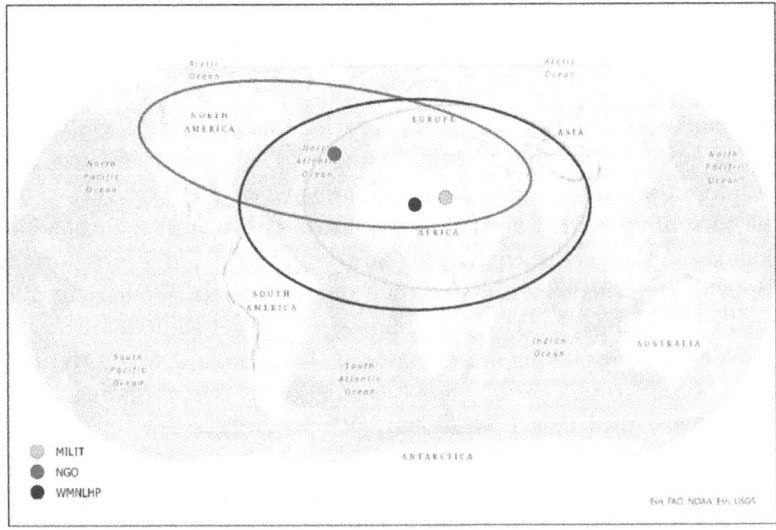

Fig. 6.14. Ellipse of the geographical average

Figure 6.14 is a chorochromatic map depicting the importance of certain factors in politics for all countries around the world: number of NGOs (the leftmost ellipse), the share of women in politics (the central ellipse) and military spending (the rightmost ellipse). We can see a significant bias in the first ellipse and its geographic average to the northwest, which reflects the predominant spatial distribution of the parameter in the North Atlantic region. The other two parameters turn out to be slightly biased relative to the centroid of the world map, and their distribution can be considered close to the standard distribution. This kind of analysis emphasizes the fact that, in centrography, conclusions may only be drawn when comparing geographic averages and their centroids, as they are not particularly informative when taken separately.

In electoral research, centrography is used to find the central trend in the distribution of votes for individual candidates and parties and to identify cross-temporal shifts in the central trend of support for a party.

6.5 Spatial Neighbourhood Weights

Observations in ordinary statistical analysis are not structured in space; we know nothing about their location relative to one another. And centrography only gives us an averaged idea of the spatial distribution of a phenomenon. For instance, in the previous example, we attempted to evaluate the relationship between voting for the Republican Party and the proportion of white population, ignoring where those voters lived. At the same time, in the United States, counties with a larger share of white population form stable continuum clusters in space, and it is important for us to take this factor into account in our analysis. We have formulated two hypotheses: H1 – Republicans tend to get additional votes in counties that neighbour those in which they have a consistently high support base (where the neighbourhood effect is observed); and H2 – Republicans tend to get additional votes in areas (spatial continuum clusters) where there is a large proportion of whites. To test these hypotheses, we need to use special spatial statistical analysis algorithms, that is, we need to use techniques where the location of observations in space – and not just their quality – matters. An inevitable primary objective of spatial statistical analysis is to define the nature of spatial relationships between observations. For instance, in our example, this is to formalize how counties in the United States are ordered relative to one another.

In spatial analysis, the task of structuring space is accomplished by constructing a matrix of *spatial neighbourhood weights*. Weights are given as a symmetrical square matrix, the elements of which indicate which objects are neighbours.

Let us calculate the neighbourhood matrix for six imaginary districts, A, B, C, D, E, F, located as shown in the diagram:

Spatial Neighbourhood Weights

A	B	C
D	E	F

Neighbourhoods are marked with a 1 in the table (matrix), and non-neighbourhoods are marked with a 0. We will then calculate the sum of neighbours for each row. B and E are in the middle and have five neighbours each, while the remaining districts have three neighbours.

	A	B	C	D	E	F	Σ
A	0	1	0	1	1	0	3
B	1	0	1	1	1	1	5
C	0	1	0	0	1	1	3
D	1	1	0	0	1	0	3
E	1	1	1	1	0	1	5
F	0	1	1	0	1	0	3

We need to standardize the matrix so that the sum of the values in each row is equal to 1.

	A	B	C	D	E	F	Σ
A	0	0.33	0	0.33	0.33	0	1
B	0.2	0	0.2	0.2	0.2	0.2	1
C	0	0.33	0	0	0.33	0.33	1
D	0.33	0.33	0	0	0.33	0	1
E	0.2	0.2	0.2	0.2	0	0.2	1
F	0	0.33	0.33	0	0.33	0	1

Now imagine that district C is particularly important (say, it has twice as many voters as the other districts). In this case, we would denote the weight of the neighbourhood with a 2.

	A	B	C	D	E	F	Σ
A	0	1	0	1	1	0	3
B	1	0	2	1	1	1	6
C	0	2	0	0	2	2	6
D	1	1	0	0	1	0	3
E	1	1	2	1	0	1	6
F	0	1	2	0	1	0	4

Again, we will normalize the weighted neighbourhood matrix so that the sum of each row is equal to 1.

	A	B	C	D	E	F	Σ
A	0	0.33	0	0.33	0.33	0	1
B	0.2	0	0.4	0.2	0.2	0.2	1
C	0	0.33	0	0	0.33	0.33	1
D	0.33	0.33	0	0	0.33	0	1
E	0.2	0.2	0.4	0.2	0	0.2	1
F	0	0.25	0.5	0	0.25	0	1

In its basic form, the $n \times n$ neighbourhood matrix W, where w_{ij} denotes the weight of the neighbourhood between objects i and j, looks like this:

$$W = \begin{bmatrix} w_{11} & w_{12} & \cdots & w_{1n} \\ w_{21} & w_{22} & \cdots & w_{2n} \\ \vdots & \vdots & \ddots & \vdots \\ w_{n1} & w_{n2} & \cdots & w_{nn} \end{bmatrix}$$

Spatial Neighbourhood Weights

If i and j are neighbours, the spatial weights of w_{ij} will be greater than zero. If they are not neighbours, then $w_{ij} = 0$. In contrast, it is impossible to be one's own neighbour, so $w_{ii} = 0$.

Since we assume that the sum of each row of the spatial neighbourhood weights matrix should be equal to 1, the general standardization formula for calculating the neighbourhood measure $w_{ij(s)}$ of a single cell wil look like this:

$$w_{ij(s)} = w_{ij} / \sum_{j} w_{ij}$$

The sum of all spatial weights will be equal to the number of observations $S_0 = \sum_{i}\sum_{j} w_{ij}$.

Dual standardization is also possible for $S_0 = 1$. To do this, we transform he matrix according to the formula:

$$w_{ij(ds)} = w_{ij} / \sum_{i}\sum_{j} w_{ij}$$

An object with no neighbours (say, one located on a distant island), that is, one where $w_{ij} = 0 \,\forall j$, is called an *isolate*.

The concept of neighbourhood is not strictly formalized. Consequently, it is defined in two fundamentally different ways: by adjacency and by metric.

The first principle of determining neighbourhood – *adjacency* – is based on topological relationships between objects and is applied in the analysis of data confined to areal units. Objects are considered adjacent when their boundaries have common points or lines. In this case, there are three options for identifying the kind of neighbourhood, and they are named after chess pieces: *rook* neighbourhoods, *bishop* neighbourhoods, and *queen* neighbourhoods. The rook rule defines adjacency by the existence of a common edge between two spatial units, while bishop rule defines it by the existence of a common point, and the queen rule by the existence of both (Fig. 6.15).

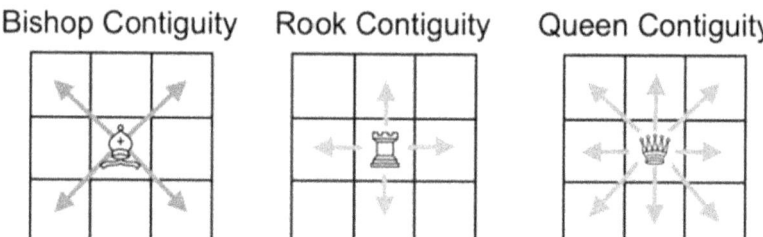

Fig. 6.15. Types of neighbourhood by adjacency
Source: L. Sedaghat, J. Hersey, and M.P. McGuire, "Detecting Spatio-Temporal Outliers in Crowdsourced Bathymetry Data," *Proceedings of the Second ACM SIGSPATIAL International Workshop on Crowdsourced and Volunteered Geographic Information*. GEOCROWD'13 (2013), p. 59.

Additionally, when determining a neighbourhood by adjacency, you can set the adjacency order that should be taken into account: 1st order neighbours only; 1st and 2nd order neighbours; or 2nd order neighbours only (Fig. 6.16).

Spatial Neighbourhood Weights

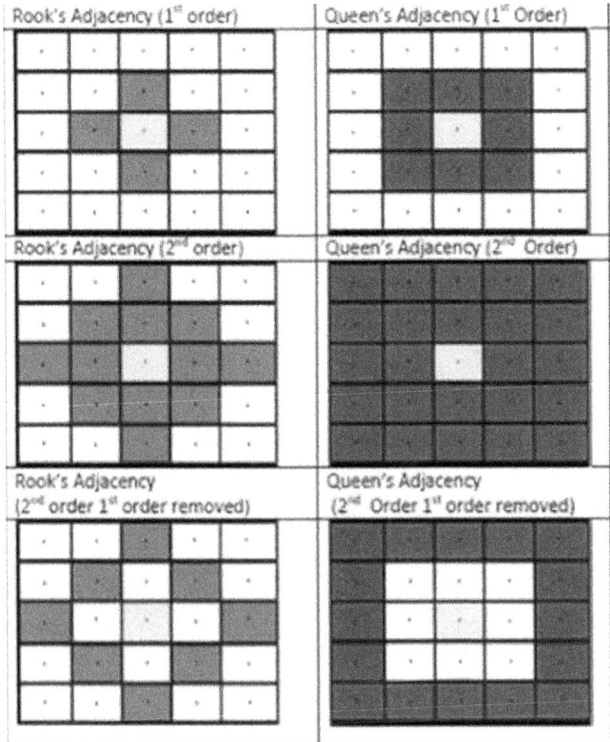

Fig. 6.16. Types of neighbour by adjacency order
Source: M. Rura, "Eigenvector Spatial Filtering for Image Analysis: An Efficient Algorithm," https://www.researchgate.net/publication/252401954_Eigenvector_spatial_filtering_for_image_analysis_An_efficient_algorithm

For most tasks, it is important for the matrix to be symmetrical, that is, for the principle $w_{ij} = w_{ji}$ to be observed for all pairs.

We will now create the neighbourhood weights for U.S. counties for further analysis: W1 according to queen's adjacency of the 1st order; and W2 according to queen's adjacency of the 1st and 2nd orders (estimates of more continuous clusters). Figure 6.17 is an adjacency graph for W1, and Table 6.3 compares their parameters.

Figure 6.17 shows the adjacency graph for W1, and Table 6.3 compares their parameters.

Fig. 6.17. Queen's adjacency graph for U.S. counties

Table 6.3. Comparison of adjacency neighbourhood matrices

Weight	W1	W2
Type	Adjacency	Adjacency
Subtype	Queen's rule – 1st order neighbourhood	Queen's rule – 1st and 2nd order neighbourhood
Minimum neighbours	0	0
Maximum neighbours	14	40
Average number of neighbours	5.84	18.30
Median number of neighbours	6.00	19.00
Symmetry	Yes	Yes

In some cases – for instance, if a state is located on islands or, as the case of the United States, the compactness of the spatial distribution of counties is extremely uneven – it is better to use a different method of calculating spatial neighbourhood weight matrices, namely, according to *metric* (distance). We should bear in mind here that two main measurements of distance are used in geography: Euclidean (d_e) and Manhattan (d_m). Euclidean distance is defined as the length of a straight line connecting two points, while Manhattan distance is the difference between the vertical and horizontal coordinates of the points (it is impossible to walk in a straight line in urban areas, so you need to walk across street intersections). For both, the formulas for calculating the distance

Spatial Neighbourhood Weights

between points i and j with coordinates (x_i, y_i) and (x_j, y_j) will be as follows:

$$d_{ij} = \sqrt{(x_i - x_j)^2 + (y_i - y_j)^2}$$

$$d_{ij}^m = |x_i - x_j| + |y_i - y_j|$$

To calculate the distance between objects, we first need to determine the point from which the distance will be calculated. For example, you can use the coordinates of the centroid (average or median centre), the administrative centre of the region, or the territorial election commission.

For spatial data in latitudes and longitudes, *arc distance* – the shortest line between two points along a section of a curve – is used. First, the *latitude* and *longitude* are transformed into radians:

$$Lat_r = (Lat_d - 90) * \pi / 180$$

$$Lon_r = Lon_d * \pi / 180$$

With $\Delta Lon = Lon_{r(j)} - Lon_{r(j)}$, the distance along the great circle for the Earth's radius $R = 6371 km$ will be:

$$d_{ij} = R * arccos\left[cos(\Delta Lon) * sin Lat_{r(i)} * sin Lat_{r(j)}) + cos Lat_{r(i)} * cos Lat_{r(j)}\right]$$

$$d_{ij} = R * arccos\left[cos(\Delta Lon) * cos Lat_{r(i)} * cos Lat_{r(j)}) + sin Lat_{r(i)} * sin Lat_{r(j)}\right]$$

Now that we know how to calculate the distances between centroids, we can create a metric spatial weights matrix. To do this, we need to set a conditional range of distances, and districts that fall within this range can be considered neighbours. In other words, i and j are considered neighbours if j falls within a given range of distances δ from centroid i, such that:

$$w_{ij} = 1, \text{if } d_{ij} \leq \delta$$

$$w_{ij} = 0, \text{if } d_{ij} > \delta$$

The proper calculation of a metric spatial weights matrix requires at least one neighbour to fall into the given range of distances for each object. In other words, we need to use the so-called maximum–minimum range selection principle, which states that the maximum distance to the nearest region should be the uppermost figure in the range. However, the existence of remote enclaves or islands often makes this figure too high.

Another algorithm for calculating weights by metric is the k-nearest neighbours method. If we move away from the axiom that we need symmetric neighbourhood weights only, this allows us to find a simpler solution to the problem of isolates. In a simplified model, this method means that the researcher specifies which single number of nearest neighbours is to be specified for each element being analysed.

An important advantage of metric-based neighbourhood is that it allows us to determine the weight of the neighbourhood in the matrix using the *inverse distance weighting (IDW)* method. In other words, the weight of a district located closer to the centroid of the district being analysed will be higher than that of a district whose centroid is located farther from the average or median centre of the district being analysed. This operation allows us to reduce the effect of including remote or isolated territories in our analysis.

Now let us create two matrices for U.S. counties by metric: W3 according to the maximum–minimum principle for selecting the range of distances from the median centre with inverse distance weighting; and W4 for $k = 6$ (which is the average and median number of neighbours according to the queen's adjacency rule) from the median centre with equal neighbourhood weight. Figure 6.18 shows the adjacency graph for W4, and both matrices are compared in Table 6.4.

Spatial Neighbourhood Weights 211

Fig. 6.18. Adjacency graph for U.S. counties using the k-nearest neighbours method

Table 6.4. Comparison of metric neighbourhood matrices

Weights	W3	W4	W5
Type	Metric	Metric	Conditional
Subtype	maximum–minimum principle, from the median centre, with IDW	$k = 6$, from the median centre, without IDW	By state affiliation
Minimum neighbours	1	6	0
Maximum neighbours	107	6	253
Average number of neighbours	48.91	6.00	96.18
Median number of neighbours	48.00	6.00	87.00
Symmetry	Yes	No	Yes

We can also create complex weights by combining several structural principles. One example would be to take the queen's adjacency rule for 1st order regions into account for counties with fewer than six neighbours, adding them according to the k-nearest neighbours principle. Finally, we can estimate the level of spatial homogeneity within states. For this, we will set conditional weights W5, in which all counties in a single state will be neighbours.

As we can see from the comparison of the range and average number of neighbours, the W1/W4 and W2/W3 matrices are extremely similar to each other with this particular dataset. Thus, for future analysis, we will leave matrix W1 (for standard spatial analysis), W2 (for finer scale analysis with larger clusters), and W5 (for analysis in a conditional space, where each U.S. state represents an isolated sub-region). This allows us to test two additional hypotheses: H3 – spatial clustering in U.S. elections is stronger at the regional level than at the local level; and H4 – spatial clustering in the United States mirrors the administrative and territorial structure of the country.

6.6 Spatial Lag

Neighbourhood spatial weights matrices allow you to perform statistical operations with observations that are structured in space. The basic vector variable of spatial analysis is *spatial lag*, which is the sum or averaged (for normalized weights) value of indicators in adjacent cells. Accordingly, median spatial lag is the middle number in a series of numbers describing the indicators in neighbouring cells.

For object i, the spatial lag $\tilde{y}$ for variable y will be calculated using the formula:

$$\tilde{y}_i = w_{i1} y_1 + w_{i2} y_2 + \ldots + w_{in} y_n$$

$$\text{or } \tilde{y}_i = \sum_{i=1}^{n} w_{ij} y_j,$$

where weights w_{ij} contain indicators in the i-row of matrix W, which coincide with the corresponding indicators in vector y.

For standardized weights, where the sum of each row is equal to 1 – $\sum_j w_{ij} = 1$, the standardized spatial lag is the weighted average of the indicators in adjacent cells. In some cases, the value of the cell being analysed is also included in the calculation of the spatial lag.

Spatial Lag

$$\tilde{y}_i = y_i + \sum_{i=1}^{n} w_{ij} y_j$$

When calculating the spatial lag with a neighbourhood matrix obtained using the inverse distance weighting method, the values will be divided by the distance between observations, while the coefficient α may be equal to 1 for standard situations, or 2 if the task is to show an exponential increase in the weight of distances.

$$\tilde{y}_i = \sum_{i=1}^{n} \frac{w_{ij} y_j}{d_{ij}^{\alpha}}$$

$$\tilde{y}_i = y_i + \sum_{i=1}^{n} \frac{w_{ij} y_j}{d_{ij}^{\alpha}}$$

Now let us consider the spatial lag indicators for Trump's share of the votes in Manhattan in the 2016 presidential elections using our three neighbourhood spatial weight matrices (Table 6.5). Note that he only won 9.9 % of the vote on the island. Let us analyse the results. In actual fact, the lag, as a sum, tells us little about the spatial distribution of the phenomenon, because different weights contain different numbers of observations. It would thus be better to focus on the standardized spatial lag as the average. We can see that, as we move away from Manhattan – that is, as we start to expand the set of neighbours (initially only the island's immediate neighbours, then 2nd order neighbours, and finally all the counties in the state) – Trump's indicators grow. Calculating the median spatial lag with account for the value in the cell being analysed (Manhattan itself) produced almost exactly the same results in all three neighbourhood weights, which can be considered confirmation of the validity of our conclusions.

Table 6.5. Calculation of spatial lag indicators

	W1	W2	W5
Non-row standardized spatial lag	1.14	3.85	32.1
Standardized spatial lag	0.22	0.32	0.52
Standardized median spatial lag	0.22	0.31	0.54
Standardized spatial lag with account for the value in the cell under analysis	0.21	0.31	0.52

Now we will map the distribution of the spatial lag using the natural interval position algorithm (Fig. 6.19). To do this, we used the neighbourhood matrix W1 and the standardized spatial lag, taking the value in the cell we are analysing. It is this that allows us to compare the data with a simple chorochromatic map of the distribution of votes for Trump (Fig. 6.4). The spatial lag allowed us to average the values across the territory and present the voting results in the form of sequential isolines.

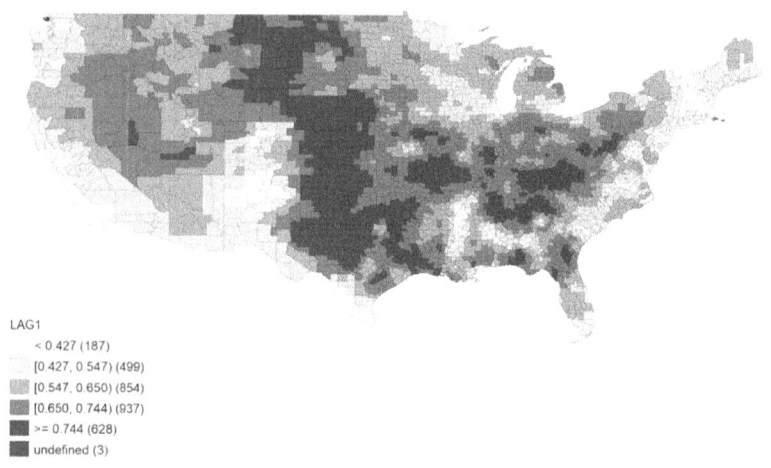

Fig. 6.19. Spatial lag chorochromatic map

Figure 6.20 depicts the dynamics of change in the spatial lag of support for the Republican candidate by neighbourhood order. Support for Trump in Manhattan is plotted at the 0 on the x-axis and, as we move along the x-axis, we see the spatial lag of his support in 1,2, ..., 10 neighbourhood orders, calculated in order of distance from the island,

Spatial Lag 215

ignoring the lower orders. We can see support trending upwards to the 3rd order, at which point it stabilizes.

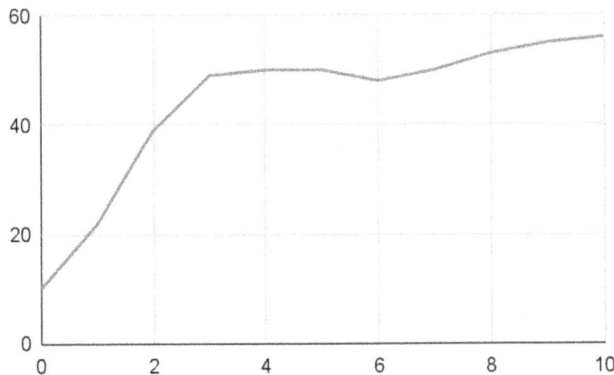

Fig. 6.20. Diagram of spatial lag changes by neighbourhood order

Spatial lag gives us an idea of relative indicators in space. Let us take a closer look at them. An analysis of the chorochromatic map allows us to conclude that the most difficult situation has developed in the Ganges Delta. However, this conclusion is incorrect, as we used a spatially extensive variable, namely, the number of deaths per $1 km^2$. This region is actually one of the most densely populated in the world, which means that we need to use a spatially intensive variable, such as the ratio of deaths per 1000 people per $1 km^2$, if we want our analysis to produce reliable results. Calculating spatially intensive indicators allows us to evaluate the situation at one point in space in relation the situation at other points in space.

In electoral geography, we typically deal with spatially intensive quantities, that is, the ratio of the number of people who voted for a candidate or party (O) to the total number of people who voted (P). Say that for region i, the share of votes for a candidate is calculated as $r_i = O_i / P_i$. The share of votes for this candidate at the national level can be represented both as O / P, and as the ratio of the level of support in all regions to the total number of voters in all regions.

$$R = \frac{\sum_{i=1}^{n} O_i}{\sum_{i=1}^{n} P_i}$$

The spatial weighting ratio for region i can be calculated by following the same logic, that is, using the neighbourhood matrix w_{ij} and knowing how many votes the candidate received and how many votes were cast in total in each of the neighbouring regions. The values in the cell itself are usually taken into account too, so $w_{ii} \neq 0$.

$$sr_i = \frac{\sum_{j=1}^{n} w_{ij} O_j}{\sum_{j=1}^{n} w_{ij} P_j}$$

Figure 6.21 is a natural breaks map that shows the ratio of the number of votes for the Republican candidate in 2012 – not to the total number of votes in the country, but to the total estimated population. This allows us to draw conclusions not only about the level of support for the candidate, but also about voter turnout.

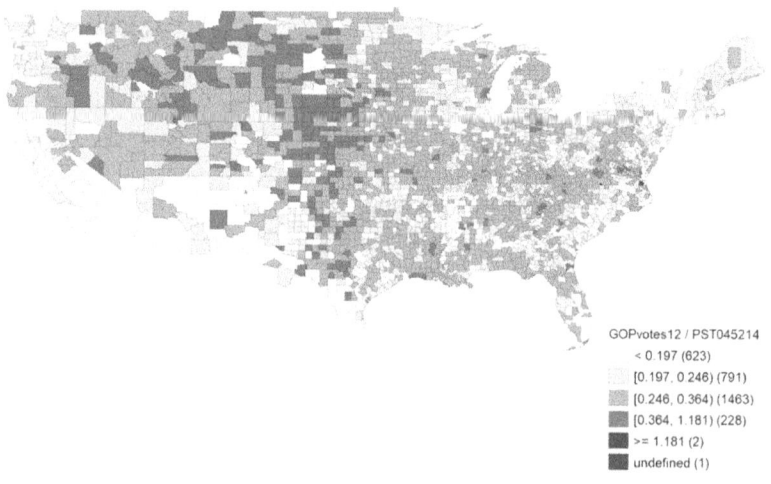

Fig. 6.21. Ratio chorochromatic map

The natural breaks map in Fig. 6.22 shows the same ratio, but spatially weighted (with W_1 weights).

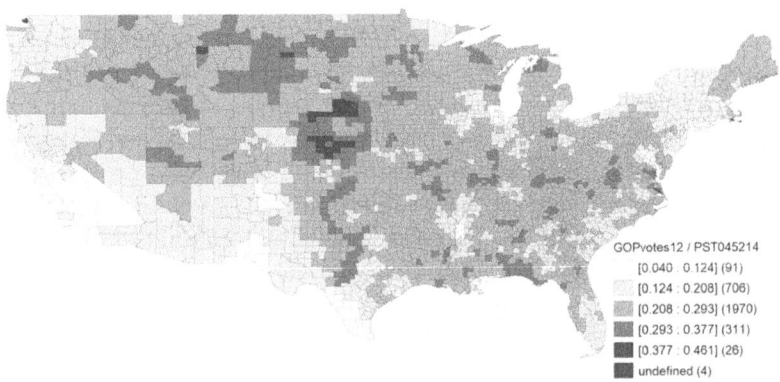

Fig. 6.22. Spatially weighted ratio chorochromatic map

An important consequence of the spatial lag concept is the notion of a spatial window of opportunity: the set of values of a variable for neighbours can be used to make a probabilistic estimate of the distribution of a phenomenon.

6.7 Spatial Autocorrection

In traditional statistics, correlation analysis tests hypotheses about the dependence of the distribution of two variables, regardless of how they are located in space. What is more, this approach assumes that the observed values of the variables are independent of their location in space. Take the following example: we are examining the correlation between household income in a city (y) and the tallness of the building they live in (x). However, each building houses a different number of households: one family might live in a one-storey box room, while another might live in a mansion, and thousands more in high-rise buildings. If we take the data of each family living in a high-rise building, then this data will be internally identical (it will be autocorrelated), and this will distort the results: we need to calculate the average income in each building to eliminate spatial autocorrelation and test the hypothesis further.

However, there are hypotheses for which *spatial autocorrelation* may be necessary as a tool for testing them, and these hypotheses are characteristic of geographical research [242–244]. For example, our hypothesis is that the higher the income level of households, the higher the income level of neighbouring households. To test this hypothesis, we will also need to standardize the data by type of house in order to not be dependent on the number of apartments in the building and thus eliminate this kind of spatial autocorrelation. However, the clustering of buildings in space – the fact that upscale houses tend to be located in specific areas of the city, and poorer houses tend to be located in other areas – is also a kind of spatial autocorrelation, but this will be the metric we are looking for to test our hypothesis. Traditional statistics are not up to this task. And there are two reasons for this. First, traditional statistics does not structure data in space, we have no idea who is located next to whom. Second, it deals with two variables, and we are looking for a correlation of the dependent variable (y) with it, but weighted in space (Wy) . In spatial statistics, the neighbourhood matrix gives us the structure of space, and spatial lag gives us the spatial weight of a variable.

However, spatial correlation analysis can also determine the relationship between two variables, and this kind of *bivariate spatial correlation* can be used to test the hypothesis, for example, that higher income level means taller buildings in their area. Once again, we will need to standardize the data for each building to eliminate the spatial autocorrelation that distorts the results, but now we will identify the relationship of the independent variable (y) with clustering in the space of the independent variable (Wx), for which we will need to specify the structure of space in the form of a neighbourhood matrix and define this variable in terms of its spatial lag.

Finally, it is *possible to correlate two spatially weighted variables* (Wy, Wx), say, in the following hypothesis: the higher the income level of households in the area, the taller the buildings there. Table 6.6 summarizes the difference between the types of correlation.

Table 6.6. Comparison of types of correlation

Correlation type	Hypothesis tested	Null hypothesis	Example hypothesis
Correlation	$y_i = f(x_i) + \varepsilon$	$y_i \neq f(x_i)$	The higher the income level of households, the taller the buildings they live in.
Spatial autocorrelation	$y_i = f(Wy_i) + \varepsilon$	$y_i \neq f(Wy_i)$	The higher the income level of households, the higher the income level of households living next to them.
Bivariate spatial correlation	$y_i = f(Wx_i) + \varepsilon$ or $Wy_i = f(x_i) + \varepsilon$	$y_i \neq f(Wx_i)$ or $Wy_i \neq f(x_i)$	The higher the income level of households, the taller the buildings in the area they live in.
Correlation of two spatially weighted variables	$Wy_i = f(Wx_i) + \varepsilon$	$Wy_i \neq f(Wx_i)$	The higher the income level of households, the taller the buildings in that area.

Spatial autocorrelation is assessed using the Moran and Geary indices of spatial autocorrelation.

The most commonly used method for estimating spatial dependence in the distribution of a variable is *Moran's I*, named after the Australian statistician Patrick Moran (1917–1988) [395]. In terms of its idea, it is almost identical to the Pearson correlation coefficient. However, in it, the arithmetic mean $\underline{y}$, the deviation from the arithmetic mean $y_i - \underline{y}$ and the variance in the denominator $\sum_{i=1}^{n}(y_i - \underline{y})^2$ are calculated for indicator y only, and the coefficient of spatial weights of the neighbourhood w_{ij} is introduced:

$$I_Y = \frac{n}{\sum_{i=1}^{n}\sum_{j=1}^{n} w_{ij}} \cdot \frac{\sum_{i=1}^{n}\sum_{j=1}^{n} w_{ij}(y_i - \underline{y})(y_j - \underline{y})}{\sum_{i=1}^{n}(y_i - \underline{y})^2}$$

with $i \neq j$, where n is the number of spatial objects, y_i and y_j are the values of indicator y for the i-th and j-th objects, $\underline{y}$ is the average value of the indicator, w_{ij} is the spatial weight of the neighbourhood between the i-th and j-th objects, and $\sum_{i=1}^{n}\sum_{j=1}^{n} w_{ij}$ is the sum of all spatial weights.

For standardized neighbourhood weights, the sum of all spatial weights is equal to the number of observations $\left(n = \sum_{i=1}^{n}\sum_{j=1}^{n} w_{ij}\right)$ and thus the Moran's I is simply calculated as:

$$I_Y = \frac{\sum_{i=1}^{n}\sum_{j=1}^{n} w_{ij}(y_i - \underline{y})(y_j - \underline{y})}{\sum_{i=1}^{n}(y_i - \underline{y})^2} = \frac{\sum_{i=1}^{n}\left((y_i - \underline{y}) \times \sum_{j=1}^{n} w_{ij}(y_j - \underline{y})\right)}{\sum_{i=1}^{n}(y_i - \underline{y})^2}$$

Note than Moran's spatial autocorrelation index is very close to the Pearson correlation coefficient between variable y and its spatial lag $\tilde{y}_i = \sum_{j=1}^{n} w_{ij} y_j$, which is especially noticeable if you enter the value of the lag into the formula and perform identical transformations.

$$r_{Y,\tilde{Y}} = \frac{\sum_{i=1}^{n}(y_i - \underline{y})(\tilde{y}_i - \underline{\tilde{y}})}{\sqrt{\sum_{i=1}^{n}(y_i - \underline{y})^2}\sqrt{\sum_{i=1}^{n}(\tilde{y}_i - \underline{\tilde{y}})^2}}$$

$$I_Y = \frac{\sum_{i=1}^{n}(y_i - \underline{y})(\tilde{y}_i - \underline{y})}{\sqrt{\sum_{i=1}^{n}(y_i - \underline{y})^2}\sqrt{\sum_{i=1}^{n}(y_i - \underline{y})^2}} = \sqrt{\frac{\sum_{i=1}^{n}(\tilde{y}_i - \underline{\tilde{y}})^2}{\sum_{i=1}^{n}(y_i - \underline{y})^2}} \cdot r_{Y,\tilde{Y}}$$

Since the ratio under the square root in the last formula is almost equal to one, we can say that $I_Y \approx r_{Y,\tilde{Y}}$ [366].

Moran's *I* values range from −1 to 1, where 1 is absolute positive spatial autocorrelation (observations are perfectly clustered; picture a in Fig. 6.23), −1 is absolute negative spatial autocorrelation (observations are perfectly spaced so that each has neighbours with the opposite value; picture b in Fig. 6.23), and 0 is the lack of any connection (observations are distributed randomly; picture c in Fig. 6.23).

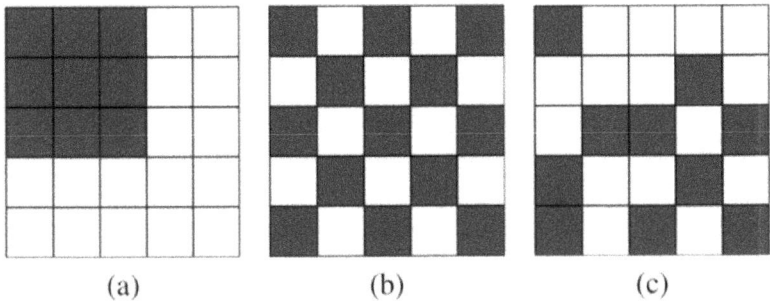

(a) (b) (c)

Fig. 6.23. Examples of spatial autocorrelation
Source: L. Anselin, and G. Piras, "Approaches towards the Identification of Patterns in Violent Events Baghdad, Iraq" (2009), http://hdl.handle.net/11681/13859

Moran's *I* allows us to test the proposed hypothesis H1 – Republicans tend to get additional votes in counties that neighbour those in which they have a consistently high support base (where the neighbourhood effect is observed). To do this, we will build a Moran scatterplot for the neighbourhood matrix *W*1 (Fig. 6.24).

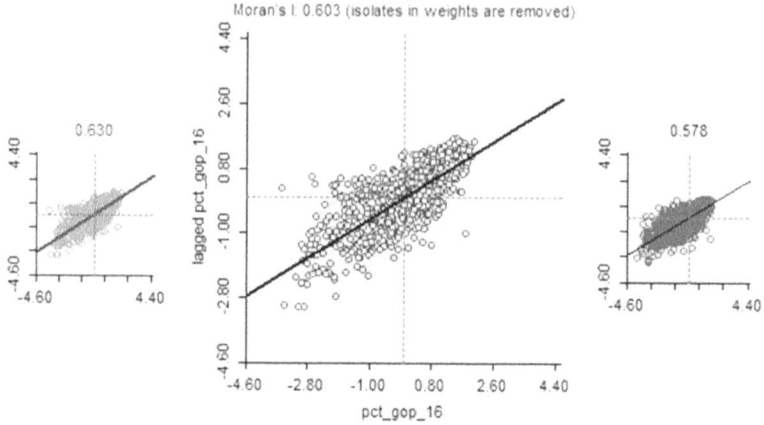

Fig. 6.24. Moran scatterplot for spatial autocorrelation

On the diagram, the x-axis plots the value in each territorial unit, and the y-axis shows the spatial lag, which is the weighted average of all neighbours, ignoring the value of the cell being analysed. Isolates are excluded from the sample during analysis. The dotted lines in the Moran scatterplot represent the mean values along both axes, and the slanted line represents the linear regression of these values. Values to the right of the dotted line $y_i - \underline{y} > 0$ are called high (above average), and values to the left of the dotted line $y_i - \underline{y} < 0$ are called low (below average). The same applies to the y-axis with a variable in the spatial lag, $w_{ij}(y_j - \underline{y}) > 0$ on the right, and $w_{ij}(y_j - \underline{y}) < 0$ on the left. The Moran chart is then divided into four quadrants: upper right (high–high) and lower left (low–low) denote positive spatial autocorrelation. Conversely, the upper left (low–high) and lower right (high–low) quadrants reflect negative spatial autocorrelation (Fig. 6.25).

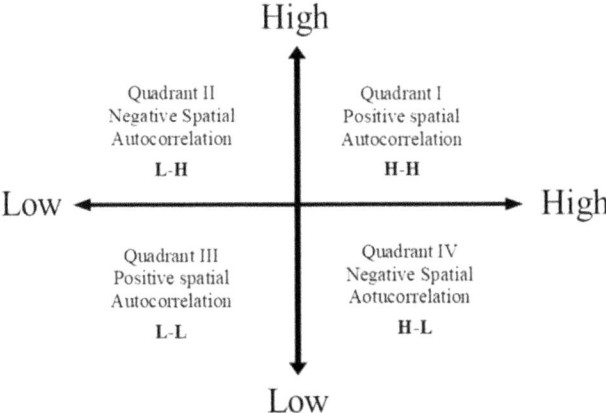

Fig. 6.25. Quadrants of the Moran scatterplot

The Moran's *I* for our hypothesis is 0.603, and the degree of its determination is calculated in the same way as for the Pearson correlation coefficient (Table 6.1). In this case, we can state that there is significant average degree of determination. This confirms our hypothesis that votes for the Republican candidate demonstrate a significant degree of clustering in space. At the same time, the average degree of determination for slightly higher for the western half of the country (0.630) than for the eastern half (0.578).

The validity of the Moran's *I* is tested using the significance level (*p-value*) and z-score. The level of significance for Moran's spatial autocorrelation is estimated by testing the null hypothesis that the observations are distributed in space completely randomly. The significance level (called the pseudo level of significance in this case) is calculated by permuting the observations multiple times and determining the probability that the permutations will support the null hypothesis. Figure 6.26 shows a test of the Moran's *I* we calculated earlier for the U.S. elections.

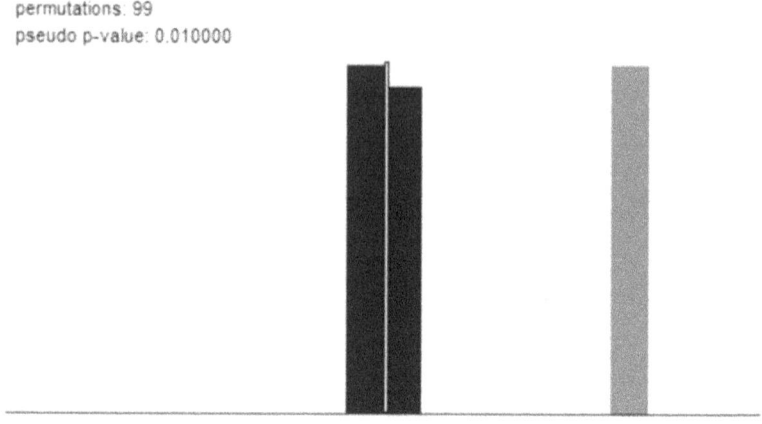

Fig. 6.26. Significance test for Moran's *I*

A total of 99 permutations were performed, producing a minimum significance level of 0.01, which is perfectly acceptable for the social sciences. The black bars show the values of the Moran's indices calculated for 99 cases of data permutation. As you can see, they are significantly smaller everywhere than the grey bar, which shows the real level of significance of the Moran's *I*, which means that we can reject the null hypothesis. Below we have additional statistics: Moran's *I*; the expected value for the random pattern $E(I) = -1/n - 1$, i.e. the value that we would get if the values of the parameter we are looking at were the result of absolute spatial randomness; and the mean and standard deviation of the expected value. Positive or negative z-scores indicate whether the spatial autocorrelation is positive or negative when the null hypothesis is wrong. In our case, the positive z-score confirmed the validity of the positive Moran's *I*.

Let us calculate the Moran's *I* for Trump for the three neighbourhood weights we have identified (Table 6.7). The null hypothesis is confidently rejected in all three cases, and we can notice that the assessment of the level of clustering of the American electoral space decreases as the scale of analysis increases: 0.603 when analysing first-order neighbours only, which drops to 0.506 when adding second-order neighbours, and 0.302 when analysing state by state.

Table 6.7. Moran's *I* for different neighbourhood matrices

Parameters	W1	W2	W5
Moran's *I*	0.603	0.506	0.302
Significant level (at 99 permutations)	0.01	0.01	0.01
Z-score	47.1	74.7	95.1

Moran's bivariate spatial autocorrelation index measures the association between the dependent variable y and the spatial lag of another variable $\tilde{x}$. Accordingly, it is expressed by the following formula:

$$I_{Y,\tilde{X}} = \frac{n}{\sum_{i=1}^{n}\sum_{j=1}^{n}w_{ij}} \cdot \frac{\sum_{i=1}^{n}\sum_{j=1}^{n}w_{ij}\left(y_i - \underline{y}\right)\left(x_j - \underline{x}\right)}{\sqrt{\sum_{i=1}^{n}\left(y_i - \underline{y}\right)^2}\sqrt{\sum_{i=1}^{n}\left(x_i - \underline{x}\right)^2}}$$

For standardized spatial weights and spatial lag $\tilde{x}$, it will be approximately equal to the Pearson coefficient between y and $\tilde{x}$.

$$I_{Y,X} = \frac{\sum_{i=1}^{n}\left(y_i - \underline{y}\right)(\tilde{x}_i - \underline{x})}{\sqrt{\sum_{i=1}^{n}\left(y_i - \underline{y}\right)^2}\sqrt{\sum_{i=1}^{n}\left(x_i - \underline{x}\right)^2}} \approx$$

$$\approx r_{Y,\tilde{X}} = \frac{\sum_{i=1}^{n}\left(y_i - \underline{y}\right)(\tilde{x}_i - \underline{x})}{\sqrt{\sum_{i=1}^{n}\left(y_i - \underline{y}\right)^2}\sqrt{\sum_{i=1}^{n}(\tilde{x}_i - \underline{x})^2}}$$

To calculate the correlation of two spatially weighted variables, it is easier to simply determine the Pearson coefficient for the spatial lag $\tilde{y}$ and $\tilde{x}$.

$$I_{\tilde{Y},\tilde{X}} = \frac{n}{\sum_{i=1}^{n}\left(\sum_{j=1}^{n}w_{ij}\right)^2} \cdot \frac{\sum_{i=1}^{n}\left[\left(\sum_{j=1}^{n}w_{ij}\left(y_i - \underline{y}\right)\right) \cdot \left(\sum_{j=1}^{n}w_{ij}\left(x_i - \underline{x}\right)\right)\right]}{\sqrt{\sum_{i=1}^{n}\left(y_i - \underline{y}\right)^2}\sqrt{\sum_{i=1}^{n}\left(x_i - \underline{x}\right)^2}}$$

$$I_{\tilde{Y},\tilde{X}} = \frac{\sum_{i=1}^{n}(\tilde{y}_i - \underline{y})(\tilde{x}_i - \underline{x})}{\sqrt{\sum_{i=1}^{n}(y_i - \underline{y})^2}\sqrt{\sum_{i=1}^{n}(x_i - \underline{x})^2}} \approx$$

$$\approx r_{\tilde{Y},\tilde{X}} = \frac{\sum_{i=1}^{n}(\tilde{y}_i - \underline{\tilde{y}})(\tilde{x}_i - \underline{\tilde{x}})}{\sqrt{\sum_{i=1}^{n}(\tilde{y}_i - \underline{\tilde{y}})^2}\sqrt{\sum_{i=1}^{n}(\tilde{x}_i - \underline{\tilde{x}})^2}}$$

Bivariate spatial correlation allows us to test our second hypothesis – that Republicans tend to get additional votes in areas (spatially continuous clusters) where a large proportion of white population lives (Fig. 6.27). For $W1$ weights that demonstrate a satisfactory significance level and z-score, the Moran's I for the country as a whole was weakly significant (0.273). What is more, it was insignificant for the west of the country (0.077) and slightly more significant for the east (0.370).

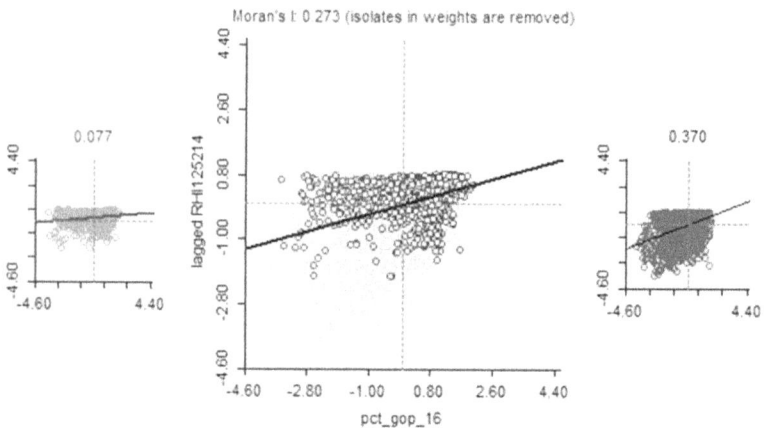

Fig. 6.27. Moran scatterplot for bivariate spatial correlation

The second basic method for assessing the degree of clustering is the *Geary's C spatial autocorrelation index*, named after the Irish statistician Roy Geary (1896–1983) [298]. When $i \neq j$, the following formula is used for calculation:

$$C_Y = \frac{(n-1)}{2\sum_{i=1}^{n}\sum_{j=1}^{n}w_{ij}} \cdot \frac{\sum_{i=1}^{n}\sum_{j=1}^{n}w_{ij}(y_i - y_j)^2}{\sum_{i=1}^{n}(y_i - \underline{y})^2}$$

Accordingly, the formula is converted to the following for bivariate spatial correlation and the correlation of two spatially weighted variables:

$$C_{Y,\tilde{X}} = \frac{(n-1)}{2\sum_{i=1}^{n}\sum_{j=1}^{n}w_{ij}} \cdot \frac{\sum_{i=1}^{n}\sum_{j=1}^{n}w_{ij}\left(\sqrt{\sum_{i=1}^{n}(y_i - y_j)^2}\sqrt{\sum_{i=1}^{n}(x_i - x_j)^2}\right)}{\sum_{i=1}^{n}\left(\sqrt{\sum_{i=1}^{n}(y_i - \underline{y})^2}\sqrt{\sum_{i=1}^{n}(x_i - \underline{x})^2}\right)}$$

$$C_{\tilde{Y},\tilde{X}} = \frac{(n-1)}{2\sum_{i=1}^{n}\sum_{j=1}^{n}w_{ij}} \cdot$$

$$\cdot \frac{\sum_{i=1}^{n}\left(\sum_{j=1}^{n}w_{ij}\sqrt{\sum_{i=1}^{n}(y_i - y_j)^2} \cdot \sum_{j=1}^{n}w_{ij}\sqrt{\sum_{i=1}^{n}(x_i - x_j)^2}\right)}{\sum_{i=1}^{n}\left(\sqrt{\sum_{i=1}^{n}(y_i - \underline{y})^2}\sqrt{\sum_{i=1}^{n}(x_i - \underline{x})^2}\right)}$$

Geary's C ranges from 0 to 2, where 1 indicates no spatial autocorrelation, values significantly less than 1 indicate positive (direct) spatial correlation, and values significantly greater than 1 indicate negative (inverse) correlation. At the same time, it refutes another null hypothesis: not that the phenomenon is randomly distributed in space, but that there is no spatial differentiation in its distribution, that is, that neighbouring phenomena are not different.

Table 6.8. Levels of spatial autocorrelation links

Degree	Moran's I	Geary's C
Absolute positive	1	0
Strong positive	1.0⟨...⟩0.7	0.3⟨...⟩0.0

(Continued)

Table 6.8. Continued

Degree	Moran's I	Geary's C
Average positive	$0.7\langle...\rangle 0.5$	$0.5\langle...\rangle 0.3$
Weak positive	$0.5\langle...\rangle 0.3$	$0.7\langle...\rangle 0.5$
No connection	0	1
Weak negative	$-0.3\langle...\rangle -0.5$	$1.5\langle...\rangle 1.3$
Average negative	$-0.5\langle...\rangle -0.7$	$1.7\langle...\rangle 1.5$
Strong negative	$-0.7\langle...\rangle 1.0$	$2.0\langle...\rangle 1.7$
Absolute negative	-1	2

Geary's C can be used to cross-check the result of the Moran's I (or the Pearson correlation coefficient for a variable and its spatial lag) with a normal spatial distribution of the data, for which it should be represented as $I \approx 1 - C$ or, conversely, $C \approx -I + 1$. Data on approximate comparisons between the levels of spatial autocorrelation relationships are presented in Table 6.8.

6.8 Local Spatial Autocorrelation

Earlier, we looked at methods for calculating spatial autocorrelation indicators for an entire data sample (i.e. the entire space being analysed). Such indices are called *global*. However, sometimes we need to measure *local* space autocorrelation, which determines the variability of spatial heterogeneity at the local (individual) level. Local correlation is revealed by calculating another key vector variable (in addition to spatial lag) in space called that we call the *local indicator of spatial association* (*LISA*) [194]. LISA is calculated for each observation individually (as opposed to global indices, which evaluate the entire population) and answers the question of how spatially correlated (clustered) the data around the observation is.

To continue our example, while global autocorrelation gives us an answer to the hypothesis that the higher the income level of households,

Local Spatial Autocorrelation

the higher the income of households living next to them, local autocorrelation allows us to answer more specific questions – for instance, in which areas of the city is this correlation is confirmed, and in which areas it is not confirmed.

The basic global and local spatial autocorrelation formulas can be found in Table 6.9.

Table 6.9. Comparison of types of spatial autocorrelation

Type of spatial autocorrelation	Basic formula
Global spatial autocorrelation	$\Gamma = \sum_{i=1}^{n}\sum_{j=1}^{n} w_{ij} f(y_i, y_j)$
Local spatial autocorrelation (local indicators of spatial autocorrelation)	$\Gamma_i = \sum_{j=1}^{n} w_{ij} f(y_i, y_j)$

The Local Moran and Geary indices, as well as the Getis–Ord index, are used to calculate local spatial autocorrelation.

The *Local Moran's I* [395] looks like this:

$$I_{Y_i} = \frac{n}{\sum_{i=1}^{n}\sum_{j=1}^{n} w_{ij}} \cdot \frac{\sum_{j=1}^{n} w_{ij}(y_i - \underline{y})(y_j - \underline{y})}{\sum_{i=1}^{n}(y_i - \underline{y})^2}$$

for $i \neq j$, where n is the number of spatial objects, y_i and y_j are the values of indicator y for the i-th and j-th objects, respectively, $\underline{y}$ is the average value of the indicator, w_{ij} is the spatial weight of the neighbourhood between the i-th and j-th objects, and $\sum_{i=1}^{n}\sum_{j=1}^{n} w_{ij}$ is the sum of all spatial weights.

The formula for standardized neighbourhood weights, where $n = \sum_{i=1}^{n}\sum_{j=1}^{n} w_{ij}$, will look like this:

$$I_{Y_i} = \frac{\sum_{j=1}^{n} w_{ij}(y_i - \underline{y})(y_j - \underline{y})}{\sum_{i=1}^{n}(y_i - \underline{y})^2}$$

The method for calculating the Local Moran's *I* allows us to identify the cores and peripheries of spatial clusters, as well as to divide them into four types:

- *high–high* – a local spatial autocorrelation cluster with high parameter values;
- *low–low* – a local spatial autocorrelation cluster with low parameter values;
- *high–low* – cells that are spatial outliers in which high parameter values contrast with low parameter values in adjacent cells;
- *low–high* – cells that are spatial outliers in which low parameter values contrast with high parameter values in adjacent cells.

The validity of Local Moran's *I* is tested using significance level (*p-value*) and z-score. A pseudo-significance level is obtained by carrying out recalculations with a random distribution of data. The expected value for a random pattern that would be obtained if the values of the parameter being analysed were the result of absolute spatial correlation is determined using the following formula:

$$E(I_i) = -1 \sum_{j=1}^{n} w_{ij} / n - 1$$

Now we will create a Local Moran's chorochromatic map for the share of support for Donald Trump in the 2016 U.S. presidential elections (Fig. 6.28) with the neighbourhood matrix *W*1. Unlike the spatial lag chorochromatic map we worked with earlier, the breaks for clusters of high and low support were more accurate because we were now able to estimate the level of significance for each observation (Fig. 6.29) and ignored only those observations on the chorochromatic map for which it was below 0.05. We can also see spatial outliers that were imperceptible on the previous chorochromatic map.

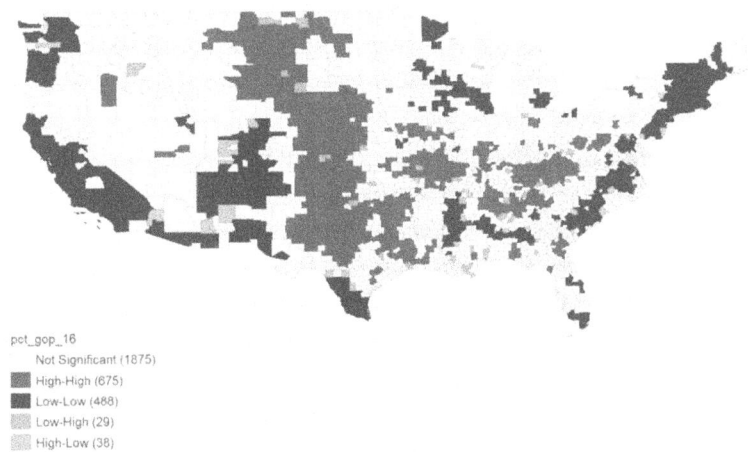

Fig. 6.28. Local Moran chorochromatic map

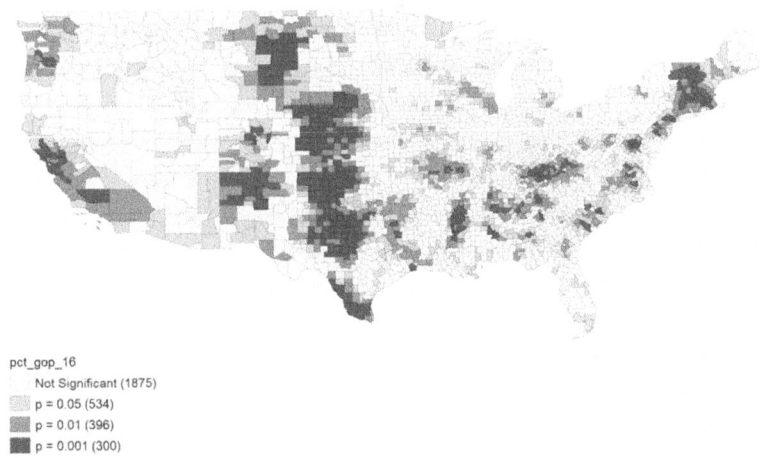

Fig. 6.29. Local Moran significance chorochromatic map

For bivariate spatial autocorrelation, the Local Moran's *I* will be calculated using the following formula:

$$I_{Y,X^-} = \frac{n}{\sum_{i=1}^{n}\sum_{j=1}^{n}w_{ij}} \cdot \frac{\sum_{j=1}^{n}w_{ij}\left(y_i - \underline{y}\right)\left(x_j - \underline{x}\right)}{\sqrt{\sum_{i=1}^{n}\left(y_i - \underline{y}\right)^2}\sqrt{\sum_{i=1}^{n}\left(x_i - \underline{x}\right)^2}}$$

Let us construct a bivariate Local Moran's *I* chorochromatic map for our second hypothesis – that Republicans tend to receive additional votes in regions of the country where more white people live (Fig. 6.30). Recall that the Global Moran's *I* for this spatial dependence turned out to be weak, which simply means that there are regions where this value is significant, and regions where it is not. It is in the latter case that the Local Moran's *I* can be useful.

Fig. 6.30. Bivariate Local Moran's *I* chorochromatic map

The chorochromatic map shows outlier regions that skew the results so that our hypothesis is not supported: in the South, there are regions where support for Trump is high, despite the lower proportion of white men; and in the Midwest, there are counties where support for Trump is low, even though the proportion of whites there is higher.

Another way to assess local spatial correlation in Geary's local spatial autocorrelation index (*local Geary's C*) [298], which, in its generalized form, looks like this:

$$C_{Y_i} = \sum_{j=1}^{n} w_{ij} \left(y_i - y_j \right)^2$$

Fig. 6.31. Local Geary's C chorochromatic map

Calculating and mapping the local Geary's C can be used to test the null hypothesis about the absence of differentiation of the space surrounding the observation, as well as to check the results of the Local Moran's I. As we can see from Fig. 6.31, constructed for the same variable as the Moran's I, local Geary's C can separate out "high–high" and "low–low" clusters, but does not see a difference between the "high–low" and "low–high" types of spatial outliers of negative spatial autocorrelation.

The formula for local Geary's C allows us to calculate the indicator for several variables at once. We call this the multivariate local Geary index. For k variables with index h, it will be calculated as the sum of the local Geary indices for each variable:

$$C_{Y_i}^M = \sum_{h=1}^{k} \sum_{j=1}^{n} w_{ij} \left(y_{hi} - y_{hj} \right)^2$$

A third and final way to measure LISA indicators is the relatively easy-to-calculate *Getis–Ord General G*, named after the American geographer Arthur Getis and his English colleague Keith Ord [301, 418]. It can be calculated either for neighbouring values only (G_i), or for all values, including the object being analysed (G_i^*). The formulas will look like this:

$$G_i = \frac{\sum_{j \neq 1}^{n} w_{ij} x_j}{\sum_{j \neq 1}^{n} w_{ij} x_j}, \quad G_i^* = \frac{\sum_{j=1}^{n} w_{ij} x_j}{\sum_{j=1}^{n} w_{ij} x_j}$$

As Fig. 6.32 shows, a Getis–Ord spatial autocorrelation chorochromatic map calculated for the same indicator as the univariate Local Moran's and Geary's indices confirms the existence of a high–high spatial cluster (called "hot spots" in Getis–Ord statistics) and a low–low spatial cluster ("cold spots"). This method does not identify spatial outliers.

Fig. 6.32. Getis–Ord general G chorochromatic map

6.9 Spatio-Temporal Autocorrelation

Since elections take place at regular intervals, electoral geography deals with relatively regular time series data. Therefore, an important

Spatio-Temporal Autocorrelation

element of the analysis is to check the continuity of electoral behaviour between cycles. In statistics, temporal autocorrelation methods are used to test the hypothesis about the connectedness of the distribution of voting results on neighbouring cycles $y_{i,t} N y_{i,t-1}$. However, a more accurate measure is spatio-temporal autocorrelation, defined as the degree to which data for the current cycle $y_{i,t}$ for each district is connected to the average values for neighbouring districts in the past cycle $W(y_{i,t-1})$; or spatio-temporal autocorrelation of two spatially weighted variables, which is determined by comparing spatial data lags for both electoral cycles $(Wy_{i,t}, Wy_{i,t-1})$. Another type of measurement we can use is differential spatio-temporal autocorrelation, when the difference in the results in neighbouring cycles $y_{i,t} - y_{i,t-1}$ is compared with the spatial lag of these values $W(y_{i,t} - y_{i,t-1})$. Finally, spatio-temporal regression is used to build a model in which both parameters are independent variables: the value at the previous cycle $y_{i,t-1}$ and its spatial lag $W(y_{i,t-1})$. The basic types of temporal and spatio-temporal autocorrelation are listed in Table 6.10.

Table 6.10. Comparison of spatial autocorrelation types

Correlation type	Verifiable hypothesis
Spatial autocorrelation	$y_{i,t} = f(y_{i,t-1}) + \varepsilon$
Spatio-temporal autocorrelation	$y_{i,t} = f\left[W(y_{i,t-1})\right] + \varepsilon$
Spatio-temporal autocorrelation of two spatially weighted variables	$Wy_{i,t} = f\left[W(y_{i,t-1})\right] + \varepsilon$
Differential spatio-temporal autocorrelation	$y_{i,t} - y_{i,t-1} = f\left[W(y_{i,t} - y_{i,t-1})\right] + \varepsilon$

For simple temporal autocorrelation, you can use the Pearson correlation coefficient for variables Y_t, Y_{t-1} in neighbouring cycles:

$$r_{Y_t, Y_{t-1}} = \frac{\sum_{i=1}^{n}(y_{i,t} - \underline{y_t})(y_{i,t-1} - \underline{y_{t-1}})}{\sqrt{\sum_{i=1}^{n}(y_{i,t} - \underline{y_t})^2} \sqrt{\sum_{i=1}^{n}(y_{i,t-1} - \underline{y_{t-1}})^2}}$$

Spatio-temporal autocorrelation is similar to bivariate spatial correlation, only the spatial lag of values in the previous electoral cycle $\tilde{Y}_{t-1}$ acts as an independent variable. Below, we have the three formulas for spatio-temporal autocorrelation: (1) Global Moran's I; (2) Local Moran's I (both of these are for non-standardized weights); and (3) Global Moran's I calculated for standardized weights (i.e. where $n = \sum_{i=1}^{n}\sum_{j=1}^{n} w_{ij}$) and when replacing weights with a spatial lag $\tilde{y}$ compared to the Pearson correlation coefficient.

$$I_{Y_t,\tilde{Y}_{t-1}} = \frac{n}{\sum_{i=1}^{n}\sum_{j=1}^{n} w_{ij}} \cdot \frac{\sum_{i=1}^{n}\sum_{j=1}^{n} w_{ij}\left(y_{i,t} - \underline{y_t}\right)\left(\tilde{y}_{j,t-1} - \underline{\tilde{y}_{t-1}}\right)}{\sqrt{\sum_{i=1}^{n}\left(y_{i,t} - \underline{y_t}\right)^2}\sqrt{\sum_{i=1}^{n}\left(\tilde{y}_{j,t-1} - \underline{\tilde{y}_{t-1}}\right)^2}}$$

$$I_{Y_{t_i},\tilde{Y}_{t-1}} = \frac{n}{\sum_{i=1}^{n}\sum_{j=1}^{n} w_{ij}} \cdot \frac{\sum_{j=1}^{n} w_{ij}\left(y_{i,t} - \underline{y_t}\right)\left(\tilde{y}_{j,t-1} - \underline{\tilde{y}_{t-1}}\right)}{\sqrt{\sum_{i=1}^{n}\left(y_{i,t} - \underline{y_t}\right)^2}\sqrt{\sum_{i=1}^{n}\left(\tilde{y}_{j,t-1} - \underline{\tilde{y}_{t-1}}\right)^2}}$$

$$I_{Y_t,\tilde{Y}_{t-1}} = \frac{\sum_{i=1}^{n}\left(y_{i,t} - \underline{y_t}\right)\left(\tilde{y}_{j,t-1} - \underline{\tilde{y}_{t-1}}\right)}{\sqrt{\sum_{i=1}^{n}\left(y_{i,t} - \underline{y_t}\right)^2}\sqrt{\sum_{i=1}^{n}\left(\tilde{y}_{j,t-1} - \underline{\tilde{y}_{t-1}}\right)^2}} \approx$$

$$\approx r_{Y_t,\tilde{Y}_{t-1}} = \frac{\sum_{i=1}^{n}\left(y_{i,t} - \underline{y_t}\right)\left(\tilde{y}_{t-1} - \underline{y_{t-1}}\right)}{\sqrt{\sum_{i=1}^{n}\left(y_{i,t} - \underline{y_t}\right)^2}\sqrt{\sum_{i=1}^{n}\left(\tilde{y}_{t-1} - \underline{y_{t-1}}\right)^2}}$$

The spatio-temporal autocorrelation of two spatially weighted variables is similar to the correlation of two spatially weighted variables, but it compares the spatial lag of a variable in the current cycle $\tilde{Y}_t$ to the spatial lag of the same variable in the previous cycle $\tilde{Y}_{t-1}$. The formulas for Global Moran's I (for non-standardized weights), as well as for the Global Moran's I when replacing lag $\tilde{y}$ and the Pearson correlation coefficient look like this:

Spatio-Temporal Autocorrelation

$$I_{\tilde{Y}_t \tilde{Y}_{t-1}} = \frac{n}{\sum_{i=1}^{n}\left(\sum_{j=1}^{n} w_{ij}\right)^2} \cdot$$

$$\cdot \frac{\sum_{i=1}^{n}\left[\left(\sum_{j=1}^{n} w_{ij}\left(y_{i,t} - \underline{y_t}\right)\right) \cdot \left(\sum_{j=1}^{n} w_{ij}\left(y_{i,t-1} - \underline{y_{t-1}}\right)\right)\right]}{\sqrt{\sum_{i=1}^{n}\left(y_{i,t} - \underline{y_t}\right)^2} \sqrt{\sum_{i=1}^{n}\left(y_{i,t-1} - \underline{y_{t-1}}\right)^2}}$$

$$I_{I_{\tilde{Y}_t \tilde{Y}_{t-1}}} = \frac{\sum_{i=1}^{n}\left(\tilde{y}_{i,t} - \underline{y_t}\right)\left(\tilde{y}_{i,t-1} - \underline{y_{t-1}}\right)}{\sqrt{\sum_{i=1}^{n}\left(y_{i,t} - \underline{y_t}\right)^2} \sqrt{\sum_{i=1}^{n}\left(y_{i,t-1} - \underline{y_{t-1}}\right)^2}} \approx$$

$$\approx r_{\tilde{Y}_t, \tilde{Y}_{t-1}} = \frac{\sum_{i=1}^{n}\left(\tilde{y}_{i,t} - \underline{\tilde{y}_t}\right)\left(\tilde{y}_{i,t-1} - \underline{\tilde{y}_{t-1}}\right)}{\sqrt{\sum_{i=1}^{n}\left(\tilde{y}_{i,t} - \underline{\tilde{y}_t}\right)^2} \sqrt{\sum_{i=1}^{n}\left(\tilde{y}_{i,t-1} - \underline{\tilde{y}_{t-1}}\right)^2}}$$

Differential (i.e. difference-based) spatio-temporal autocorrelation is based on the idea of spatial autocorrelation $Y_t - Y_{t-1}$. The formulas for the global and Local Moran's I (for non-standardized weights) for calculating differential spatio-temporal autocorrelation are as follows:

$$I_{Y_t - Y_{t-1}} = \frac{n}{\sum_{i=1}^{n}\sum_{j=1}^{n} w_{ij}} \cdot$$

$$\cdot \frac{\sum_{i=1}^{n}\sum_{j=1}^{n} w_{ij}\left[\left(y_{i,t} - y_{i,t-1}\right) - \underline{y_{i,t} - y_{i,t-1}}\right] \cdot \left[\left(y_{j,t} - y_{j,t-1}\right) - \underline{y_{i,t} - y_{i,t-1}}\right]}{\sum_{i=1}^{n}\left[\left(y_{i,t} - y_{i,t-1}\right) - \underline{y_{i,t} - y_{i,t-1}}\right]^2}$$

$$I_{(Y_t - Y_{t-1})_i} = \frac{n}{\sum_{i=1}^{n}\sum_{j=1}^{n} w_{ij}} \cdot$$

$$\frac{\sum_{j=1}^{n} w_{ij}\left[\left(y_{i,t}-y_{i,t-1}\right)-\overline{y_{i,t}-y_{i,t-1}}\right]\cdot\left[\left(y_{j,t}-y_{j,t-1}\right)-\overline{y_{i,t}-y_{i,t-1}}\right]}{\sum_{i=1}^{n}\left[\left(y_{i,t}-y_{i,t-1}\right)-\overline{y_{i,t}-y_{i,t-1}}\right]^{2}}$$

Now let us transform the Global Moran's I calculated for standardized weights $\left(\text{i.e. where } n = \sum_{i=1}^{n}\sum_{j=1}^{n} w_{ij}\right)$ by replacing the weights with the spatial lag $\tilde{y}$ compared to the Pearson correlation coefficient for estimating differential spatio-temporal autocorrelation.

$$I_{Y_t - Y_{t-1}} \approx r_{Y_t - Y_{t-1}, \tilde{Y}_t - \tilde{Y}_{t-1}} =$$

$$= \frac{\sum_{i=1}^{n}\left[\left(y_{i,t}-y_{i,t-1}\right)-\overline{y_{i,t}-y_{i,t-1}}\right]\cdot\left[\left(\tilde{y}_{i,t}-\tilde{y}_{i,t-1}\right)-\overline{y_{i,t}-y_{i,t-1}}\right]}{\sqrt{\sum_{i=1}^{n}\left[\left(y_{i,t}-y_{i,t-1}\right)-\overline{y_{i,t}-y_{i,t-1}}\right]^{2}}\sqrt{\sum_{i=1}^{n}\left[\left(y_{i,t}-y_{i,t-1}\right)-\overline{y_{i,t}-y_{i,t-1}}\right]^{2}}} \approx$$

$$\approx \frac{\sum_{i=1}^{n}\left[\left(y_{i,t}-y_{i,t-1}\right)-\overline{y_{i,t}-y_{i,t-1}}\right]\cdot\left[\left(\tilde{y}_{i,t}-\tilde{y}_{i,t-1}\right)-\overline{\tilde{y}_{i,t}-\tilde{y}_{i,t-1}}\right]}{\sqrt{\sum_{i=1}^{n}\left[\left(y_{i,t}-y_{i,t-1}\right)-\overline{y_{i,t}-y_{i,t-1}}\right]^{2}}\sqrt{\sum_{i=1}^{n}\left[\left(\tilde{y}_{i,t}-\tilde{y}_{i,t-1}\right)-\overline{\tilde{y}_{i,t}-\tilde{y}_{i,t-1}}\right]^{2}}}$$

Now we will estimate the temporal autocorrelation of voting for Republican candidates in the U.S. presidential elections in 2012 and 2016. First, we will build a scatterplot based on the Pearson correlation coefficient (Fig. 6.33).

Spatio-Temporal Autocorrelation

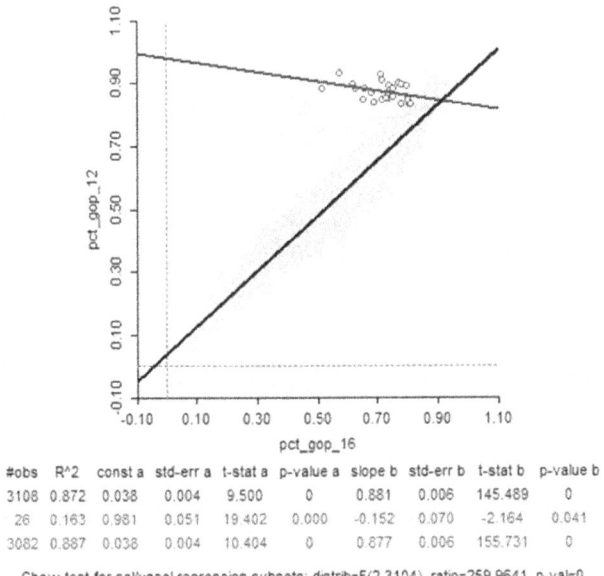

Fig. 6.33. Scatterplot for temporal autocorrelation

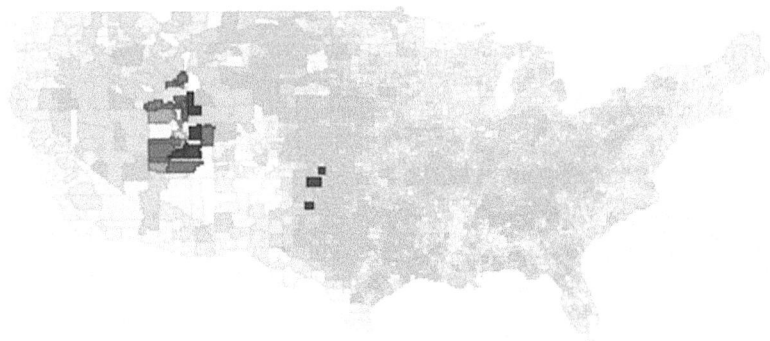

Fig. 6.34. Natural breaks map marked with a zone of anomalous temporal autocorrelation

On the whole, the electoral support for candidates shows a confident continuity ($R^2 = 0,872$), although there is an obvious group of counties in which this regularity is not clearly traced ($R^2 = 0{,}163$). Figure 6.34 shows that these outliers are predominantly located in Utah.

The differential Local Moran's I chorochromatic map (Fig. 6.35) shows a clear difference between the two clusters of changes: a decrease in the level of support for the candidate relative to the previous election cycle was observed in the "low–low" cluster in the southwest; while the opposite is true of the "high–high" cluster in the northeast, where support increased compared to the previous cycle.

Fig. 6.35. Differential Local Moran's I chorochromatic map

6.10 Spatial Regression

Regression is the statistical dependence of the mean value of a random variable on the values of one or more other random variables. Unlike correlation, where a relationship is established between two variables, regression, by increasing the number of independent variables, allows us to determine the direction of the relationship. The formula for simple two-dimensional linear regression is:

$$y = \beta X + \varepsilon$$

Spatial Regression

where y is the dependent variable, X is a group of independent variables $x_1,...,x_n$, β is the coefficient that determines the slope of the regression line (i.e. the expected change in the dependent variable when the independent variable changes), and ε is the error (disparity, residual) of the model, that is, the random deviation from the deterministic function. Geographically weighted variables take part in spatial regression, and the following main models exist (Table 6.11).

Table 6.11. Spatial regression models

Model	Formula	Principle
Spatial Durbin Model (SDM)	$y = \beta X + \theta WX + \varepsilon$	y depends on the values of the explanatory variables in the region under observation and the weighted values of the same explanatory variables in neighbouring regions
Spatial Autoregressive Model (SAR)	$y = \beta X + \rho W y + \varepsilon$	y depends on the values of the explanatory variables in the region under observation and the weighted values of y in neighbouring regions
Spatial Error Model (SEM)	$y = \beta X + (1 - \lambda W)^{-1} \varepsilon$	y depends on the values of the explanatory variables in the region under observation and on the neighbour-weighted error component

There are three main spatial regression models: the *Spatial Durbin Model (SDM)*, the *Spatial Autoregressive Model (SAR)*, and the *Spatial Error Model (SEM)* [195]. In the *Spatial Durbin Model*, geographically weighted parameters (the spatial lag of the phenomenon) are included in the number of independent variables – this is what sets it apart from ordinary regression. The *Spatial Autoregressive Model* is used to test the hypothesis that the outcome (dependent variable) is affected not only by the independent variables, but also by neighbouring values (spatial lag) of the dependent variable (the diagram on the right in Fig. 6.36). In other words, this model allows us to identify the neighbourhood effect in social processes. With the *Spatial Error Model*, it is not the dependent variable itself that is considered autoregressive, but the error component (disparity, residual) of the entire model (the diagram on the left in Fig. 6.36). It is also possible to use models that combine the spatially weighted explanatory variables from the Durbin model with those from the spatial autoregressive or spatial error models.

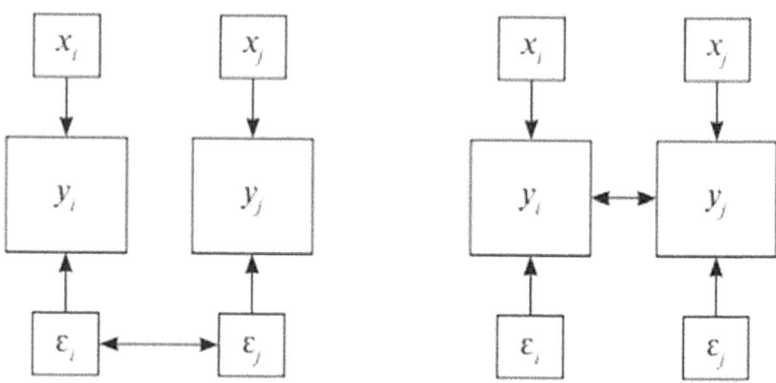

Fig. 6.36. Spatial regression models

Let us use regression analysis to evaluate the dependence of voting for the Republican candidate in the U.S. presidential election on two factors: the proportion of white men (RHI125214) and the proportion of citizens with a higher education (EDU685213). We will use the ordinary least squares (OLS) method for multiple linear regression (MLR) – that is, estimating more than one independent variable to determine a linear relationship.

```
SUMMARY OF OUTPUT: ORDINARY LEAST SQUARES
ESTIMATION
Data set: election
Dependent Variable  : pct_gop_16  Number of Observations: 3108
Mean dependent var  : 0.636607    Number of Variables    : 3
S.D. dependent var  : 0.156059    Degrees of Freedom     : 3105
R-squared           : 0.504309    F-statistic            : 1579.49
Adjusted R-squared  : 0.503990    Prob(F-statistic)      : 0
Sum squared residual: 37.5204     Log likelihood         : 2453.72
Sigma-square        : 0.0120839   Akaike info criterion  : -4901.45
S.E. of regression  : 0.109927    Schwarz criterion      : -4883.32
Sigma-square ML     : 0.0120722
S.E of regression ML: 0.109874
----------------------------------------------------------------
Variable       Coefficient   Std.Error      t-Statistic   Probability
----------------------------------------------------------------
CONSTANT    0.364052      0.0117393      31.0113       0.00000
RHI125214   0.00516359    0.000125296    41.2113       0.00000
EDU685213  -0.00854768    0.000223303   -38.2784       0.00000
----------------------------------------------------------------
REGRESSION DIAGNOSTICS
MULTICOLLINEARITY CONDITION NUMBER 13.463841
TEST ON NORMALITY OF ERRORS
TEST         DF          VALUE        PROB
Jarque-Bera  2           934.4029     0.00000
DIAGNOSTICS FOR HETEROSKEDASTICITY
RANDOM COEFFICIENTS
TEST                DF          VALUE        PROB
Breusch-Pagan test  2           29.2685      0.00000
Koenker-Bassett test 2          13.5011      0.00117
SPECIFICATION ROBUST TEST
TEST         DF          VALUE        PROB
White        5           40.7076      0.00000
```

When analysing the resulting regression model, special attention should be paid to the following indicators:

1. R^2 (*R-squared*) – the coefficient of determination, the percentage of the variance explained by the model. It is generally accepted that

a model with a coefficient of determination of less than 0.5 cannot be used, whereas a coefficient of greater than 0.8 is desirable. In our case, $R^2 = 0.504$, which can be interpreted as relationship between the variables being weak, although this is not enough for us to draw any conclusions about the explanatory power of the model. The fact that the result does not differ significantly from the *adjusted R-squared* figure indicates a high accuracy of the calculation.

The coefficient of determination is the absolute value of the correlation between observation y and valuation function $\check{y}$ predicted by the regression equation (what the observation would be if there was no fractional error ε in the regression model). In this case, the coefficient of determination is calculated on the basis of the total variance (*sum of squares total, SST*), which is the sum of the explained variance (*sum of squared regression, SSR*) and the unexplained variance (*sum squared error, SSE*).

$$SST = SSR + SSE$$

$$SST = \sum_{i=1}^{n}(y_1 - \underline{y})^2, SSR = \sum_{i=1}^{n}(\widehat{y}_i - \underline{y})^2, SSE = \sum_{i=1}^{n}(y_1 - \check{y})^2$$

$$R^2 = \frac{SSR}{SST} = \frac{\sum_{i=1}^{n}(\widehat{y}_i - \underline{y})^2}{\sum_{i=1}^{n}(y_1 - \underline{y})^2} = 1 - \frac{SSE}{SST} = 1 - \frac{\sum_{i=1}^{n}(y_1 - \check{y})^2}{\sum_{i=1}^{n}(y_1 - \underline{y})^2}$$

The adjusted coefficient of determination also takes the number of observations n and parameters p (regression coefficients), including the intersection point of constant α, into account. The difference between n and p is called the *degrees of freedom for error* (*DFE*), and the difference $p-1$ is the *degrees of freedom for model* (*DFM*).

$$DFE = n - p, DFM = p - 1$$

$$R_{adj}^2 = 1 - \left(\frac{n-1}{DFE}\right)\frac{SSE}{SST} = 1 - \left(\frac{n-1}{n-p}\right)\frac{\sum_{i=1}^{n}(y_1 - \check{y})^2}{\sum_{i=1}^{n}(y_1 - \underline{y})^2}$$

2. The F-statistic – the overall evaluation of the quality of the resulting model. Its significance level (Prob(F-statistic)) should be below 0.05.

The F-statistic tests the plausibility of the null hypothesis that the coefficient values are zero.

$$F = \frac{SSR / DFM}{SSE / DFE} = \frac{\sum_{i=1}^{n}(\widehat{y}_i - \underline{y})^2 \div (p-1)}{\sum_{i=1}^{n}(y_1 - \check{y})^2 \div (n-p)}$$

3. The *Akaike info criterion (AIC), Schwarz criterion (SC)*, and *log likelihood* are used to compare different regression models with the same dataset. The Akaike info criterion is more suitable for comparing models of the same type with one another (say, spatial models with other spatial models), while the other two are better for comparing a non-spatial model with a spatial model. The lower the Akaike and Schwarz criteria and the higher the log-likelihood, the more robust the model.
4. The level of significance (*probability, p-value*) of the *t-statistic* of all variables, except for the constant, should be lower than 0.05. This confirms the validity of using the variable in this model. The t-statistic is calculated as the ratio of the coefficient to its standard error $t_\beta = \beta / SE_\beta$. The sign (plus or minus) of the coefficient of the variable indicates the direction of the relationship. In our case, voting Republican has a negative relationship with the proportion of people with a higher education: the fewer people with a university degree, the more votes for Trump.
5. The *Multicollinearity condition number/index (CI)* should be below 30. *Multicollinearity* speaks to a strong correlation between the individual variables of the model. If the value is above the threshold, then we need to calculate the *variance inflation factor (VIF)*, which estimates the degree to which the variance of each individual coefficient is inflated as a direct result of multicollinearity. For example, a variance inflation factor of 3 means that the variance of the coefficient estimate for a variable in the regression equation is increased by a factor of 3 due to the strong correlation of this independent variable with one or more other independent variables

included in the model. Table 6.12 shows the data for carrying our diagnostics for the multicollinearity of the model.

Table 6.12. Diagnostics for multicollinearity

Multicollinearity	CI	VIF
Weak	<30	<4
Moderate	$30\langle ... \rangle 100$	$4\langle ... \rangle 10$
Strong	>100	>10

6. The *Jarque–Bera test on normality of errors*, which tests the normality of the distribution of residuals as evidence of the absence of bias, should be greater than 0.1 (i.e. there should be no such normality). However, this test is valid for a large number of observations, so it can be ignored in this case.

7. *Heteroskedasticity* refers to the heterogeneity of observations, that is, the nonuniformity of the variance of random errors in the model. It is verified using the Breusch–Pagan, Koenker–Bassett and White tests, the latter of which is more accurate for spatially distributed data. If there is a significant difference between their values, or their significance level is below 0.1, then the probability of the heteroskedasticity of the model is confirmed. The data needs to be brought to a normal distribution, for example, by carrying out a logarithmic transformation.

The heteroskedasticity obtained in the regression may indirectly indicate spatial dependence in our model that was not accounted for. Now we will carry out diagnostics for spatial dependence and the possibility of using spatial regression. To do this, we will add a standardized neighbourhood matrix $W1$.

DIAGNOSTICS FOR SPATIAL DEPENDENCE			
FOR WEIGHT MATRIX: W1 (row-standardized weights)			
TEST	MI/DF	VALUE	PROB
Moran's I (error)	0.6232	58.1053	0.00000
Lagrange Multiplier (lag)	1	2055.4246	0.00000
Robust LM (lag)	1	9.2562	0.00235
Lagrange Multiplier (error)	1	3356.6808	0.00000
Robust LM (error)	1	1310.5123	0.00000
Lagrange Multiplier (SARMA)	2	3365.9369	0.00000

Moran's I tells us whether there is any spatial autocorrelation among the variables. Since it has a low p-value, we can conclude that there is such autocorrelation, which means that we need to use spatial regression models. Spatial autoregressive and spatial error models cannot always be used equally, and spatial tests are carried out to determine an acceptable model using the *Lagrange multiplier (LM)*, which estimates the degree of spatial autocorrelation of the dependent variable (for the spatial autoregressive model) or residuals (for the spatial error model). Note the last column in the Diagnostics for Spatial Dependence section, "PROB (p-value)," and select the model (*Lagrange Multiplier (lag)* or *Lagrange Multiplier (error)*) for which the p-value is less than or equal to 0.05. If both qualify, then you need to repeat the process, but this time using the robust Lagrange multiplier (*Robust LM (lag)* and *Robust LM (error)*). And if both these give us p-values that are less than or equal to 0.05, then we select the one with the lowest p-value. In our case, both models can be applied, although the spatial error model is expected to be slightly more valid. We will apply the spatial autoregressive model first.

```
SUMMARY OF OUTPUT: SPATIAL LAG MODEL –
MAXIMUM LIKELIHOOD ESTIMATION
Data set              : election
Spatial Weight        : W1
Dependent Variable    : pct_gop_16  Number of Observations: 3108
Mean dependent var    : 0.636607    Number of Variables   : 4
S.D. dependent var    : 0.156059    Degrees of Freedom    : 3104
Lag coeff. (Rho)      : 0.608959
R-squared             : 0.727308    Log likelihood        : 3257.41
Sq. Correlation       : –           Akaike info criterion : -6506.82
Sigma-square          : 0.00664122  Schwarz criterion     : -6482.66
S.E of regression     : 0.0814937
-----------------------------------------------------------------
Variable         Coefficient   Std.Error    z-value    Probability
-----------------------------------------------------------------
W_pct_gop_16     0.608959      0.0131619    46.267     0.00000
CONSTANT         0.0509801     0.00932695   5.46589    0.00000
RHI125214        0.00353854    0.000119279  29.666     0.00000
EDU685213       -0.00529236    0.000179716 -29.4485    0.00000
-----------------------------------------------------------------
REGRESSION DIAGNOSTICS
DIAGNOSTICS FOR HETEROSKEDASTICITY
RANDOM COEFFICIENTS
TEST                        DF       VALUE       PROB
Breusch-Pagan test          2        190.2566    0.00000
DIAGNOSTICS FOR SPATIAL DEPENDENCE
SPATIAL LAG DEPENDENCE FOR WEIGHT MATRIX: W1
TEST                        DF       VALUE       PROB
Likelihood Ratio Test       1        1607.3794   0.00000
```

For spatial regressions, it is best to use the *maximum likelihood estimation (MLE)* method, so comparing the coefficient of determination between the two models is not entirely correct. Nevertheless, we see that the degree of determination increased significantly when adding a new variable – namely, the spatial lag of the dependent variable. The spatial model again proves to be more valid here, which is confirmed by the lower Schwarz criterion and the higher log likelihood. From this, we can conclude that the neighbourhood effect is significant for the U.S. electoral space.

Now let us look at the spatial error model.

```
SUMMARY OF OUTPUT: SPATIAL ERROR MODEL –
MAXIMUM LIKELIHOOD ESTIMATION
Data set            : election
Spatial Weight      : W1
Dependent Variable  :  pct_gop_16   Number of Observations : 3108
Mean dependent var  :  0.636607     Number of Variables    : 3
S.D. dependent var  :  0.156059     Degrees of Freedom     : 3105
Lag coeff. (Lambda) :  0.857272
R-squared           :  0.834040     R-squared (BUSE)       : -
Sq. Correlation     :  -            Log likelihood         : 3846.676040
Sigma-square        :  0.00404183   Akaike info criterion  : -7687.35
S.E of regression   :  0.0635754    Schwarz criterion      : -7669.23
------------------------------------------------------------------------
   Variable     Coefficient    Std.Error      z-value    Probability
------------------------------------------------------------------------
CONSTANT        0.0983935      0.0135416      7.26602    0.00000
RHI125214       0.00732343     0.000121679    60.1864    0.00000
EDU685213      -0.00534314     0.000165543   -32.2765    0.00000
LAMBDA          0.857272       0.0104785      81.8124    0.00000
------------------------------------------------------------------------

REGRESSION DIAGNOSTICS
DIAGNOSTICS FOR HETEROSKEDASTICITY
RANDOM COEFFICIENTS
TEST                          DF         VALUE       PROB
Breusch-Pagan test            2          242.7414    0.00000
DIAGNOSTICS FOR SPATIAL DEPENDENCE
SPATIAL ERROR DEPENDENCE FOR WEIGHT MATRIX : W1
TEST                          DF         VALUE       PROB
Likelihood Ratio Test         1          2785.9065   0.00000
```

The lambda variable has not been added to the regression model to reflect the spatially correlated component of the error. The coefficient of determination in such a model confidently indicates the presence of a statistical dependence, while the Akaike info and Schwarz criteria are lower once again, while the log likelihood is higher, thus increasing the validity of the model.

The following conclusion can be drawn from our study: the level of support for Trump in 2016 was closely connected to three variables: (1) the share of the white population; (2) the share of the population who do not have a higher education; and (3) the neighbourhood effect, that is, proximity to regions where support for Trump was high. At the same time, regions in which such dependence was not observed also had significant clustering in space; in other words, there was a reactive neighbourhood effect that increased the level of support for Trump's opponent.

Thus, the use of spatial statistical analytical methods has allowed us to confirm the basic hypothesis of electoral geography – the neighbourhood effect – namely, that the structure of space has an independent (separate from other factors) influence on the electoral process and voting results.

<u>Key Terms</u>

- map, chorochromatic map, quantile, isolate, orthodrome, polygon
- chorochromatic map: equal intervals, equal classes, natural intervals
- the modifiable areal unit problem, the boundary problem, the multiple comparisons problem, the distance decay effect, spatial inversion
- variation, outlier, variance, validity, correlation (covariation), causation, null hypothesis, robustness
- median, arithmetic mean, arithmetic weighted mean, standard deviation, normal distribution, standard error, *p*-value
- centrography, geographical average (geographic midpoint), geographical weighted average, median centre, standard distance, standard deviation ellipse, spatial outlier, bivariate normal distribution, spatial neighbourhood weights, spatial lag
- the Pearson coefficient, the Moran, Geary, and Getis–Ord indices, the local indicator of spatial association (LISA)
- spatial autocorrelation, bivariate spatial correlation, correlation of two spatially weighted variables, global and local spatial autocorrelation
- spatial regression models: spatial Durbin model, spatial autoregressive model, spatial error model
- multicollinearity, heteroskedasticity

Questions and Exercises

Choose an electoral campaign and analyse it according to the following scheme.

1. Select a basic variable for analysis, for example, the level of support for one party or candidate.
2. Evaluate the data distribution of the variable you are analysing using a histogram and a box and whisker plot.
3. Build a chorochromatic map of natural intervals for the variable you are analysing.
4. Calculate the Pearson correlation coefficient for the variable you are analysing and one independent variable that reflects socio-economic statistics. Assess the level of validity and determination of the correlation.
5. Construct the best matrix of spatial neighbourhood weights for the space you are analysing.
6. Calculate and map the spatial lag for the variable you are analysing and compare the result with the natural interval chorochromatic map.
7. Calculate the Global Moran's I for the variable you are analysing. Evaluate the level of validity and determination of the resulting variable.
8. Calculate the bivariate Global Moran's I for the variable you are analysing and the socioeconomic statistical variable used in the calculation of the Pearson correlation coefficient.
9. Calculate and map the Local Moran's I for the variable you are analysing, and the bivariate Local Moran's I for the two variables obtained at the previous stage. What conclusions can you draw about the space clustering?
10. Calculate a regression model with at least two independent variables. Check its validity and determination levels.
11. Use the Lagrange multiplier to determine whether spatial regression can be applied to the same model. If it can, carry out the necessary calculations, check its validity and determination levels, and compare spatial and non-spatial models.
12. Draw a general conclusion about the differentiation of the electoral space you are analysing and the level of influence of the spatial factor on electoral results.

Appendix. Spatial Statistical Analysis in Political and Electoral Geography: Methodological Guide

1. Geoinformation Mapping

Objectives

✓ Become proficient in the QGIS software package
✓ Find a suitable shapefile
✓ Select a map projection
✓ Create a chorochromatic map with typological zoning
✓ Add captions and labels to a chorochromatic map

Geographic Information Systems

A geographic map is a generalized representation of the Earth's surface on a plane. It is based on a mathematical formula that transforms the spherical shape of the Earth into a plane. At the same time, unlike photographs, maps generalize and systematize the image of reality. In other words, it is a mental, rather than a natural, product. Accordingly, since a map is only a model of reality, the process of mapping itself is an analytical procedure for perceiving reality. Moreover, the study and comparison of maps – which can be seen as nonverbal narratives of the people who created them – can be considered a kind of discourse analysis.

The discursive potential of maps can be illustrated using the maps of Valga/Valka (twin cities separated by the border between Latvia and Estonia). As the bigger of the two, the Estonian city of Valga dominates the pair and sees Valka merely as a suburb. Meanwhile, its Latvian sister, Valka, desperate to assert its own identity and fearing that it will get swallowed up by the stronger Valga, positions itself as a separate city altogether. The narrative in Valga is thus one of Valga/Valka as a single entity, while in Valka it is one of the city's uniqueness. Estonian maps depict Valga/Valka as one city separated by a barely visible border, while Latvian maps show their part of the city only, cut off at the borderline – the Estonian part of the city is replaced by shades of grey.

Spatial analysis does not involve creating and studying maps themselves. What we want to produce is *chorochromatic maps* – simplified geographic models that reflect the necessary spatial patterns. They will not include the environment of the object of study or its grid coordinates. See Fig. A2.1 to get an understanding of what a chorochromatic map looks like. Special geographic information system (GIS) software is used to create such chorochromatic maps. We will call the creation of chorochromatic maps in GIS software packages geoinformation mapping.

There are many professional GIS packages on the market today – ArcView GIS, MapInfo, GeoGraph/GeoDraw, etc. We will give examples of spatial analysis carried out primarily in the QGIS GIS package, with the use of some other programs for specific tasks, later in this guide.

Quantum GIS (QGIS) is a completely free, intuitive, open-source geographic information system that works on all major operating systems, including Linux and Windows. It can be downloaded from the QGIS website at www.qgis.org.

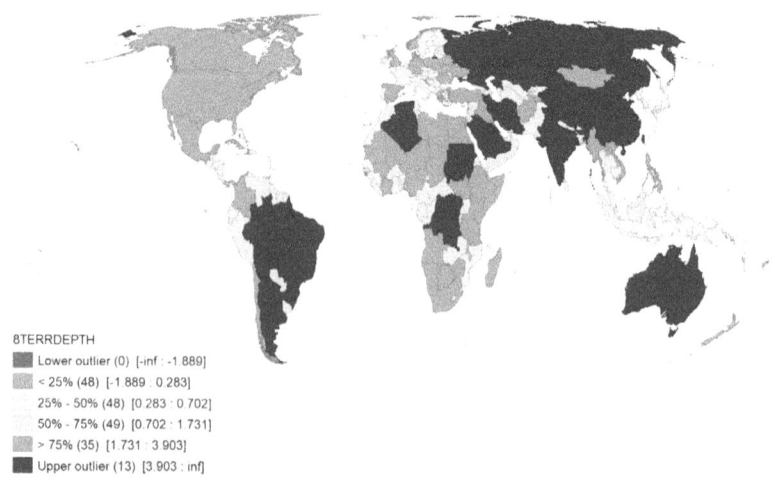

Fig. A2.1. Chorochromatic map

Chorochromatic Map Data Formats

In order to work in the QGIS app, we need images that have a spatial reference – information about the coordinate system and projection and geometric correction parameters. This means files with the .shp extension for vector chorochromatic maps and .geotiff for raster chorochromatic maps. The most common format for working with vector chorochromatic maps is also called *shapefile*. Shapefiles are not images but digital vector-based storage files.

Major shapefile repositories by country include GADM (https://gadm.org/index.html), Geofabrik (http://download.geofabrik.de) and DIVA-GIS (http://www.diva-gis.org/data/). However, the easiest way to find a shapefile is simply to type what you need into your preferred search engine. Data for Russia, including its administrative, territorial and municipal divisions, can be found at http://gis-lab.info/qa/rusbounds-rosreestr.html. The project monitors the placement of administrative boundary data for all countries at https://index.okfn.org/dataset/boundaries/ and includes links to individual shapefile databases by country.

For this task, open the shape file "Russia" in QGIS. The dataset is available on our Google Drive (https://drive.google.com/file/d/1nsjJshlCUiq7nUbDnNuos_ybYejS9LO/view?usp=drive_link) as well as on our website (https://mgimo.ru/upload/2023/05/russia-geoinformaczionnoekartografirovanie-geoinformation-mapping.zip). To open the file in QGIS, you need to load it as a separate layer. This works in a similar way to photo or video editors. There are several ways to add layers to QGIS.

Method 1: On the tab bar, select Layer, then Add Layer and Add Vector Layer. As the program tells us, this can also be done by pressing Ctrl+Shift+V (assuming you're on a Windows computer). Try to use shortcuts whenever possible – it will save you time and energy. In the pop-up window, click on the ellipsis (...) sign, select the required .shp file and click Open. Click Add, and then close the dialog box after the layer is displayed in the workspace at the bottom left of the screen.

Method 2: Drag the .shp file from the folder to the QGIS workspace.

To delete a layer, right-click on the layer name in the workspace and select Remove Layer. Click OK to confirm that you want to delete the layer.

Map Projection Transformation

The first procedure in spatial analysis is always to establish the projection of the chorochromatic map you are analysing and to transform it if necessary.

A map projection is a mathematically defined representation of the spherical surface of the Earth on a plane. It is a one-to-one representation of the geodetic coordinates of points (latitude B and longitude L) and their rectangular coordinates (X and Y) on the map:

$$X = f_1(B,L); \ Y = f_2(B,L).$$

Obviously, the spherical surface of the globe cannot be unfolded onto a plane without distortion. To look for yourself, take a tangerine peel and then try to stretch it out on a plane.

Fig. A2.2. Chorochromatic map of administrative borders of the Russian Federation

When we flatten the Earth's surface into a plane, distances, areas or directions (angles) can be distorted. Compare, for example, the conformal *Mercator projection* created for navigation, which distorts areas, with the equal-area *Gall–Peters projection*, which correctly reflects areas but distorts the shape of the polar regions. Countries in the northern hemisphere have traditionally preferred to use the Mercator projection for school maps, as it increases the area of these countries and reduces that of countries in the southern hemisphere.

What is more, the Earth is not a perfect sphere. It is both flattened at the poles (an *ellipsoid*) and has an uneven distribution of mass plus ongoing global tectonic deformations (which is why the Earth's unique shape is referred to as a *geoid*). However, despite the irregularity and variability of the Earth's shape, what we need to do if we want to build cartographic projections is set (and regularly update) the parameters of a geometric *reference ellipsoid*. Such an ellipsoid will contain three basic parameters: semi-major axis *a*, semi-minor axis *b*, and polar contraction index α, calculated using the formula:

$$\alpha = (a - b) / a$$

There are many standards of reference ellipsoids. Table 2.1 lists the basic parameters of two reference ellipsoids that we will use as a basis for calculating widely used cartographic projections: the World Geodetic System 1984 and the Russian "Surface of the Earth" 1990.

Table 2.1 Basic Parameters of International and Russian Reference Ellipsoids

Reference Ellipsoid	Code	*a* axis	*b axis*	Contraction α
World Geodetic System	WGS 84	6,378,137	6,356,752	1:298.257
Surface of the Earth	PZ-90	6,378,136	6,356,751	1:298.358

The map projection menu in QGIS is located in the bottom righthand corner. You can use it to search for a map projection from the included list, or import a user map projection. Most map projections that exist are easy to find by typing in a unified EPSG code. The easiest map projections for working with maps of Russia with that come with the program are listed in Table 2.2.

Table 2.2 Map Projections for Chorochromatic maps of Russia

EPSG Code	Map Projection Name
EPSG:3576	WGS 84 / North Pole LAEA Russia
EPSG:5940	WGS 84 / EPSG Russia Polar Stereographic
EPSG:32643	WGS 84 / UTM Zone 43N
EPSG:32648	WGS 84 / UTM Zone 48N
EPSG:3388	Pulkovo 1942 / Caspian Sea Mercator

Note that distances in the WGS 84 projection are measured in degrees and, say, if we want to measure distances later on, we will need a projection that uses the metric system. You can find the necessary projection, as well as more detailed information about it, at www.epsg.io.

Let's load the EPSG:5940 projection so that we can create a chorochromatic map of Russia. When adding new chorochromatic map layers, you need to make sure that they are all in the same map projection, otherwise the results of our analysis will be incorrect. You may need to convert the projection of each layer to the one selected for rendering.

Zoning with Typological Differentiation

Now let's look at the information that is stored in our shapefile. To do this, right click on the layer you need and select the Open Attribute Table menu (Fig. A2.3).

Some parameters in the attribute table may be unreadable. All you need to do here is fix their encoding. Right-click on the layer in the table in the bottom left and select the Layer Properties tab. Select Source and change the encoding. UTF-8 or Windows-1252 will usually do the trick.

Methodological Guide

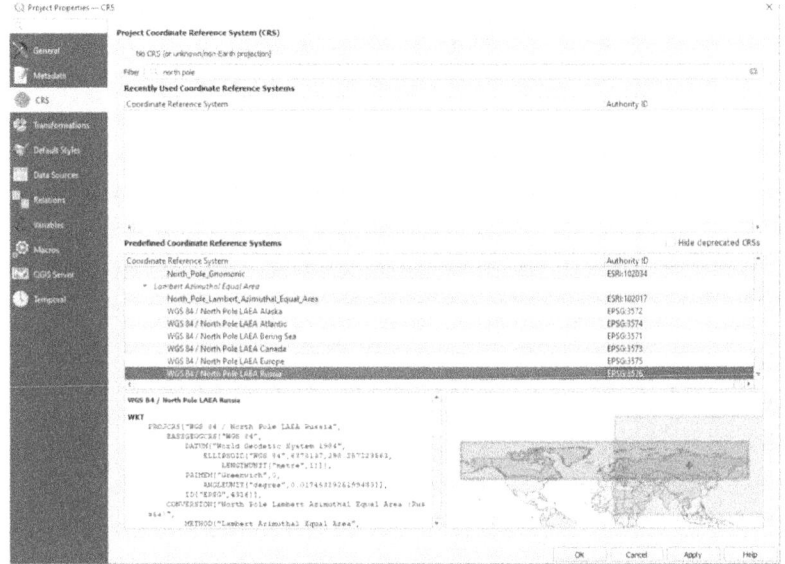

Fig. A2.3. Map projection transformation

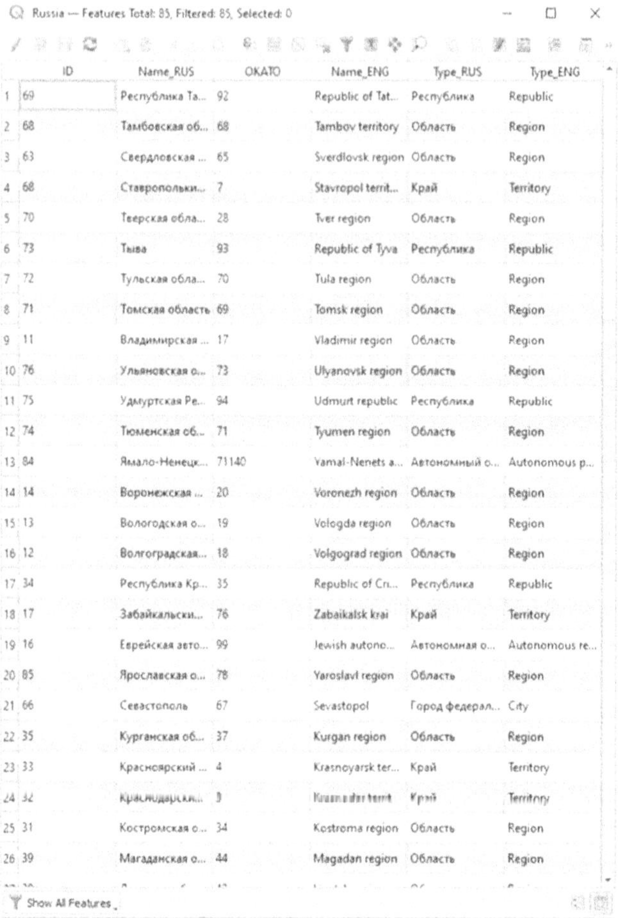

Russia — Features Total: 85, Filtered: 85, Selected: 0

	ID	Name_RUS	OKATO	Name_ENG	Type_RUS	Type_ENG
1	69	Республика Та...	92	Republic of Tat...	Республика	Republic
2	68	Тамбовская об...	68	Tambov territory	Область	Region
3	63	Свердловская ...	65	Sverdlovsk region	Область	Region
4	68	Ставропольки...	7	Stavropol territ...	Край	Territory
5	70	Тверская обла...	28	Tver region	Область	Region
6	73	Тыва	93	Republic of Tyva	Республика	Republic
7	72	Тульская обла...	70	Tula region	Область	Region
8	71	Томская область	69	Tomsk region	Область	Region
9	11	Владимирская ...	17	Vladimir region	Область	Region
10	76	Ульяновская о...	73	Ulyanovsk region	Область	Region
11	75	Удмуртская Ре...	94	Udmurt republic	Республика	Republic
12	74	Тюменская об...	71	Tyumen region	Область	Region
13	84	Ямало-Ненецк...	71140	Yamal-Nenets a...	Автономный о...	Autonomous p...
14	14	Воронежская ...	20	Voronezh region	Область	Region
15	13	Вологодская о...	19	Vologda region	Область	Region
16	12	Волгоградская...	18	Volgograd region	Область	Region
17	34	Республика Кр...	35	Republic of Cr...	Республика	Republic
18	17	Забайкальски...	76	Zabaikalsk krai	Край	Territory
19	16	Еврейская авто...	99	Jewish autono...	Автономная о...	Autonomous re...
20	85	Ярославская о...	78	Yaroslavl region	Область	Region
21	66	Севастополь	67	Sevastopol	Город федерал...	City
22	35	Курганская об...	37	Kurgan region	Область	Region
23	33	Красноярский ...	4	Krasnoyarsk ter...	Край	Territory
24	32	Краснодарски...	3	Krasnodar territ...	Край	Territory
25	31	Костромская о...	34	Kostroma region	Область	Region
26	39	Магаданская о...	44	Magadan region	Область	Region

Show All Features

Fig. A2.4. Shapefile attribute table

Our file contains information about the types of regions in the Russian Federation. We will use this data to create a chorochromatic map that displays the spatial distribution of the constituent entities of the Russian Federation by type of region. To do this, right-click on the layer we need and select the Layer Properties menu (Fig. A2.4). If Single Symbol is selected in the window that opens at the top, then our chorochromatic map is evenly painted over with a single colour. Here, we can choose the fill colour, or fill it with hatching. We want the regions to have different colours depending on their type. This procedure is known as *zoning with typological differentiation* and is associated with the classification subdivision of territories (their differentiation) according to a qualitatively distinguishable feature (type).

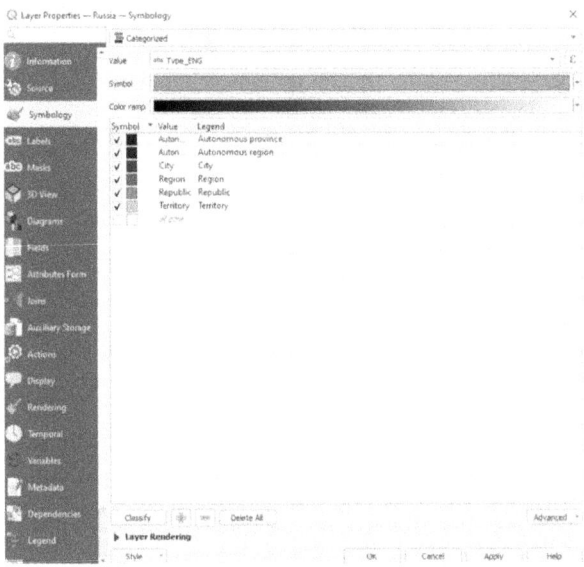

Fig. A2.5. Adjusting symbology settings to perform zoning with typological differentiation

Fig. A2.6. Spatial Distribution of the Constituent Entities of the Russian Federation by Type of Region

To do this, we will select the Unique Values tab at the top instead of Single Symbol. In the Field column, we will set a qualitative attribute from the shapefile attribute table we are using as a basis for classifying the space – in this case, "Region Type."

In the Color Ramp field, we can choose a colour scheme for the different types of regions. Click Classify at the bottom and the program will suggest a shade for each type of region. We can manually add and remove types of classified regions, rename them, and change their shade. Click Apply and we have our results (Fig. A2.6). If there is something we don't like in the render, we can go back to the Style field and experiment with other settings.

As you can see from the dialog box on the result, we have created separate layers for each of the classified types of regions. We can choose whether to show all or just some of them.

Modelling Cartographic Semiotics

The chorochromatic map we have created is still unreadable. We need to feed additional data into it, starting with name, symbols, scale bar, compass sign and copyright notice. These components of the map's language create a unique geographical semiotic system that allows anyone to understand what it means.

To model a chorochromatic map and add symbols, mark the layers you need in the window on the left and select the New Print Layout function (Ctrl + P). The program will ask you if you want to create a new .qgz or .qgs file and a new dialog box will open (Fig. A2.7).

To add our newly created chorochromatic map to the layout, click Add a Map a select the area on the layout where you want to place it. The position and dimensions of the map can then be corrected using the Move Item and Move Item Content functions on the left.

We will use the Add New Scalebar function to place this object of cartographic semiotics onto the map. In the window on the left, you can configure the scale bar parameters: select the style, unit of measurement, number and length of segments, etc. Scale can also be expressed numerically, as a fraction with a "1" in the numerator. It shows how many times the lengths on the map are less than the lengths on the ground (for example, 1:1,000,000).

We will use the Add Legend function to insert symbols that correspond to the layers in our chorochromatic map into the image. You can also format this area in the window on the left – change its name, add labels for each layer, etc.

We will use the Add Image function to place a schematic compass that points north and replaces the coordinate grid of meridians and parallels on our chorochromatic map. There are several compass images to choose from in the image search on the left.

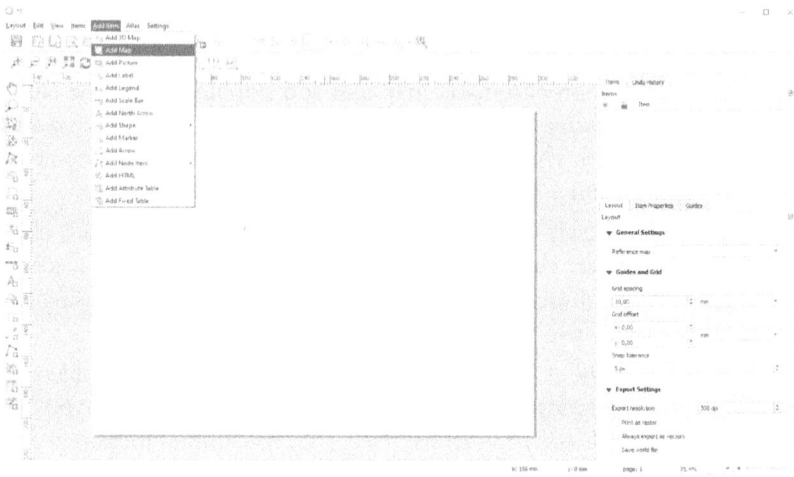

Fig. A2.7. Modelling chorochromatic maps

Fig. A2.8. Chorochromatic map of the Spatial Distribution of Constituent Entities of the Russian Federation

Methodological Guide

Finally, using the Label tool, we will insert the name of the chorochromatic map and a copyright notice in the image. Choose the font and enter the text in the properties window on the left.

Now the chorochromatic map is ready (Fig. A2.8). You can use the Composition tab to save or export it as an image in any file format, or as a .pdf file.

Questions and Tasks

1. Find two or more maps of the same object that convey its characteristics in different ways. What narratives are the authors of these maps trying to convey? Analyse the conditions in which the maps were created and the goals they are trying to achieve.
2. Explain the factors that contribute to making Russia a difficult country to map.
3. Analyse the chorochromatic map of the spatial distribution of the constituent entities of the Russian Federation by type of region. Are there any geographical patterns in the distribution of certain types of region in Russia?
4. Remove layers of the chorochromatic map depicting constituent entities that are built upon the national-territorial principle.
5. Select a new projection for the chorochromatic map. What purposes does the projection serve?
6. Load the ESPG:4940 projection for the chorochromatic map. What changes need to be made to the legend now?
7. What other chorochromatic maps can be created using zoning with typological differentiation?

2. Geoinformation Zoning

Objectives

✓ Find the most appropriate nonspatial data
✓ Attach nonspatial data to a chorochromatic map
✓ Convert text to numbers
✓ Create a chorochromatic map with a data scale

Nonspatial Statistics

The shapefile we analysed, like most other files of this format, contains almost no statistical data for analysis. Moreover, almost all statistical data (including demographic, socioeconomic and electoral indicators) are nonspatial in nature, that is, they are not tied to any specific geographic coordinates, which means that it is impossible to analyse their spatial distribution. Our new task is to analyse statistical data that does not have a spatial dimension and is not contained in the original shapefile.

Statistical data that can be mathematically processed is contained in comma-separated values (CSV) files or files that can be converted to CSV format. Unfortunately, most statistical services do not provide data in a machine-readable format, so it has to be transformed manually, for example, from .xls format.

For this task, download the .xsl file from the "World" archive available on our Google Drive (https://drive.google.com/file/d/1BGfb_7 3KvAvgw7dtgI1myf3WTz3c3-av/view?usp=drive_link) as well as on the website (https://mgimo.ru/upload/2023/05/world-geoinformation-zoning.zip). It is provided in a .zip format, so we will first need to unzip it – which can be done online using a service like https://extract.me/ru/. Open the .xls file in Excel and study the contents: the first column contains the country name, the second the country code, the third the total area of the country, and the fourth the share of arable land as part of the total area of the country. The statistical data in the .xls file needs to be converted to CSV format in order to "link" it with the spatial data in the .shp file.

A number of checks have to be made to ensure the data is converted correctly. First, the format in which the data is stored in the cells. The format of the cells containing statistical data typically has to be "Number". You can do this by selecting the column and right-clicking, and then selecting Format Cells. Second, whether or not a period is used as the decimal separator. To do this, open Options in Excel (under the File menu) and change the separator to a period if it is not already set. Finally, in the File menu, select Save As and set the format to "CSV (Comma separated values)". The leading global data repositories that use CSV format include the World Bank Open Data portal (https://data.worldbank.org/) and the Global Open Data Index (https://index.okfn.org), which monitors the publication of statistical information by countries in open and parsable formats. Russian repositories include the government's Open Data Portal (https://data.gov.ru) and the open data portals of individual regions (for example, https://data.mos.ru for Moscow and http://data.gov.spb.ru for St. Petersburg). Project Center Infometer (https://read.infometer.org) monitors the openness of the data contained in Russian federal and regional data portals. Our task is to create a chorochromatic map of the distribution of arable land in various countries. To start, let's open the shapefile in QGIS (Fig. A3.1).

Fig. A3.1. QGIS interface

Transforming Nonspatial Data into Spatial Data

The statistical data we have found needs to be attached to geographic coordinates, which will allow us to analyse their spatial distribution. To do this, you will need to load the .csv file as well as the World shapefile into QGIS, then open the shapefile's properties, select Links and click on the green icon with the + sign at the bottom of the dialog box.

Set the required parameters in the pop-up window (Fig. A3.2). Join Layer is a layer that contains statistical data. Join Field is the field in a file containing statistical data that is used for linking. Target Field is the field in the shapefile used for the link. Joined Fields are the fields (indicators, columns) that should be linked to the spatial data. Custom Field Name Prefix is the prefix to the name of the joined fields.

Fig. A3.2. Joining layers

Pay special attention to Join Field and Target Field. These are fundamental for successful data merging, so there must be strict correspondence between them. For the purposes of our practical task, we set up these fields in advance: each country was assigned a code that was contained in the shapefile in the ISO_GIS column. The same codes were assigned to the countries in the CSV file, in the column with the same name. The program identified the correspondence between the values of the cells in these columns and combined the data. The easiest way to mark correspondence is to use a numeric code, but alphabetical characters can also be used. When the Join Layer appears in the Joins tab, click OK.

Transforming Text into Numbers

All the data in a CSV file is stored in text format. But if we want to create a quantitative scale reflecting the distribution of the given indicator by region, we need to transform it into numeric values. This function is called Text to Float and is found in the Toolbox, which can be opened from the Processing menu.

In the Input Layer field, specify the layer that contains the data we want to convert, and then specify the field whose format you want to change in the Text Attribute to Convert to Float field. Then click Run and wait for the process to finish (Fig. A3.3).

Fig. A3.3. Converting cell formats

After it is done, the dialog box will take you to the Log window, and the program will create a new layer where the format of the field we specified will be different, while the rest will stay the same. You will need to change the format of two fields, so switch back from Log to the Parameters tab in the dialog box. Select the layer we have just created as the input layer and the field whose format we want to change and click Run. Once you have converted all the required fields, click Close. To continue with fields that have been converted into numbers, use the last layer you created. You can also rename it by right-clicking.

Zoning with Dasymetric Differentiation

In geography, zoning with dasymetric differentiation is used to convey quantitative differences between phenomena with a continuous distribution. Unlike zoning with typological differentiation, which classifies objects qualitatively, zoning with dasymetric differentiation uses quantitative indicators.

Scaled colouring is used, meaning that the colour intensity increases or decreases depending on the quantitative change of a given attribute. To apply colouring, open the properties of the layer we have just created and select the Style tab (also called Appearance in other versions of QGIS). Then select Graduated in the top drop-down box. In Value, select the field on which we want to carry out zoning, in this case, 7ArLand. In Color Ramp, select the colour scale you want to use. Classify objects by selecting scaling formats (located above the Classify button) and the number of classes (on the right).

To better understand the data, we can plot a histogram reflecting the distribution of a feature based on the scaling parameters. To do this, select the format and the number of classes, open the Histogram tab, click Load Values and evaluate their distribution (Fig. A3.4). You can also specify the calculation of the mean μ and standard deviation σ by checking the corresponding boxes in the dialog box. Now the chorochromatic map is ready for modelling (Fig. A3.5).

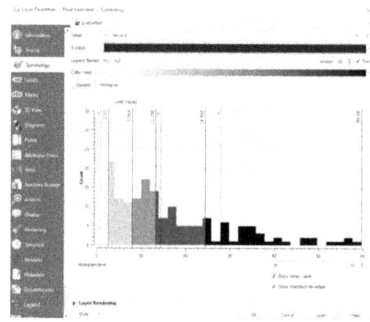

Fig. A3.4. The distribution of the feature based on the scaling parameters

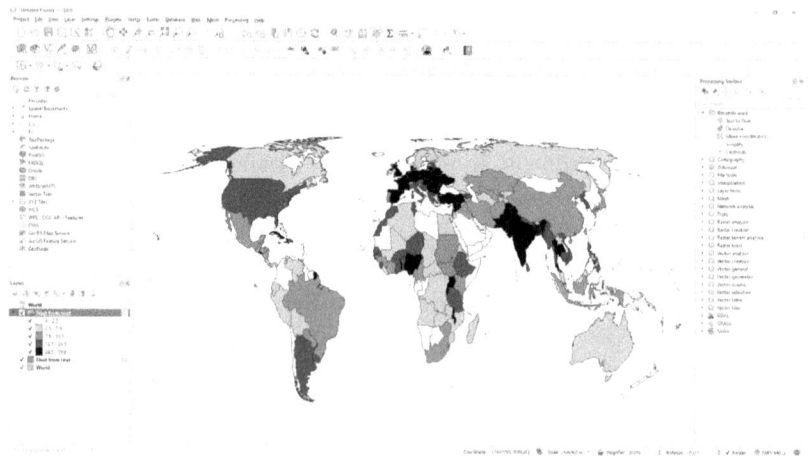

Fig. A3.5. Final chorochromatic map of the arable land area distribution by country

Methodological Guide

Common mistakes	Ways to fix them
You forgot to convert formulas (which make reference to other cells) to values when closing your Excel session.	Select and copy the column with formulas, then highlight an empty column, right-click, choose Paste Special, check the Values box and then click OK. Then make sure to remove the formula column completely.
You have not deleted columns and rows that you used in Excel but that are not needed for joining – this is often the case when carrying out calculations in Excel.	Delete unnecessary columns and rows.
When working in QGIS, you somehow linked fields that you didn't want to link.	Return to Excel, select a few columns to the right of the one you want to link to and right-click to delete them, even if they appear to be empty. Then do the same with the lines below the last line you need. You can also delete the columns directly in QGIS using the Edit mode in the table.
You made a mistake or a typo when assigning codes to be used for linking, so the program does not see the field as having a match in the shapefile.	Recheck and correct your error or typo.
You are using the same code for multiple fields.	Assign a unique code to each individual observation.

Common mistakes	Ways to fix them
You are using numeric codes for linking and some values start with a zero, for example, 0041, 0052, etc.	Don't use zeros at the beginning of codes. You can use 41, 52, etc., instead. However, make sure that the new codes have not been used in other fields.
You have set the wrong column format.	Fields usually take the following formats: Columns with names or letters – General The columns with codes for linking – General Columns with indicator values – Number. Make sure you put the correct number of decimal places in numeric columns. If all the figures in a numeric column are right-aligned, but one cell is left-aligned, then the format of that cell is likely General. Retype the value in this cell manually.
File format with statistical data required for linking.	"CSV (Comma delimited)" Be careful, there are different CSV formats.
Also, when working in QGIS, make sure that Join Field and Target Field are exactly the same. They can be named differently, but they have to have the same codes/symbols for linking.	
When converting text to numbers in QGIS, select Number from Text layer (at the bottom) for each new conversion.	

Questions and Tasks

1. Find statistics for your home country that can be dasymetrically zoned by administrative region.
2. What other indicators besides ISO codes can be used to merge nonspatial statistics with geographic coordinates? Will postcodes or electoral districts work for this?
3. Analyse the chorochromatic map of arable land distribution in the world by country. Can you see any spatial patterns?
4. Create and compare chorochromatic maps with different scaling formats and number of classes. Are there any among them that illustrate better the spatial distribution of arable lands?
5. Compare chorochromatic maps with single and two-colour scaling. Does one of the types visualize the data more clearly?
6. Compare a chorochromatic map with highlighted area borders to a chorochromatic map with transparent area borders. In what cases are each type of chorochromatic map more suitable?
7. Create a chorochromatic map with dasymetric zoning of a different variable given in the dataset. Compare it with the chorochromatic map created in this chapter. What conclusions can you make on the basis of this comparison?

3. Geoinformation Mapping and Zoning in R

Objectives

✓ Become proficient in RStudio
✓ Understand the basics of programming in R Markdown
✓ Connect nonspatial data to a chorochromatic map
✓ Create a chorochromatic map with dasymetric zoning

So far, we have been performing manual mapping and zoning operations in a GIS program. But the same operations can be performed using a programming language. This enables us, first of all, to formalize our commands in the form of a reproducible algorithm and, second, to control every step of analysis. Finally, by putting the algorithm of analysis into code, we can demonstrate the validity of our research to the scientific community (providing code is becoming increasingly necessary these days when publishing articles or speaking at conferences).

For spatial analysis, we will use the R programming language. A free software environment has been created for working in this language that is widely used as statistical software for data analysis and visualization. It can be downloaded from https://cran.r-project.org/, and the program for working with R (RStudio) is available at https://www.rstudio.com/produ cts/rstudio/download. Open RStudio (Fig. A4.1) and use the + sign in the menu to create a new markdown file (Fig. A4.2, markdown is a simplified computer markup language).

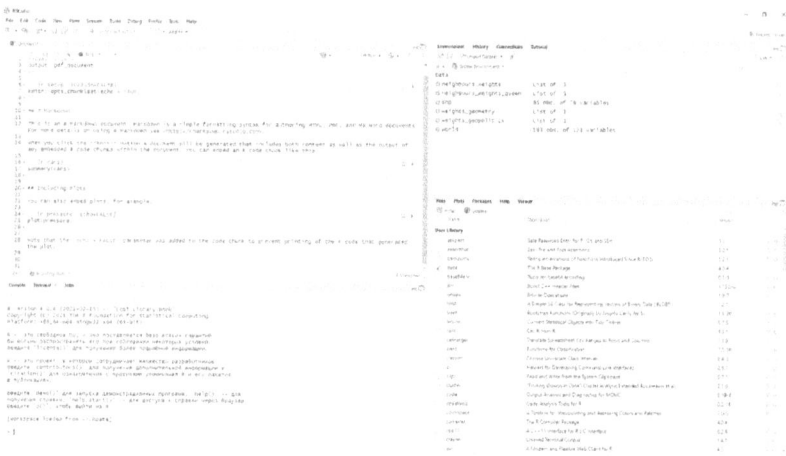

Fig. A4.1. RStudio interface

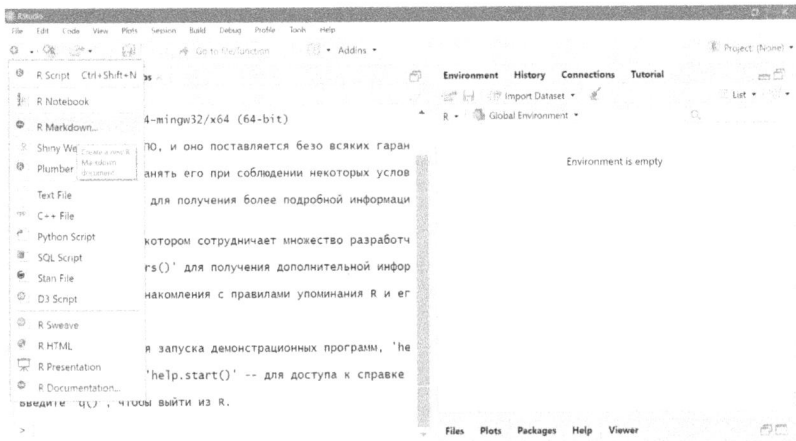

Fig. A4.2. Creating R markdown

If you do not have the markdown extension installed, select Packages in the bottom right window, then Install and enter rmarkdown in the pop-up window (Fig. A4.3). You can also type install.packages("rmarkdown") in the working console (upper left blank field with numbered lines) and press Ctrl+Enter. To download R package extensions, your computer must be connected to the internet and the user must have admin rights.

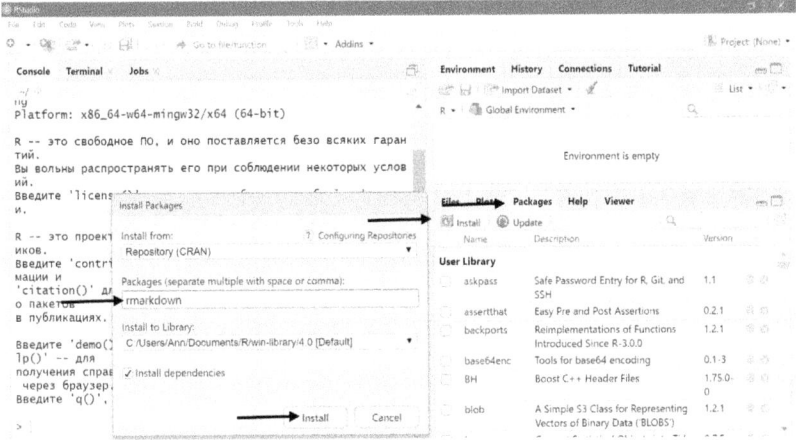

Fig. A4.3. Installing packages

Give the file a name and save it in .html, .pdf or .doc format. The markdown should be saved in the same folder as all other files that you are planning to use. Standard text will appear in the code entry window. You can delete this, but make sure to keep the information about the file itself (name, author, date, format) and the parameters for generating the final report (knit). Set the following parameters for generating the final report so that system messages do not get into it.

To insert a new chunk of code, press the green "+c" icon at the top and choose R (Fig. A4.4).

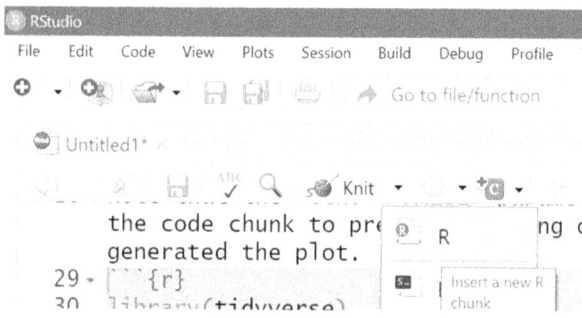

Fig. A4.4. Inserting a new R chunk

To run a command, press Ctrl+Enter after you have written the command. If the command was given correctly, the line will blink green; if it wasn't, it will blink red and an error message will appear in the Console. To run several chunks at once, select them and press Ctrl+Enter, and to run the entire code, press Ctrl+Alt+Enter.

To get started with geographic data in R, you want to run special packages. Enter some geographic data into the code, and if the message pops up that the necessary packages are not installed, download them as you did with "markdown". After you have entered each package into the console, run these commands. In this lab you will need the following packages: rgdal, maptools, shapefiles, sp, ggplot2, ggsn, scales, reshape, tidyverse. To load them into your markdown run the library(package name) command.

To create any statistical chorochromatic map, we need to load the shapefile into RStudio (we will use the same World package from Chapter 2). Note that the file must be located in the same root folder as the

Methodological Guide

markdown. Let's name this operation World. Use <- to assign a command to World. This way when you use World as an attribute to other commands, the program will know exactly what you're referring to.

```
```{r}
World <- readOGR("World.shp")
```
```

The shapefile should appear in the "Environment" tab in the upper right field. Double-click on it and take a look at the attribute table (Fig. A4.5). Running the World command will give us information about the file, including its projection (WGS-84). This may come in handy if we need to do a transformation.

Next load the CSV file containing statistical data on countries that you have created in the previous lab. Let's name it data. While we are doing this, we will also change the encoding to UTF-8 (encoding="UTF-8"). Specify if the first line in the table contains headings (head = TRUE) and specify the separator, here a semicolon (sep=";"). The file and the number of variables should also appear in the Environment tab. Double-clicking on it will invoke the contents of the file.

Fig. A4.5. Attribute Table in R

```
```{r}
data <- read.csv("World.csv", encoding="UTF-8", head=TRUE, sep=";")
```
```

Let's use ISO codes to join data from the two files. Now, just like in QGIS, we need to convert variables into a numeric format (integer). First, let's check the format of the columns in both files. The data in the CSV file was converted to numeric values automatically upon loading. However, the shapefile stores the ISO codes as text (character). Run the given command to convert these characters into numbers.

```{r}
class(World$ISO_GIS) #character
class(data$ISO_GIS) #integer
World$ISO_GIS = as.numeric(as.character(World$ISO_GIS))
```

Let's create the World_data file, in which we will join the data from both files using ISO_GIS as a joining field. The new file should also appear on the right. Double-click on it to see results. Note that if the columns containing variables that we want to use as joining fields are named differently in the shapefile and in the .csv-file, RStudio will not be able to run the merger. To fix this, you can rename columns using the rename command.

```{r}
World_data <- merge (World, data, by.x="ISO_GIS", by.y="ISO_GIS")
```

Now that you're familiar with the R language basics, try using packages given at the beginning of this chapter to perform zoning operations in R. Finally, click Knit to wrap up and save your progress.

Questions and Tasks

1. Compare the results of geoinformation zoning with dasymetric differentiation in QGIS and R. What are the pros and cons of each method?
2. Use suggested packages to perform mapping.

3. Create a chorochromatic map with dasymetric zoning using R.
4. Create the chorochromatic map we built in Chapter 2 using R.

4. Mapping Descriptive Statistics

Objectives

- ✓ Become proficient in GeoDa
- ✓ Identify the spatial distribution of a single variable using a histogram and a box plot
- ✓ Identify the spatial distribution of two variables using a scatter plot
- ✓ Analyse the relationship between more than two variables on a scatter plot matrix
- ✓ Visualize spatial clusters with a 3D scatter plot model. Plot a discrete anamorph in GeoDa
- ✓ Create a continuous anamorph in QGIS
- ✓ Determine the optimal number of iterations to make anamorphoses
- ✓ Perform a two-factor anamorph

Having linked statistical information to geographic features, we can perform many mapping operations with descriptive statistics. We will start with the simplest operation – spatial one- and two-dimensional analysis, i.e. the analysis of the distribution in space in one or two variables. We will use the free software package GeoDa to do this.

For this task, download the "World" archive available on our Google Drive (https://drive.google.com/file/d/1BGfb_73KvAvgw7dtgI1myf3WTz3c3-av/view?usp=drive_link) as well as on the website (https://mgimo.ru/upload/2023/05/world-mapping-descriptive-statitics.zip). The shapefile in the archive contains information about the borders of all countries and dependent territories in the world, their area, population size, and standard UN country and area codes, which can be found at the UN Statistics Division at https://unstats.un.org/unsd/methodology/m49/.

The GeoDa interface differs from that of most other programs. It is not a full-size window, but rather a narrow toolbar. The upper part of the panel, where commands are presented in words, offers more features than the lower part, which contains shortcuts. That said, shortcuts allow to execute frequently used commands quickly. Let's learn about the basic functionality of GeoDa.

To open a .shp file, click on the folder shortcut in the control to open a dialog box where you can open the file in several ways.

Method 1. Click on the small folder shortcut in the File tab and select the file format you want to open. Select the file you need and click Open.

Method 2. In the dialog box, drag the .shp file manually to the Drop Files Here window. To close a file, click on the folder shortcut with the X in the toolbar.

To open the attribute table, click on the table shortcut in the toolbar. Clicking on the X in the upper right-hand corner of any GeoDa window, with the exception of the main toolbar, will close the window (but not the program itself).

Now let's look at the map toolbar, which is located in the upper left-hand corner of the map window. The mouse cursor icon is the Select tool. To select a part of the map, left-click on it and stretch the selection box to the desired size. To deselect it, left-click on any part of the map. The selection is duplicated in all GeoDa windows, regardless of the type of content: maps, tables, diagrams, etc.

To make viewing the selected elements in the attribute table easier, you can move them up by right-clicking on the table and selecting Move Selected to Top. To select several objects on the map that are not located next to each other, left-click on the objects while holding Shift. To select several objects in the table that are not located next to each other, left-click on them while holding Ctrl. To select multiple objects that are located next to each other, hold down Shift and then select the first and last objects in the row.

To exclude objects from the analysis, you can invert your selection – that is, select all objects on the map that are not already selected. To do this, click on the Invert Select shortcut on the map toolbar. To invert the selected area in a table, right-click on it and choose Invert Selection.

You can zoom in or out of the chorochromatic map by selecting the magnifying glass icon with the + or – sign (whichever you need) and select the area you want to enlarge or reduce. To return to the full-view map, click on the Full Extent shortcut on the map toolbar. To move the chorochromatic map around the screen, click on the Pan icon on the map toolbar and then move it by holding down the left mouse button.

Histograms for Spatial Visualization

The simplest way to perform one-dimensional statistical analysis is to plot a histogram. We will use the histogram function in the main menu to build a graph of the distribution of countries by area (Explore → Histogram). The purpose of constructing a histogram here is to obtain an empirical estimate of the distribution density of the value. To build a histogram, the observed range of change in the required value is divided into several intervals and the proportion of all measurements that fall into each of the intervals is calculated. Right-click and select Choose Intervals to set the number of intervals you need. Then select View → Display Statistics. It is generally believed that the optimum number of intervals is the square root of the total number of measurements: $n = \lvert \sqrt{N} \rvert$. In our case, that works out as $n > 15$.

Clicking on the intervals of the histogram, or selecting several of them, will highlight the groups of countries that are of interest to us on the map – for example, those falling in the largest interval or all but those in the smallest interval (Fig. A5.3). Similarly, clicking on a country on the chorochromatic map will highlight the interval to which it belongs. You can choose multiple countries by clicking-and-dragging or selecting them individually by holding Ctrl+Alt. Using Invert Selection on the chorochromatic map allows you to select objects that you want to exclude from the analysis.

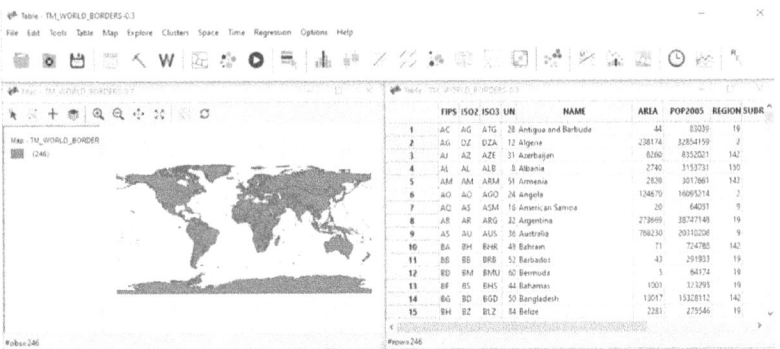

Fig. A5.1. GeoDa interface

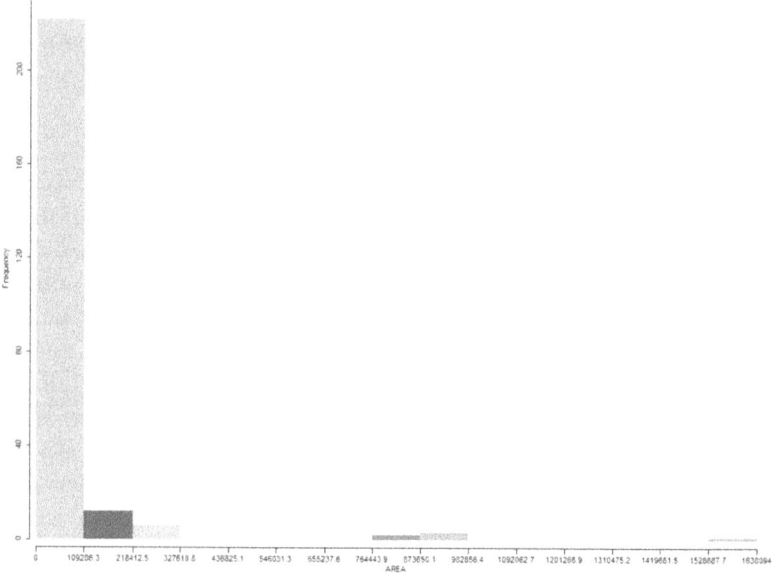

Fig. A5.2. Histogram of the Distribution of Countries by Area

Methodological Guide

Fig. A5.3. Chorochromatic maps constructed on the basis of the Histogram of the Distribution of Countries by Area

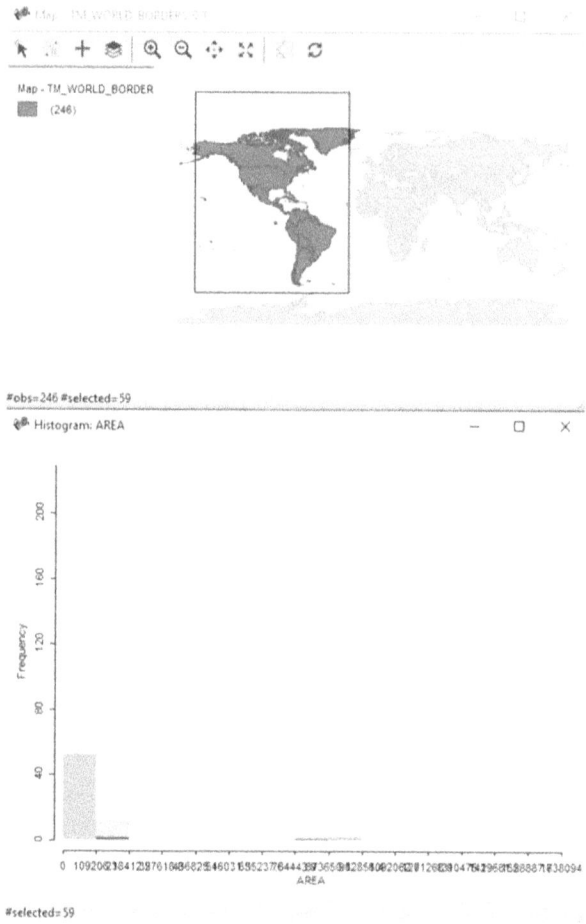

Fig. A5.4. Histogram for spatial visualization

Highlighting an area on the map, say, the countries of North and South America, will show on the histogram the proportion that these countries make up relative to other countries in the same interval (in a semi-transparent shade on the histogram) (Fig. A5.4). Right-click to save the maps and histograms.

Box Plots for Spatial Visualization

Box plots are a more advanced method of descriptive univariate statistics (Explore → Box Plot). This type of chart conveniently shows the median, lower and upper quartiles, minimum and maximum sample values, and outliers. The borders of the box are the first and third quartiles (the 25th and 75th percentiles, respectively), and the line in the middle of the box is the median (the 50th percentile). The whisker ends are the edges of a statistically significant sample.

Let's create a box plot showing the area and population of the countries of the world. The simplest way to carry out a visual two-dimensional analysis is by placing two box charts side by side. Just like with a histogram, we can visualize the parts of the diagram that are of interest to us on the map, or, conversely, see where one of the selected objects on the map is located on the diagram (for example, China on Fig. A5.5).

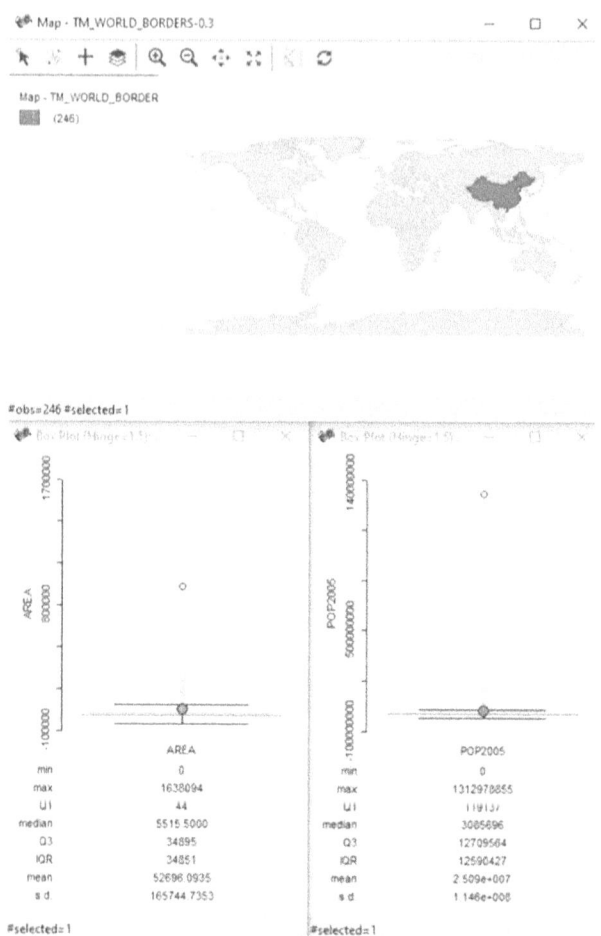

Fig. A5.5. Spatial visualization of a box plot

Scatter Plots for Spatial Visualization

Unlike histograms and box plots, *scatter plots* let you analyse the dispersion of not one, but two variables. Each object on a scatter plot corresponds to a point whose coordinates are equal to the values of two of the parameters being observed. Scatter plots are used to show if there is a correlation between two variables. Let us build a scatter plot (Explore

→ Scatter Plot) to analyse the ratio between area and population size in the countries of the world (Fig. A5.6).

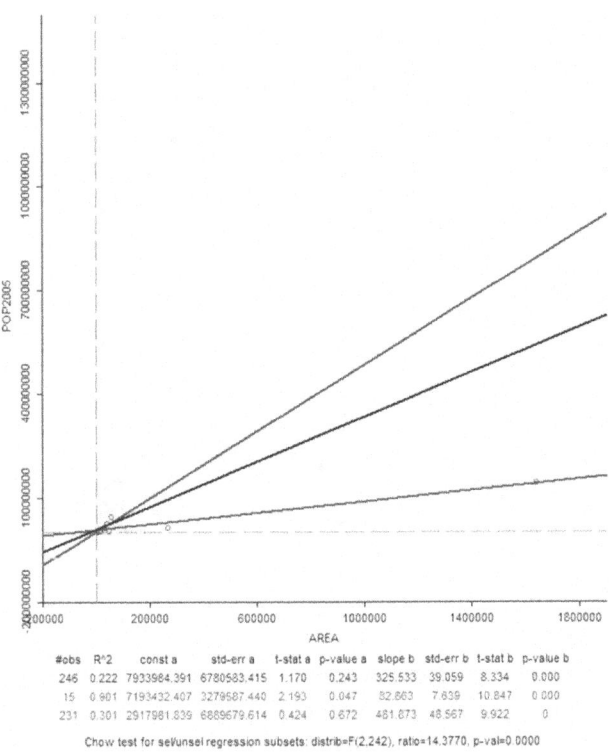

Fig. A5.6. Spatial visualization of a scatter plot

Right-click to set the properties of the graph. For example, you can use the standardized scale for variables instead of the original one. Select objects on the map that we want to calculate the correlation for, such as the former Soviet republics. The graph now shows three regressions: red (the lower line) for the population of objects we are analysing (former Soviet republics); purple (the middle line) for the entire population of objects on the map; and blue (the upper line) for the entire population of objects minus the one we are analysing (former Soviet republics).

Below the graph are statistical indicators that allow you to assess the degree of dependence between the variables in each population. The following indicators are of interest to us:

1. The *coefficient of determination (R squared)* – the proportion of the variation in the dependent variable that is explained by the dependence model we are looking at, i.e. explanatory variables. In a pairwise linear regression model, the coefficient of determination is equal to the square of the normal correlation coefficient between x and y. The coefficient takes values from 0 to 1. The closer it is to 1, the stronger the dependence. When evaluating regression models, this is interpreted as the model matching the data. A model is deemed acceptable if the coefficient of determination is at least 0.5, and good if it is above 0.8. A coefficient of determination of 1 means that there is a functional relationship between the variables.

2. The *p-value* – the value used when testing statistical hypotheses. The *p-value* is the probability of error when rejecting the null hypothesis. If the value of the test statistic calculated from the sample corresponds to p = 0.005, this indicates that the probability that the hypothesis is valid is extremely high. In other words, the smaller the p-value, the better, since it increases the "strength" of rejecting the null hypothesis and increases the expected significance of the result.

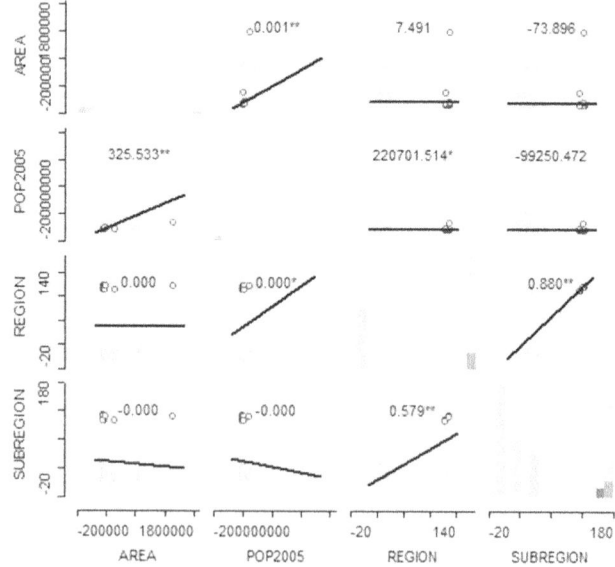

Fig. A5.7. A scatter plot matrix

Let's try and test which variables in our shapefile have the greatest dependency. To do this, we will need to create a *scatter plot matrix* (Explore → Scatter Plot Matrix) and add the region and subregion codes as parameters. As expected, there is no correlation between the numerical expression of a region and other parameters (Fig. A5.7).

GeoDa also allows you to build *three-dimensional scatter plots* (Explore → 3D Scatter Plot) to analyse the correlation between three parameters. Let's create one for the following variables: territory area, population size and subregion code (Fig. A5.8). As the model shows, the countries formed two stable clusters. Check the box that says "Select, hold CTRL for brushing" to see the spatial distribution of these two clusters (Fig. A5.9).

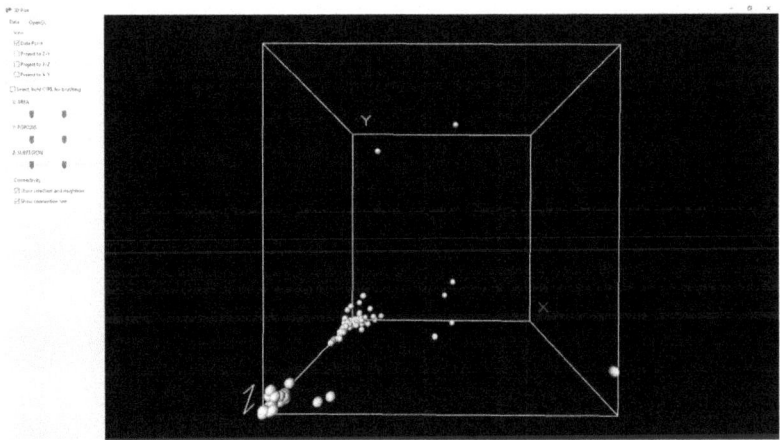

Fig. A5.8. Spatial visualization of a 3D scatter plot

Fig. A5.9. Spatial clusters in a 3D scatter plot

Questions and Tasks

1. Analyse the graphs you have obtained. What conclusions can you draw about the distribution of countries by area and population size?
2. Is it possible to say there is spatial dependence in the distribution of countries in terms of area and population size?
3. Add country GDP data to the shapefile and analyse the dependence of GDP on land area and population size in a scatter plot matrix. Use the international ISO code to include nonspatial statistics.
4. Create a 3D scatter plot depicting territory area, population size and GDP. Visualize the resulting clusters.

5. Zoning Descriptive Statistics

Objectives

- ✓ Select the required scaling type based on a histogram or box plot
- ✓ Transform dasymetric zoning into typological zoning using categorical variables
- ✓ Become proficient in multivariate zoning
- ✓ Create a discrete cartogram in GeoDa
- ✓ Create a continuous cartogram in QGIS
- ✓ Determine the optimal number of iterations when creating anamorphoses
- ✓ Perform a two-factor cartogram

Dasymetric Zoning Using Histograms

In this chapter, we are going to continue working in GeoDa. Knowledge of the methods of descriptive statistics will help us create more accurate maps and analyse the dependence of several parameters on them. For our analysis, we are going to download a shapefile containing the moral statistics developed by André-Michel Guerry in France in the 19th century. The dataset contains an abundance of variables for the regions of France and is available in the "Guerry" archive on our Google

Drive (https://drive.google.com/file/d/1ymjuxfFSwtayhgOQYG1dD90Q_4m0tDth/view?usp=drive_link) as well as on our website (https://mgimo.ru/upload/2023/05/guerry-zoning-descriptive-statistics.zip). Original source: https://geodacenter.github.io/data-and-lab/Guerry/. Original source: https://geodacenter.github.io/data-and-lab/Guerry/.

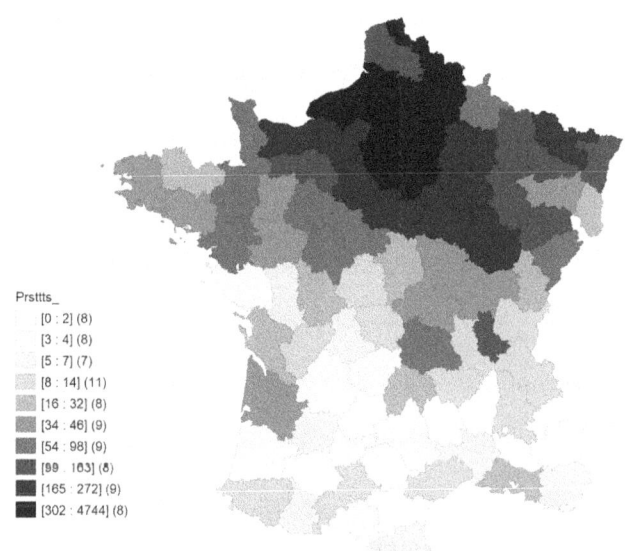

Fig. A6.1. Chorochromatic map of the moral statistics of France with equal quantiles

GeoDa can build zoning chorochromatic maps with dasymetric differentiation, which we have already plotted in QGIS. We will use the function Map → Quantile Map → 10 and select the parameter Prsttts to build *a chorochromatic map with ten classes* (that have the same number of readings and are called quantiles) reflecting the distribution of prostitutes in Paris by place of birth. Note that the map symbols show three classes for readings 0 to 7, six classes for readings 8 to 272, and only one class for readings 302 to 4744 (Fig. A6.1)

Let's build *the same chorochromatic map using a scale with natural intervals* (Map → Natural Breaks Map → 10, Fig. A6.2) and equal intervals (Map → Equal Intervals Map → 10, Fig. A6.3). Note the differences in the cartographic material depending on the choice of scaling method.

Let's build a histogram to evaluate the distribution of indicators by interval (Fig. A6.4). As we can see, the second of the three mapping methods is the most suitable in this case.

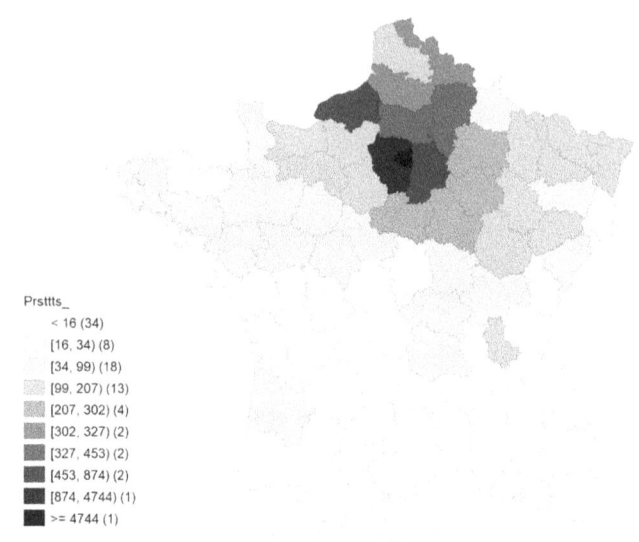

Fig. A6.2. Chorochromatic map of the moral statistics of France with natural intervals

Methodological Guide

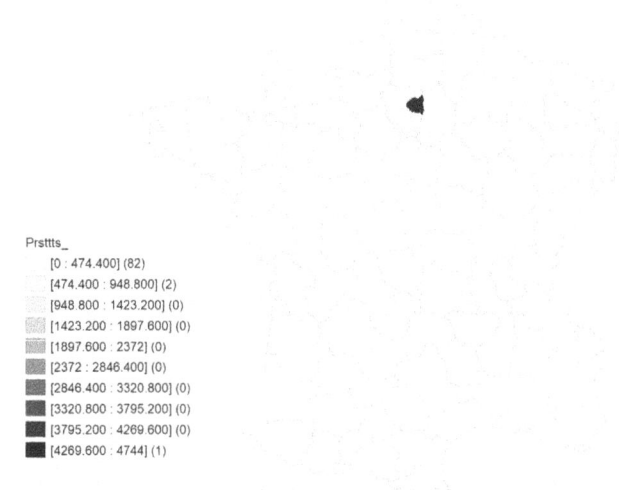

Fig. A6.3. Chorochromatic map of the moral statistics of France with equal intervals

Dasymetric Zoning Using Range Charts

Chorochromatic maps of extreme values are better suited for zoning statistics with significant outliers (as in our case): *percentile maps* (Map → Percentile Map, Fig. A6.6), *box maps* (Map → Box Map, Fig. A6.7) and *standard deviation maps* (Map → Standard Deviation Map).

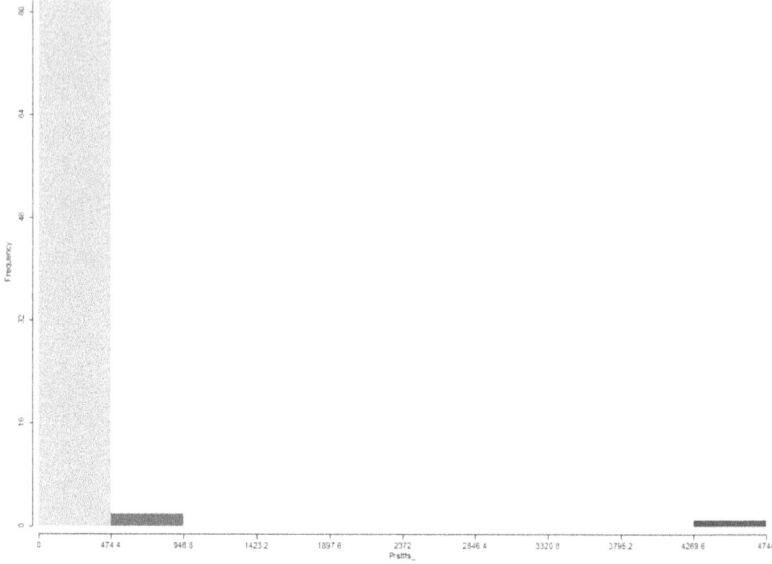

Fig. A6.4. Histogram of the distribution of the moral statistics of France

The percentile map will differ from the box map in terms of their interval boundaries: in percentile maps, 10 % of the extreme values will be allocated to separate intervals; in box maps it is 25 %. In addition, the percentile map will further separate 1 % of the extreme outliers into separate intervals, while the box map will do this based on the corresponding diagram (Fig. A6.5).

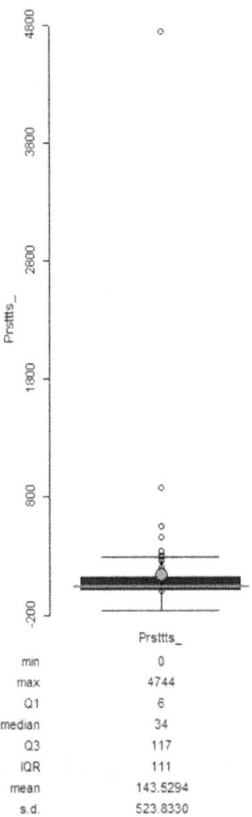

Fig. A6.5. Box plot of the moral statistics of France

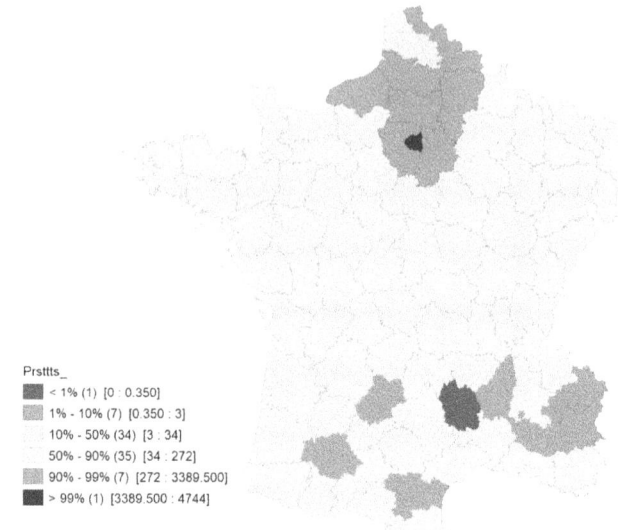

Fig. A6.6. Percentile map of the moral statistics of France

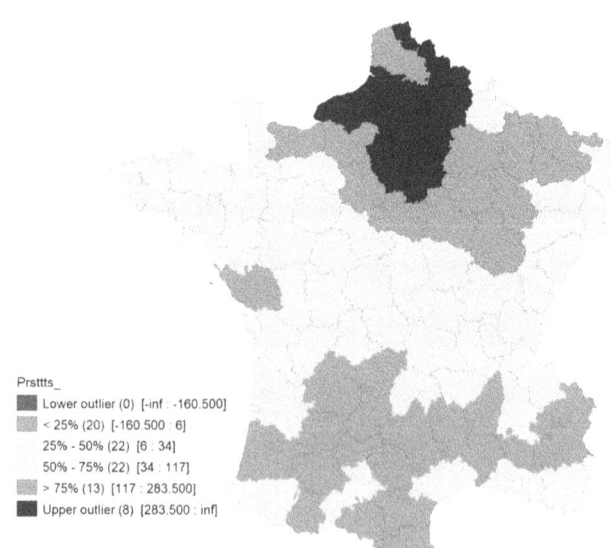

Fig. A6.7. Box map of the moral statistics of France

Typological Zoning Using Box Maps

GeoDa allows mapping and typological zoning, which is what we started with in Chapter 2. The simplest thing to do is to colour the regions of the country by assigning codes to each (alphabetic or numeric) in the attribute table (Map → Unique Values Map). Let's do this for the Region indicator (Fig. A6.8).

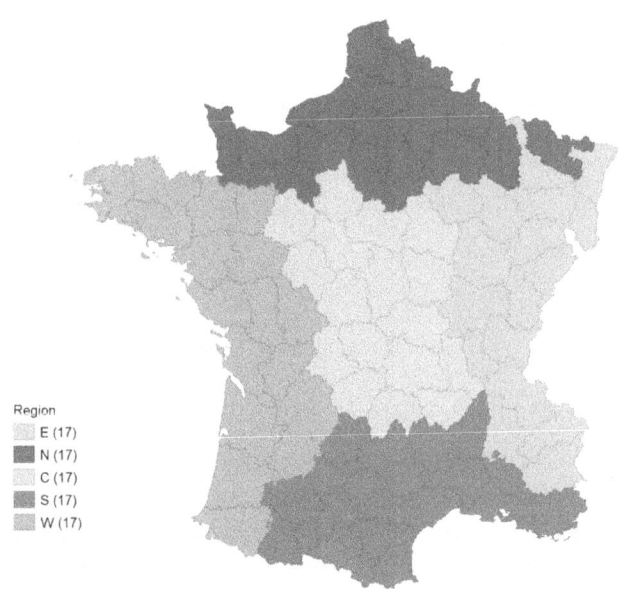

Fig. A6.8. Chorochromatic map of France with typological zoning

However, typological zoning is also possible for graded indicators. You can select a range of values on any of the statistical graphs, give it a unique numerical code and display it on a *chorochromatic map of unique values*. Let's use the box map we have already analysed. Click on the green dot representing the median and save it as an indicator (Variable Name = Median, Selected = 1, Unselected = 0) by right-clicking (Save Selection). Now let's bring up the chorochromatic map of unique values with this parameter (Fig. A6.9). Now let's display a chorochromatic map of unique values with this parameter.

Typological zoning can also be used to save the categories we created on any statistical map as unique. Open the Prostitutes (Prsttts) box map that we have already analysed. Right click on the map and save the categories in a separate column of the attribute table (Save Categories). Now we will create a chorochromatic map of unique values for this parameter (Fig. A6.10). Unlike the dasymetric zoning maps, each category has its own colour. Their configuration can be changed by dragging the categories from place to place in the symbol list.

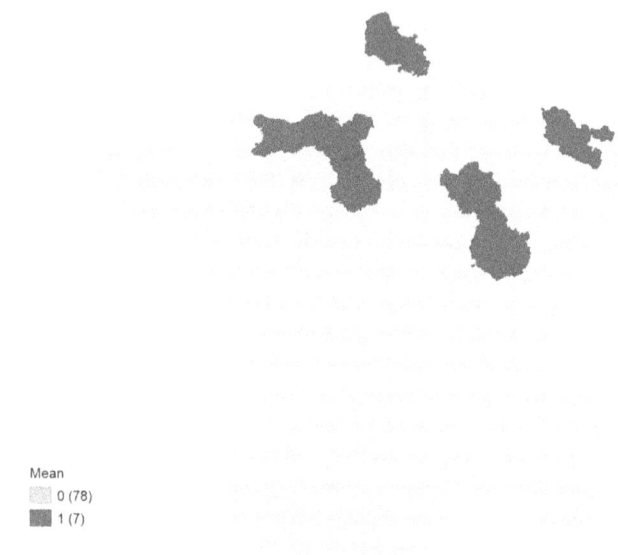

Fig. A6.9. Chorochromatic map of unique values of the moral statistics of France

Methodological Guide 299

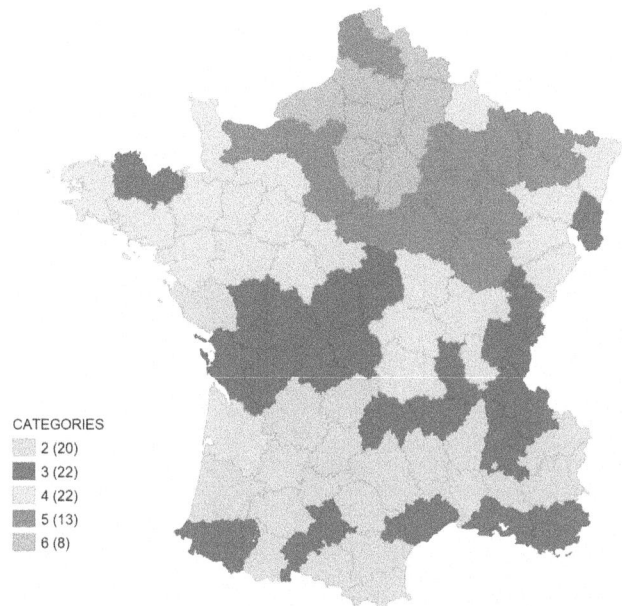

Fig. A6.10. Chorochromatic map with typological zoning of the moral statistics of France

Multivariate Typological Zoning

Up until now, we have divided the map into regions based on a single parameter. However, sometimes multivariate zoning is required. We'll start with the most basic operation: using the co-location map (Map → Co-Location Map), we will build a chorochromatic map with typological zoning according to two parameters: the number of prostitutes in Paris from the regions of France (`Prsttts`) and the number of priests in the regions of France (`clergy`). The system will bring up just one region where the indicators match. Clicking on it in the attribute table shows us that this is the Jura department. A total of 32 priests lived in the Jura department, and there were 32 prostitutes in Paris from the same region. Selecting different parameters in the co-location window allows us to see which of them match.

Fig. A6.11. Bivariate typological zoning of the moral statistics of France

However, we are probably interested in an inexact coincidence of indicators. In this case, in the `Prsttts` and `clergy` box maps, we will save the categories in the columns CATEGORIES1 and CATEGORIES2, respectively. Now we will build a co-location map according to the coincidence of these categories, making sure the type of scaling (colour scheme) at the bottom is the same as in the box plot that we used to save the categories. We now have a chorochromatic map that shows which regions had a coincidence of both priests and prostitutes (Fig. A6.11). Similarly, we can increase the number of factors for analysing in co-location maps.

Bivariate Zoning Using Conditional Maps

Now let's move on to bivariate dasymetric zoning. To do this, will use a *conditional map* (Map → Conditional Map) to create chorochromatic maps of the distribution of criminals (`crm _ prs` in Map Theme) relative to the number of prostitutes (`Prsttts` in Horizontal Cells) and the number of priests (`clergy` in Vertical Cells) in the regions of France (Fig. A6.12). On the chorochromatic map, the numbers of prostitutes and priests are divided into three groups – low, medium and high – and the number of criminals is only shown in those cells where these two groups coincide. For example, the map in the upper right shows the number of crimes in regions where there are a lot of priests and prostitutes, while the map in the lower left shows the number of crimes in regions where there are few of both.

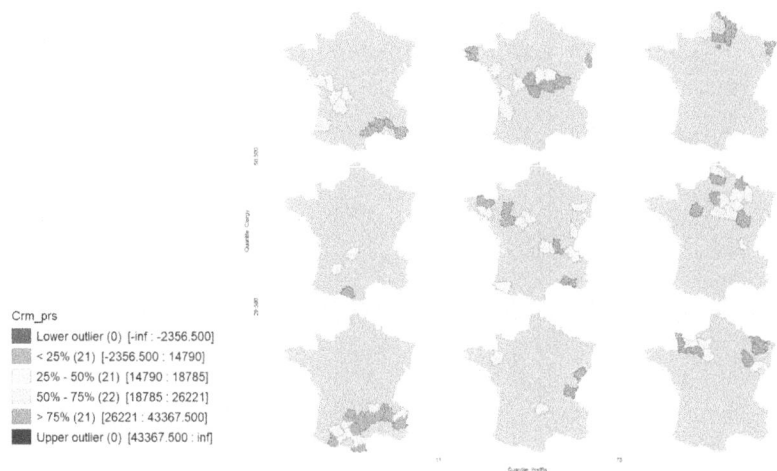

Fig. A6.12. Bivariate zoning of the moral statistics of France using a conditional map

Trivariate Zoning Using Bubble Charts

What if we want to test a hypothesis about the dependency of three variables: the further from Paris a region is, the larger its area, and the smaller the population and number of prostitutes coming from it are. To do this, we will build a *bubble chart* (Explore → Bubble Chart). We will mark the distance to Paris (Distance) on the x-axis, the area of the region (Area) on the y-axis; population size (Pop1831) is represented by the size of the circles; and the number of prostitutes in Paris from the respective regions is designated by colour (Fig. A6.13). The size of the bubbles can be changed in map properties (Adjust Bubble Size).

As we can see from the bubble chart. France is divided into three clusters: Paris, regions closest to it with a large number of prostitutes, and regions that are far from Paris and have a small number of prostitutes.

The size of the region and population density have almost no effect on the number of prostitutes.

Now we will select points on the diagram that are further than 250km from Paris, and they will be highlighted in our box map. As we can see, two departments do not fit in with the dependence we have established (Fig. A6.14).

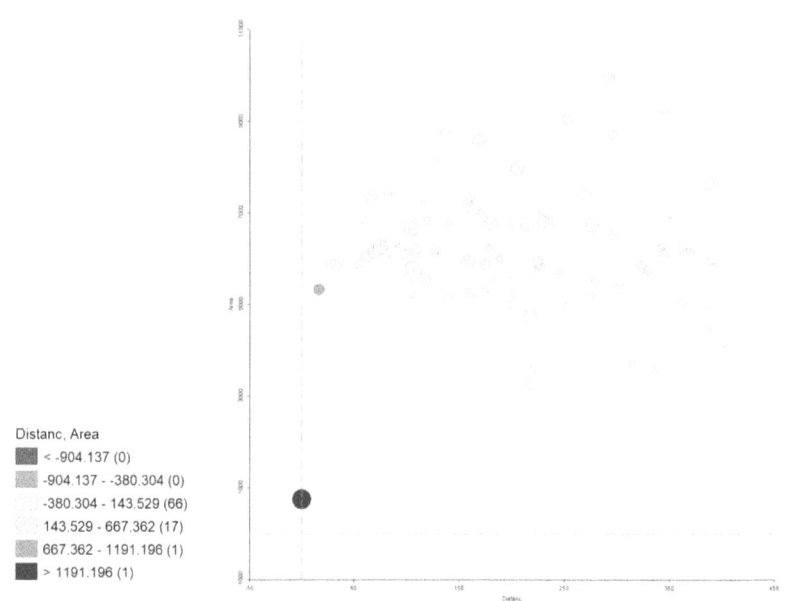

Fig. A6.13. Bubble chart of the moral statistics of France

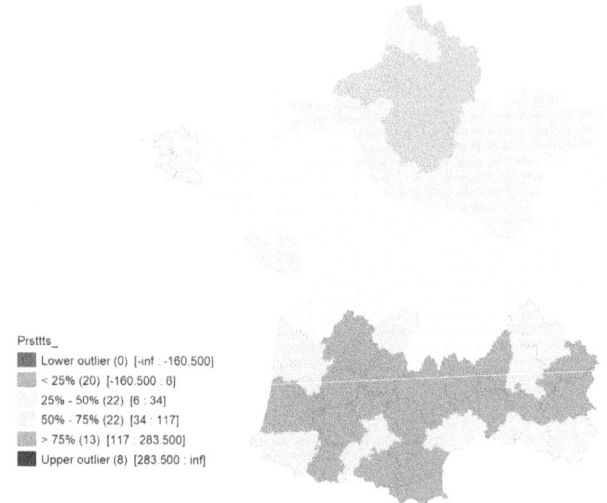

Fig. A6.14. Chorochromatic map of the moral statistics of France in the regions farthest from Paris

Multivariate Zoning Using Parallel Coordinate Plots

Bubble charts allow you to analyse the influence of only three factors on the dependent variable. Now we will look at *parallel coordinate plots*, which allow you to explore an unlimited number of variables (Explore → Parallel Coordinate Plot). We will set all the parameters we have studied and change the type of range mapping in the properties (Fig. A6.15).

On the right, we will mark the regions where these factors correlate most strongly and plot them on the box map, where they will be highlighted (Fig. A6.16).

304 Appendix

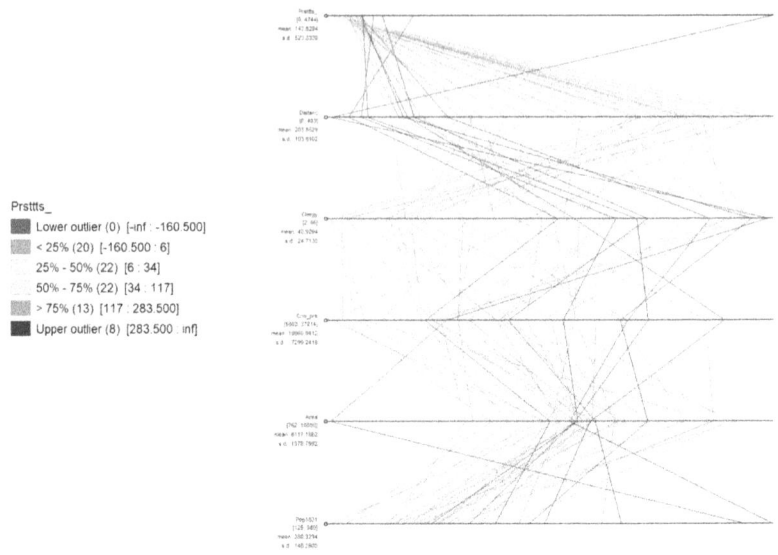

Fig. A6.15. A parallel coordinate plot of the moral statistics of France

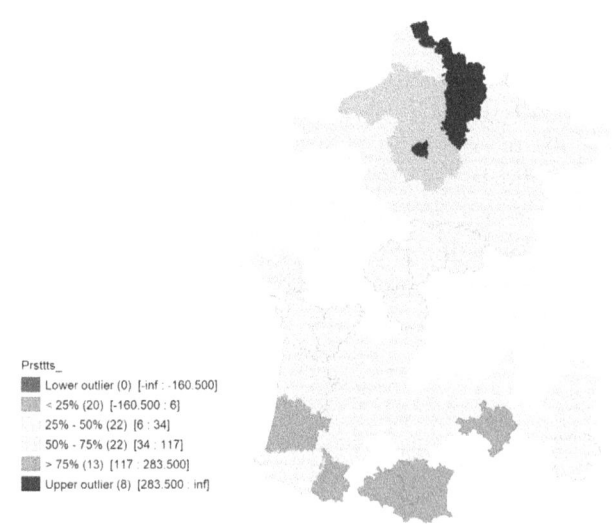

Fig. A6.16. Box map with sampled values of the moral statistics of France

Methodological Guide

Discrete Anamorphoses

Classical geographic maps are based on the Euclidean metric. In other words, distances and areas on the map correspond to real distances and areas in absolute space, measured in metres and square metres. However, for the purposes of socio-geographic analysis, absolute distances and areas may not be as important. In this case, we can build chorochromatic maps in which the topographic metric is replaced with the metric of the phenomenon we are mapping. For example, the area of a territory may not correspond to square kilometres, but to the size of the population in millions of people. These kinds of chorochromatic map are called anamorphoses. They are useful because (1) they allow phenomena to be mapped more clearly, and (2) they expand the possibilities of bivariate zoning.

The anamorphic process can be discrete, when the geographic coordinates of each individual object are transformed, or continuous, when the entire space is transformed. GeoDa is capable of performing discrete transformations.

We will use the New York city housing market shapefile available in the "NYC" archive on the website (https://mgimo.ru/upload/2023/05/nyc-zoning-descriptive-statistics.zip). Original source: https://geodacenter.github.io/data-and-lab//nyc/. We will create an anamorphose of the average monthly rental cost in 2008 (rent2008). After opening the shapefile, run the function for creating anamorphoses (Map → Chorochromatic map) and set the colour and size of the circle for the rent2008 indicator.

Fig. A6.17. Chorochromatic map of the standard deviation of housing prices in New York City neighbourhoods

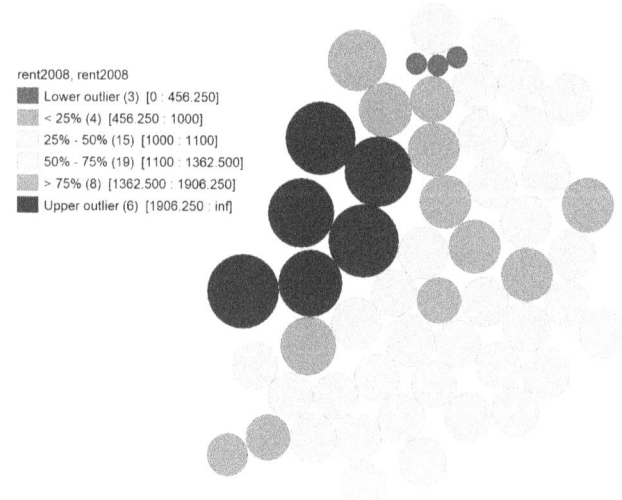

Fig. A6.18. Discrete Cartogram of housing costs in New York City neighbourhoods

Methodological Guide

Let's compare the chorochromatic map we have obtained from the cartogram (Fig. A6.18) with the standard deviation (Fig. A6.17). As we can see, the areas in the centre (Manhattan) differ significantly from the rest of the city. The GeoDa program tries to place the centre of the circle as close as possible to the region's centroid without the circles overlapping. If that proves too tricky, you can ask the program to recalculate (iterate) by right clicking and selecting Improve Chorochromatic Map.

Continuous Anamorphoses

We will be using QGIS to create continuous anamorphoses. In the program, open the anamorphic layer (Fig. A6.19). First, you will have to install a special extension that creates anamorphoses. To do this, select Modules → Manage Modules, type `chorochromatic map` in the search field and install the extension. Before you start, check in the attribute table that the indicator we need does not have values equal to or less than zero, and that each row contains a value, otherwise this will lead to an error when constructing the cartogram. Now click Vector → Chorochromatic map → Compute Chorochromatic map, select `rent2008`, the number of iterations (5, 10 or 15 are recommended) and start the process. Compare the resulting cartogram (Fig. A6.20) – the program has created a separate layer for it – with the original map.

Fig. A6.19. Chorochromatic map of housing costs in New York City neighbourhoods

Bivariate Anamorphoses

On the chorochromatic map, we could only zone a single feature. Anamorphoses allow us to perform the same function, but with two features. It effectively makes the object area parameter available for analysis. To do this in Geoda, you need to set a different indicator for the colour of the circle. In QGIS, you need to perform typological or dasymetric zoning in the layer properties, just as we did in Chapters 2 and 3. As an example, we have grey-scaled the average number of people per household (hhsiz2008). While Manhattan is still the largest, Queens is now the most densely populated borough.

Fig. A6.20. Continuous Cartogram of housing costs in New York City neighbourhoods

Methodological Guide 309

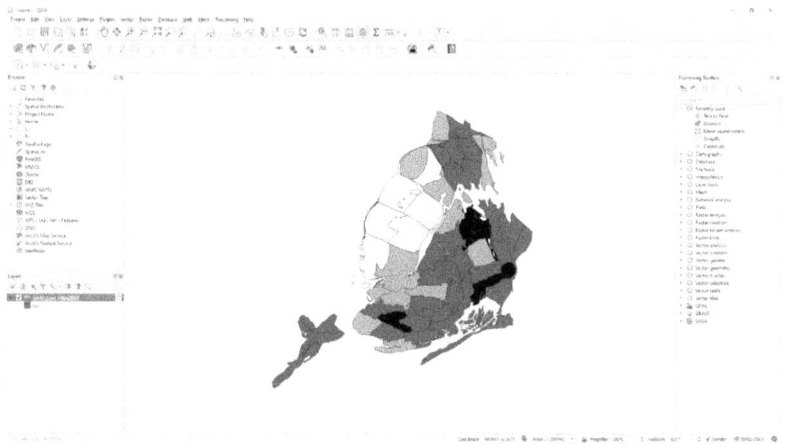

Fig. A6.21. Bivariate Cartogram of housing costs in New York City neighbourhoods

Questions and Tasks

1. What type of scaling is appropriate for data with and without significant outliers? Determine the types of data distribution that are appropriate for each type of scaling.
2. Why does dasymetric zoning need to be converted into typological zoning? It what cases are categorial variables needed?
3. Find other variables in the attribute table of the moral statistics of France suitable for multivariate zoning. Form hypotheses and test them.
4. Weigh up the advantages and disadvantages of multivariate zoning using conditional maps, bubble charts and parallel coordinate plots.
5. Suggest some analytical tasks for the multivariate zoning of Russia. What is the best method for solving each of them?
6. What kind of data is best suited for cartogram?
7. What research tasks can be solved by zoning using cartogram?
8. Perform an cartogram with various iterations. Compare your results.
9. What other parameters of the shapefile we used can be zoned using the bivariate cartogram method? What should you choose as the anamorphic parameter and what should be the scaled parameter? What conclusions can we make about the city based on this?

6. Mapping and Zoning Descriptive Statistics in R

Objectives

- ✓ Create and analyse a histogram, box plot, scatter plot and bubble chart in the ggplot package for R
- ✓ Calculate scatter plot matrices in the GGally package for R
- ✓ Plot a three-dimensional scatter plot model in the scatterplot3d package for R
- ✓ Map descriptive statistics in the tmap package for R
- ✓ Anamorphize descriptive statists in the chorochromatic map package for R

In this chapter, we are going to repeat the tasks we completed previously, but this time using the R programming language. We will start by creating a markdown, downloading the necessary packages (we will also need to install new packages – `ggthemes`, `Hmisc` and `gap`) and loading the shapefiles of the world and French statistics that we have analysed into the system. Don't forget to save the markdown in the same directory as the shapefiles.

```
```{r}
library(tidyverse)
library(sf)
library(tmap)
library(ggplot2)
library(ggthemes)
library(GGally)
library(scatterplot3d)
library(chorochromatic map)

world <- read_sf("TM_WORLD_BORDERS-0.3.shp")
france <- read_sf("Guerry.shp")
```
```

A few R commands allow us to get basic statistics for any parameter contained in the shapefile. Let's use the area of countries (AREA) as our sample and calculate the basic indicators (summary), range, variance and standard deviation (sd).

```
```{r}
summary(world$AREA)
```

Min. 1st Qu. Median Mean 3rd Qu. Max.
0.0 44.5 5515.5 52696.1 34708.8 1638094.0

```{r}
range(world$AREA)
```

[1]   0 1638094
```{r}
var(world$AREA)
```

[1] 27471317282
```{r}
sd(world$AREA)
```

[1] 165744.7
```

We will now build a histogram of the distribution of countries by area containing 15 intervals (bins). Give the graph a name and label each of the axes: x should be area, and y, the number of countries. We will also add the Stata program design theme (theme _ stata), but there is nothing stopping you from choosing more classic themes (theme _ classis, theme _ minimal, theme _ tufte, etc.), or perhaps the *Economist* theme (theme _ economist), or any other theme the program offers up when you type theme.

```
{r}
ggplot(data = world, aes(AREA))
+ geom _
histogram(bins = 15) + xlab("area")
+ ylab("number of countries")
+ ggtitle("Histogram of the
distribution of countries by area") +
theme _ stata()
```

Гистограмма распределения стран мира	Histogram of the distribution of countries by area
Количество стран	Number of countries
Площадь территории	Area

Save the histogram as a separate object (world_plot) and in the console below enter the command to generate histogram statistics by number of intervals (`plot.data`).

```
> world _ plot <- ggplot(data = world,
aes(AREA))
+ geom _ histogram(bins = 15) + xlab("area") +
ylab("number of countries") +

ggtitle("Histogram of the distribution of
countries by area") +
theme _ stata()
> layer _ data(World _ plot)

y count x xmin xmax density ncount
ndensity PANEL
1 221 221 0 -46892102 46892102 9.579161e-09
1.000000000
1.000000000 1
2 17 17 93784204 46892102 140676306 7.368585e-10
0.076923077 0.076923077 1
3 5 5 187568408 140676306 234460510 2.167231e-10
0.022624434 0.022624434 1
4 1 1 281352612 234460510 328244714 4.334462e-11
0.004524887 0.004524887 1
5 0 0 375136816 328244714 422028918 0.000000e+00
0.000000000 0.000000000 1
6 0 0 468921020 422028918 515813122 0.000000e+00
0.000000000 0.000000000 1
7 0 0 562705224 515813122 609597326 0.000000e+00
0.000000000 0.000000000 1
8 0 0 656489428 609597326 703381529 0.000000e+00
0.000000000 0.000000000 1
9 0 0 750273631 703381529 797165733 0.000000e+00
0.000000000 0.000000000 1
10 0 0 844057835 797165733 890949937 0.000000e+00
```

```
0.000000000  0.000000000  1
11  0  0  937842039  890949937  984734141  0.000000e+00
0.000000000  0.000000000  1
12  0  0  1031626243  984734141  1078518345
0.000000e+00
0.000000000  1
13  1  1  1125410447  1078518345  1172302549  4.334462e-11
0.004524887  0.004524887  1
14  0  0  1219194651  1172302549  1266086753
0.000000e+00
0.000000000  0.000000000  1
15  1  1  1312978855  1266086753  1359870957  4.334462e-11
0.004524887  0.004524887  1
```

Next, build a *box plot*, save it as a separate object (world _ boxplot) and query the distribution statistics in the console below. The words lower and upper denote the first and fourth quartiles, respectively, middle means the median value, and ymin and ymax refer not to extreme values, but to the boundaries of the so-called "whiskers" of the box plot, excluding outliers. Add a "whisker" to the box chart using the errorbar command and colour the chart – black for the frame, blue for the "box" and red for outliers. Give them a common name.

```{r}
ggplot(data=world, aes(x="",y=AREA)) + geom_
boxplot(color="black", fill="blue", outlier.
color
= "red") + stat_boxplot(geom = "errorbar") +
ggtitle("Box plot") + theme_stata()
```

[Диаграмма размаха]

```
> World_boxplot <- ggplot(data=world,
aes(x="",y=AREA)) + geom_boxplot() + theme_
stata()
> layer_data(World_boxplot)
  ymin lower middle  upper  ymax
1    0  44.5 5515.5 34708.75 82329
```

Now let's build a *scatter plot*, where the x-axis represents the area of countries and the y-axis represents the population, and ask the program to display a regression line (geom_smooth(method=lm). A bar will appear on the graph by default showing a 95 % probability that the observations are true. To remove it, you need to put the se=FALSE command in brackets after the line drawing method.

```{r}
ggplot(data = world, aes(x=AREA,y=POP2005)) +
geom _ point() + (geom _ smooth (method=lm))
+ theme _
stata()
```

To analyse the statistics – specifically, to calculate the coefficient of determination (R squared) and the p-value – we will run the `lm` function and save the data in the `regression` object. Now we can see the analysis results (`summary`).

```{r}
regression <- lm(POP2005 ~ AREA, data = world)
summary(regression)
```

```
Call:
lm(formula = POP2005 ~ AREA, data = world)
Residuals:
Min 1Q Median 3Q Max
-397234462 -7940042 -7829315 -3668936 1029682026
Coefficients:
Estimate Std. Error t value Pr(>|t|)
(Intercept) 7.934e+06 6.781e+06 1.170 0.243
AREA 3.255e+02 3.906e+01 8.334 5.7e-15 ***
---
Signif. codes: 0 '***' 0.001 '**' 0.01 '*' 0.05
'.' 0.1 ' ' 1

Residual standard error: 101300000 on 244 degrees
of freedom
Multiple R-squared: 0.2216,Adjusted R-squared: 0.2184
F-statistic: 69.46 on 1 and 244 DF, p-value: 5.701e-15
```

Now we are going to create a *bubble chart*. A bubble chart is actually a kind of scatter plot, the difference is that the size and colour of the points are also variable parameters, which means that you can analyse several variables at once.

We'll build a bubble chart in which the x-axis represents the area of the countries (AREA), the y-axis the population size (POP2005), the size of the points the region number (REGION), and the colour saturation the subregion number (SUBREGION).

```{r}
ggplot(data = world,
aes(x=AREA,y=POP2005,size=R
EGION, color=SUBREGION)) + geom_ point()
+ (geom_
smooth (method=lm)) + theme_ stata()
```

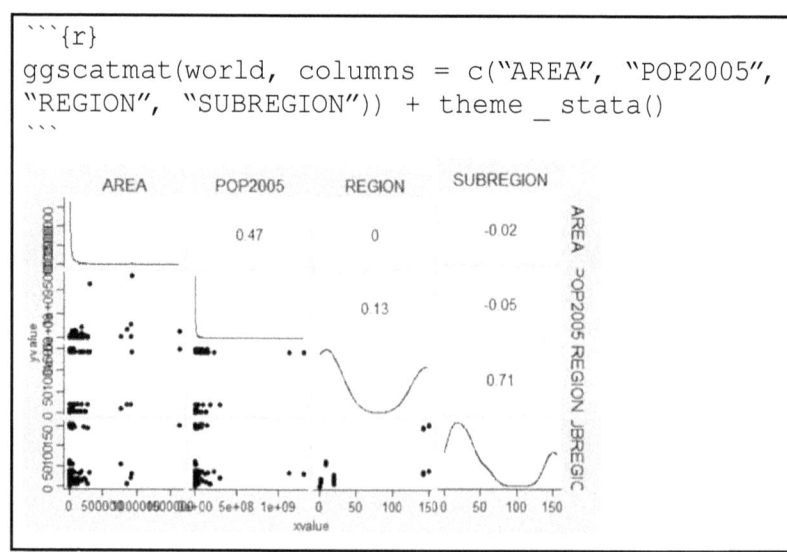

Next up is a *scatter plot matrix* for the same four parameters. To create a scatter plot matrix, use the ggscatmat function from the GGally extension.

```{r}
ggscatmat(world, columns = c("AREA", "POP2005",
"REGION", "SUBREGION")) + theme_ stata()
```

Our next task is to build a three-dimensional scatter plot model. Scatterplot3d is perfect for this.

```{r}
scatterplot3d(x = world$AREA, y = world$POP2005,
z = world$REGION, main = "Three-dimensional scatter plot model",
xlab = "Area", ylab = "Population size", zlab
= "Region code", pch = 20, color =
"grey")
```

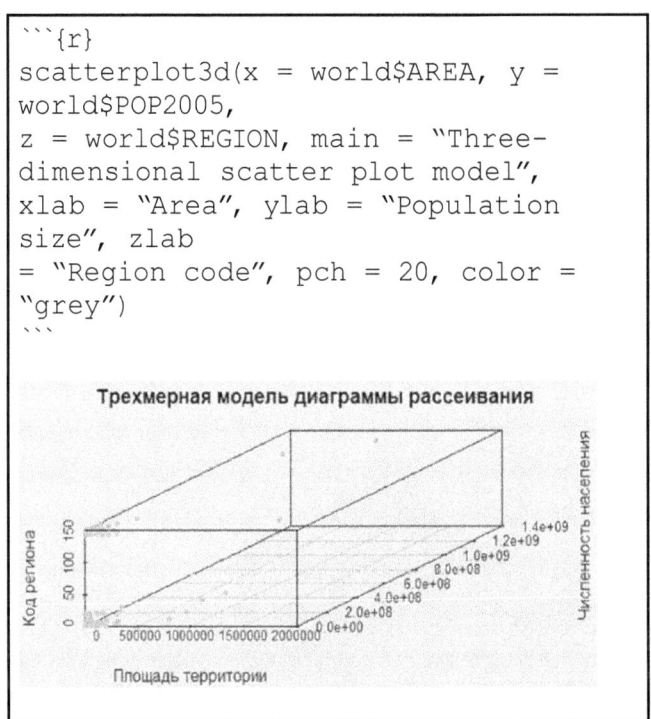

Трехмерная модель диаграммы рассеяния	Three-dimensional scatter plot model
Численность населения	Population size
Площадь территории	Area
Код региона	Region code

Now let's move on to mapping in R. In Chapter 4, we used the package `plot`. This time, we will use the more powerful `tmap`. We will be mapping the Moral Statistics of France again. For our colour scheme (`palette`), choose a monochrome grey scale (`Greys`). Of course, we could go with red (Reds), say, or polychrome – red through to white and blue (red, white, blue, or -RdBu) – or any other colours we want. We'll name the parameter in the legend "Home region," add a scale bar and compass, and place symbols outside the image. Save the file in .png format.

```{r}
tm_shape(france) + tm_fill("Prsttts", palette
= "Greys", title = "Home region") + tm_
borders() + tm_compass() + tm_
scale_bar() + tm_
layout(legend.outside = TRUE)
```

| Родной регион | Home region |

Methodological Guide

```
```{r}
tmap_save(tm = france_map,
filename = "france_map.
png")
```

Map saved to D:\france_map.
png
Resolution: 1915.625 by
1181.25 pixels
Size: 6.385416 by 3.937499
inches (300 dpi)
```

The map we have generated reflects visual intervals. Now we will build a chorochromatic map for the same parameter with equal quartiles. We have to set the number of intervals separately, in our case $n = 4$. Use `jenks` for chorochromatic maps with natural breaks, `equal` for equal intervals, and `sd` for standard deviation.

```
```{r}
tm_shape(france) + tm_fill("Prsttts",
palette =
"Greys", title = "Home region", n=4, style =
"quantile") + tm_borders() + tm_
layout(legend.
outside = TRUE)
```
```

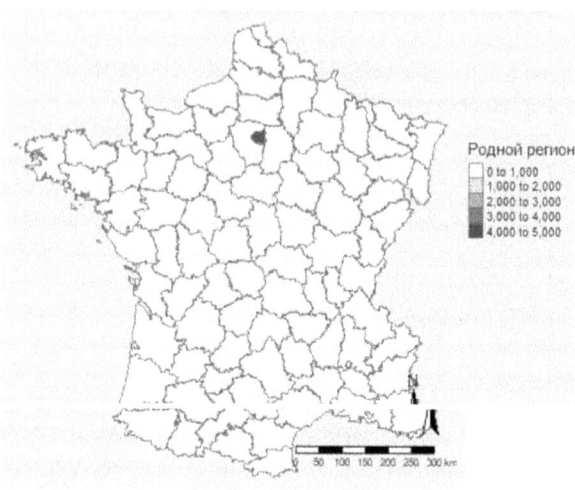

| Родной регион | Home region |

The previous chorochromatic maps we have created used dasymetric zoning. The cat style can produce typological zoning. Let's do this for the Region parameter.

```{r}
tm_shape(france) + tm_fill("Region", palette =
"Greys", style = "cat") + tm_borders() + tm_
layout(legend.outside = TRUE)
```

The next task is to create anamorphoses. We will need the chorochromatic map package to do this. We'll start with discrete cartogram using the chorochromatic_map_dorling function. The algorithm will look like this:

```{r}
anamorph <- chorochromatic 
map _ dorling(france,"Prsttts") 
class(anamorph)
```

[1] "sf" "tbl _ df" "tbl" "data.frame"
```{r}
tm _ shape(anamorph) + tm _ fi
ll("Prsttts", palette
= "Greys", title = "Home region")
+ tm _ borders()
+ tm _ compass() + tm _ scale _
bar() + tm _
layout(legend.outside = TRUE)
```

| Родной регион | Home region |

Now we will build a continuous cartogram for the same parameter. To do this, we need to use the `chorochromatic map _ cont` function. By default, the program will create 15 iterations and show the degree of error on each of them.

````{r}
anamorph_cont <- chorochromatic map_
cont(france,
"Prsttts")
````
Mean size error for iteration 1: 11.6298427664491
Mean size error for iteration 2: 10.5976014886266
Mean size error for iteration 3: 9.58145138335816
Mean size error for iteration 4: 8.58300996372125
Mean size error for iteration 5: 7.60877168420035
Mean size error for iteration 6: 6.66825389733754
Mean size error for iteration 7: 5.77682412573321
Mean size error for iteration 8: 4.94419782399732
Mean size error for iteration 9: 4.18455893496946
Mean size error for iteration 10: 3.50978820891574
Mean size error for iteration 11: 2.93059166380673
Mean size error for iteration 12: 2.4521678336134
Mean size error for iteration 13: 2.0750071702414
Mean size error for iteration 14: 1.79058672896782
Mean size error for iteration 15: 1.58441034923446

```
{r}
tm_shape(anamorph_cont) + tm_fi
ll("Prsttts",
palette = "Greys", title = "Home region") +
tm_borders() + tm_compass() + tm_scale_
bar() +
tm_layout(legend.outside = TRUE)
```

Finally, the chorochromatic map package can create a hybrid type of cartogram using the chorochromatic_map_ncont function. In the resulting image, the regions have borders that are similar to real ones, but are located separately from each other.

```
```{r}
anamorph _ ncont <- chorochromatic map _
ncont(france,
"Prsttts")
tm _ shape(anamorph _ ncont) + tm _ fi
ll("Prsttts",
palette = "Greys", title = "Home region") +
tm _ borders() + tm _ compass() + tm _ scale _
bar() +
tm _ layout(legend.outside = TRUE)
```
```

Questions and Tasks

1. Plot a histogram, box plot, scatter plot, bubble plot, scatter plot matrix and 3D scatter plot model in R for the data in the New York City population statistics shapefile we used in the previous chapter.
2. Map the data from the New York City population statistics shapefile we use in the Chapter 6 in R.
3. Anamorphize the data from the country statistics shapefile we used in Chapter 2.

4. Compare descriptive statistics analysis, mapping and anamorphization in GeoDa and R. What are the advantages and disadvantages of each software package? What tasks are they best suited for, respectively?

7. Relative Spatial Analysis

Objectives

✓ Transform an extensive variable into an intensive variable
✓ Calculate a new variable in an attribute table
✓ Create a chorochromatic map of the relative risk of the spatial distribution of a phenomenon
✓ Measure the Bayesian probability of the spatial distribution of a phenomenon
✓ Visualize the temporal dynamics of the spatial distribution of a phenomenon

Extensive and Intensive Variables

Sometimes performing proper analysis might require using spatially intensive variables, for example, the ratio of the number of deaths per 1000 people per square metre instead of just the number of deaths per square metre. Calculating spatially intensive indicators enables us to evaluate the situation at one point in space relative to the situation at other points in space. In other words, we can analyse the qualities of a phenomenon not depending on the properties of the point where it is located (vertical dependency) but depending on its location relative to other points in space (horizontal dependency). These procedures are a form of *relative spatial analysis.*

In this chapter, we will be working with a shapefile of the 1998 malaria epidemic in Colombia available in the "MalariaColomb" archive on original source: https://geodacenter.github.io/data-and-lab//colomb_malaria/.

We'll start off by building a *chorochromatic map of percentiles* for the malaria mortality rate (`MALARI98`) in GeoDa.

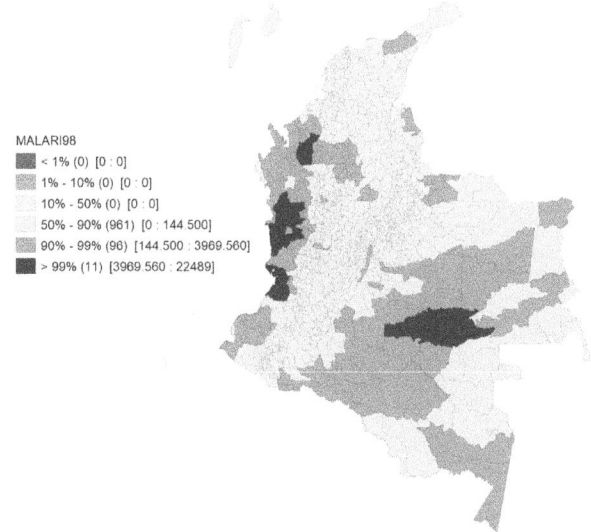

Fig. A8.1. Chorochromatic map of percentiles of the Colombian Malaria Epidemic

Now we will create a chorochromatic map (Map → Rates Calculated Map → Raw Rate) of the proportion of malaria cases in 1998 (in the numerator) relative to the population size in 1997 (TP1997). We will also select a percentile chorochromatic map to display the results (Fig. A8.1). Note how some regions that were considered significant outliers (1 % of the sample) in the previous chorochromatic map are only in the top 10 % of the sample in this one, while other regions that were not considered significant outliers now are. Right click on the map to save the indicator we have calculated to the attribute table.

The same procedure can be performed using the attribute table calculator (Table → Calculator → Bivariate). To do this, you need to create a new variable with the function RATE = MALARI98 / TP1997. Another option is to select the Rate tab and define two parameters to be analysed. It can now be mapped using the types of chorochromatic maps that we already know (Fig. A8.2).

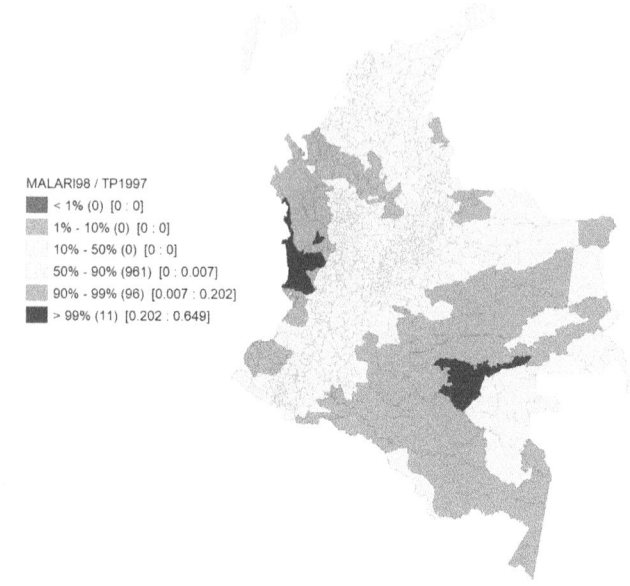

Fig. A8.2. Chorochromatic map of the proportion of Malaria cases during the epidemic in Colombia

Relative Risk of Spatial Distribution

Relative risk is the ratio of the frequency of an outcome among subjects affected by the factor under study to the frequency of outcomes among subjects not affected by that factor. Say, for example, that 50 people in a country with a population of 1000 have fallen ill, but 30 of those people are from the same region, which has a population of 100 people (Table 8.1).

Table 8.1 Relative Risk

	Number of healthy people	Number of infected people	Population size
The entire country	950 *(A)*	50 *(B)*	1000 *(A+B)*
Region	70 *(C)*	30 *(D)*	100 *(C+D)*

Methodological Guide

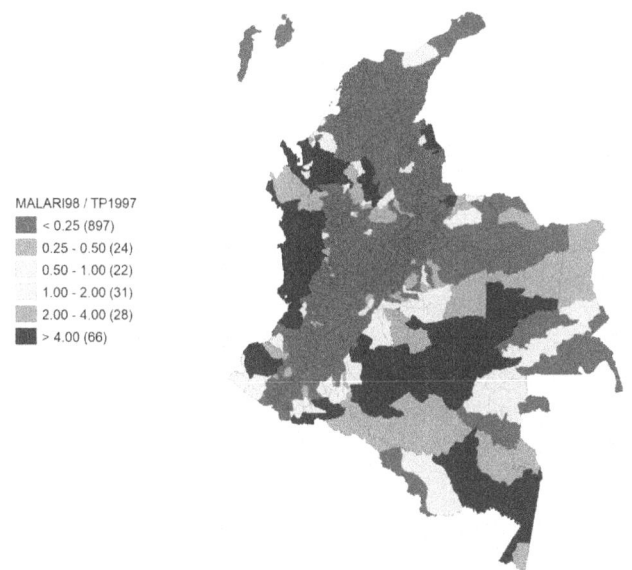

Fig. A8.3. Chorochromatic map of the relative risk of the Malaria epidemic in Colombia

In this case, the relative risk (RR) is calculated using the formula:

$$RR = \frac{\frac{A}{A+B}}{\frac{C}{C+D}} = \frac{A(C+D)}{C(A+B)}.$$

In our example, $RR \approx 1.36$. This means that the risk of infection in this is higher than the national average. If $RR = 1$, we can conclude that the factor we are studying does not affect the likelihood of the outcome (there is no relationship between the two). When $RR > 1$, the factor increases the probability of an outcome (there is a direct relationship between the two), and when $RR < 1$, the probability of an outcome decreases when exposed to the factor (feedback). Let's build chorochromatic maps of the relative risk of the malaria epidemic in Colombia (Map → Rates Calculated Map → Excess Risk). The indicator can also be calculated in the attribute table (Table → Calculator → Rate) using the Excess Rate function.

Bayesian Probability of Spatial Distribution

Instability in the parameters under analysis can lead to conclusions that are incorrect. This is why probability theory methods are sometimes used to estimate the spatial distribution of a phenomenon. One such method is based on the use of Bayes' theorem. Bayes' theorem stems from the definition of conditional probability, that is, the probability of an event occurring given that another event has already occurred. $P[A|B]$ is used to denote the probability (P) of event A when event B has already occurred. For example, the risk of contracting malaria in Colombia $P[A]$ is a fraction of one percent, but in the region where the epidemic is raging (i.e. condition B – $P[A|B]$ has occurred) the risk could be significantly higher. Bayes' theorem is generally expressed using the following formula:

$$P[AB] = P[A|B] \times P[B] = P[B|A] \times P[A],$$

where $P[A]$ is the a priori probability of A; $P[A|B]$ is the probability of A occurring given that B is true (posterior probability); $P[B|A]$ is the probability of event B occurring given that A is true; and $P[B]$ is the total probability of event B occurring.

Carrying out identical transformations will give us the following formula:

$$P[B|A] = \frac{P[B|A] \times P[A]}{P[B]}.$$

Geoda lets us calculate and map the Bayesian probability of the spatial distribution of a phenomenon based on the Poisson-Gamma distribution model (for more detail on this, see: https://geodacenter.github.io/workbook/3b_rates/lab3b.html#empirical-bayes-eb-smoothed-rate-map). To do this, select (Map → Rates Calculated Map → Empirical Bayes) and set the same parameters and percentile chorochromatic map (Fig. A8.4). The resulting chorochromatic map will provide a calculation of the incidence of malaria relative to other regions and the country as a whole (which we have already done), as well as an estimation of the probability of infection based on the size of the population in each region. In other words, the chorochromatic map takes two parameters into account: proximity to the focus of infection, and population size (which also increases the likelihood of infection).

Bayesian probability can also be calculated in the attribute table (Table → Calculator → Rate) using the Empirical Bayes function.

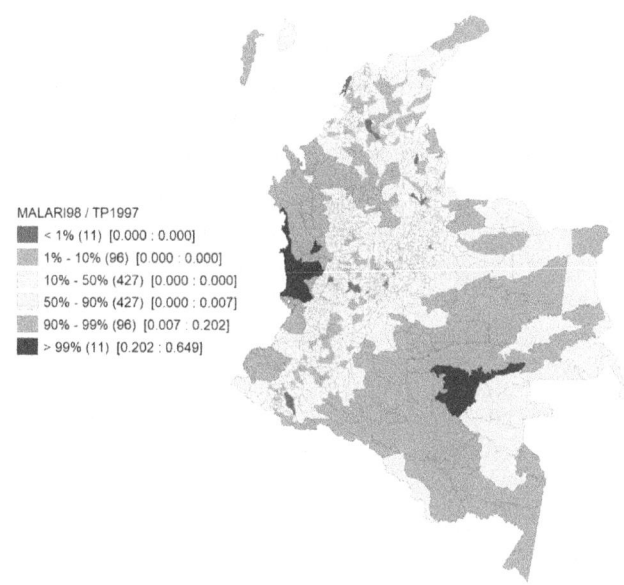

Fig. A8.4. Bayesian probability chorochromatic map of the Malaria epidemic in Colombia

Temporal Variables of Spatial Distribution

Let's try to assess whether the malaria epidemic affected population size in Colombia's regions. The easiest way to do this is to use the attribute table calculator (Table → Rates Calculator → Bivariate) to create a new variable with the function POP _ CHANGE = TP1998 - TP1997. At the same time, we can pinpoint regions with the largest outliers in terms of the number of deaths from the 1998 malaria epidemic with the help of a scatter plot. Mapping the result tells us that the epidemic did not have a significant impact (Fig. A8.5).

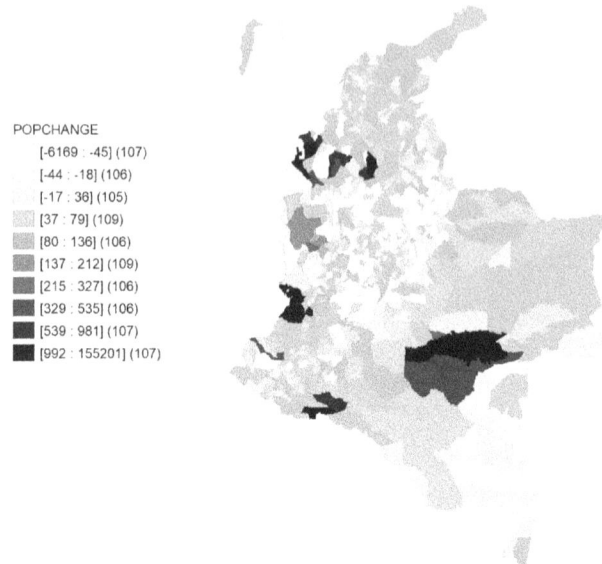

Fig. A8.5. Chorochromatic map of population change in Colombia from 1997 to 1998

However, for more reliable results, we need to study data for a longer period of time. GeoDa has a very good data visualization tool called Time Player that can help us do this. Go to Time → Time Editor, where will create a new POP _ TIMEPLAYER parameter, adding data for 1996, 1997 and 1998 from the variables on the left (Fig. A8.6).

Now open the malaria scatter plot and select the significant outliers, a chorochromatic map with ten different intervals, and Time Player. Starting Time Player will give us an animated picture of the situation as it evolved over time (Fig. A8.7). You can also save chorochromatic maps for each year separately and then combine them into a single gif file for clarity using an online converter such as https://gifmaker.me.

Methodological Guide

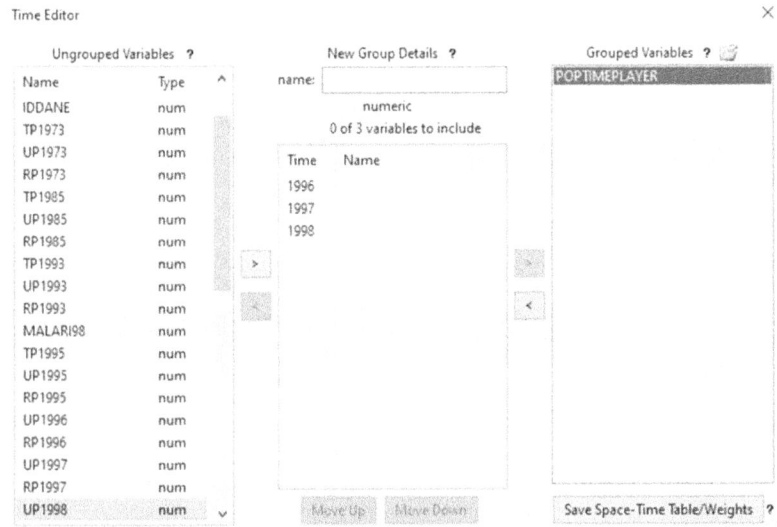

Fig. A8.6. The time editor function in GeoDa

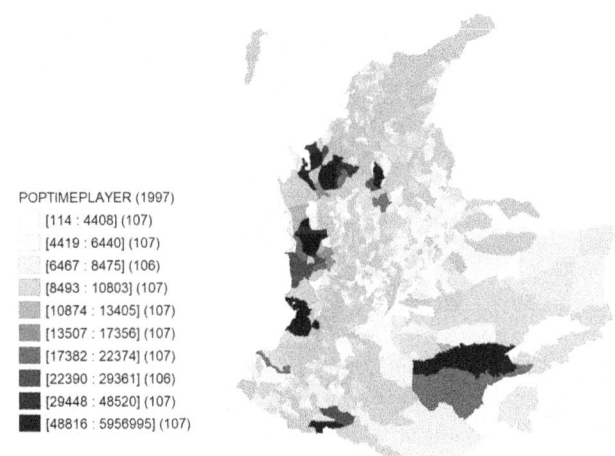

Fig. A8.7. Dynamic chorochromatic map of population change in the regions of Colombia following the epidemic

Questions and Tasks

1. Transform the extensive variables in the New York City population shapefile we used in Chapter 7 to intensive variables.
2. Select parameters and create a relative risk chorochromatic map based on the demographics of New York City we used in Chapter 6.
3. Estimate the Bayesian probability of the spatial distribution of these phenomena.
4. Compare the simple relativity, relative risk and Bayesian probability for the individual parameters of the demographics of New York City we used in Chapter 6. Which indicator was most significant in each case?
5. Show the temporal transformation of these phenomena using Time Player.

8. Neighbourhood Spatial Analysis

Objectives

- ✓ Construct a spatial weights matrix by adjacency
- ✓ Derive a spatial weights matrix by distance and calculate the arc distance
- ✓ Create a spatial weights matrix based on the k-nearest neighbours algorithm
- ✓ Derive the spatial lag in the distribution of phenomena across a territory
- ✓ Calculate indicators of relative and probabilistic spatial risk

Spatial analysis is based on the assumption that a phenomenon in one location is influenced by the properties of neighbouring locations. For example, we might hypothesize that a political regime surrounded by democracies would be more inclined towards similar policies. However, before testing such a hypothesis using geoinformation methods, we need to explain to the computer what a neighbourhood is.

We will use a neighbourhood index to carry out an integral assessment of the degree of neighbourhood:

Methodological Guide

$$\text{Neighbourhood index} = \sum_{i=1}^{n} \frac{N_i}{R_i}$$

where N_i is the number of neighbours of the *i*-th order, and R_i is the order number of the neighbourhood. In this section, we will look at how to set the neighbourhood parameters that will be used in all spatial analysis procedures.

Adjacent Neighbourhood

The first type of neighbourhood (adjacent neighbourhood) is based on topological relationships between objects and is used to analyse data confined to units of area. Adjacent objects are those whose boundaries have common points. Three variants of the neighbourhood are possible: the Rook neighbourhood; the Bishop neighbourhood; and the Queen neighbourhood.

In a Rook neighbourhood, pairs of territorial units with common sides are considered adjacent; in a Bishop neighbourhood, it is pairs of territorial units with common corners; and in a Queen neighbourhood, it is pairs of territorial units that have at least one common point on their borders, i.e. if their sides or corners touch. A Rook neighbourhood is stricter, as it only allows common sides along borders and common points are not taken into consideration.

Let's calculate the *neighbourhood matrix* (Table 12.1) for the Rook neighbourhood in the model below. First we take cells by pairs and see whether they are adjacent (putting "1" in the table on the left if they are) and calculate the sum of such "adjacent" readings in each row. Then we suppose that the sum for each row should be equal to one; to achieve that, we need to divide the 1 in each cell by the sum (see the table on the right). We now have a *spatial weights matrix*.

Table 9.1 Calculation of an Adjacent Neighbourhood

A	B	C
D	E	F

	A	B	C	D	E	F	Σ
A	0	1	0	1	0	0	2
B	1	0	1	0	1	0	3
C	0	1	0	0	0	1	2
D	1	0	0	0	1	0	2
E	0	1	0	1	0	1	3
F	0	0	1	0	1	0	2

	A	B	C	D	E	F	Σ
A	0.0	0.5	0.0	0.5	0.0	0.0	1
B	0.3	0.0	0.3	0.0	0.3	0.0	1
C	0.0	0.5	0.0	0.0	0.0	0.5	1
D	0.5	0.0	0.0	0.0	0.5	0.0	1
E	0.0	0.3	0.0	0.3	0.0	0.3	1
F	0.0	0.0	0.5	0.0	0.5	0.0	1

The short form version of the neighbourhood matrix looks like this:

$$W = \begin{bmatrix} w_{11} w_{12} \ldots w_{1n} \ w_{21} w_{22} \ldots w_{2n} \ \ldots \ \ldots \ \ldots \ \ldots w_{n1} w_{n2} \ldots w_{nn} \end{bmatrix}$$

If i and j are neighbours, then the spatial weights w_{ij} will be greater than zero. Conversely, self-neighbourhood is impossible, so $w_{ij} \neq 0$. Given our assumption that the sum of each row of the matrix of spatial weights should be equal to one, the standardized formula for calculating the neighbourhood degree of a single cell will generally look like this:

$$w_{ij(s)} = \frac{w_{ij}}{\sum_j w_{ij}}$$

The sum of the spatial weights matrix will be equal to the number of observations:

$$S_0 = \sum_i \sum_j w_{ij}$$

Methodological Guide 339

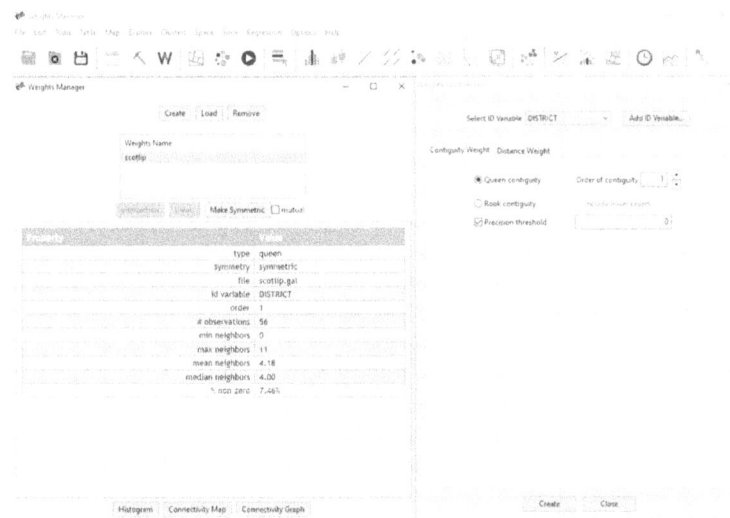

Fig. A9.1. Calculating a spatial weights matrix in GeoDa

In this lab session, we will be working in GeoDa with a shapefile of Scottish regions available in the "scotlip" archive on our Google Drive (https://drive.google.com/file/d/11XRf5h7PCYV1yVGVu3AWW7N D6daR5p-A/view?usp=drive_link) as well as on our website (https://mgimo.ru/upload/2023/05/neighbourhood-spatial-analysis.zip) Original source: https://geodacenter.github.io/data-and-lab/scotlip/.

To calculate a spatial weights matrix in GeoDa, go to Tools → Weights Manager (Fig. A9.1). Click on Create New Weights and in the window that opens set a parameter from the attribute table that reflects the number or code of the cells, or create your own variable where we will record this data. Select Contiguity Weight → Queen contiguity, adjacency order 1. Check the Precision threshold checkbox for a more accurate calculation and save the weights as a file with a .gal extension. Our weights will have appeared in the list on the left, where you can see information about them: type, symmetry, parameter, number of orders and observations, minimum, maximum, and the mean and median numbers of neighbours.

Click on Histogram at the bottom, then right-click to add statistics to it and analyse the distribution of objects by the number of neighbours (Fig. A9.2). Note that some regions do not have neighbours (islands); we will not be able to explore them further using this type of neighbourhood.

Right-click on the histogram once more and save the number of neighbours as a separate parameter in the attribute table.

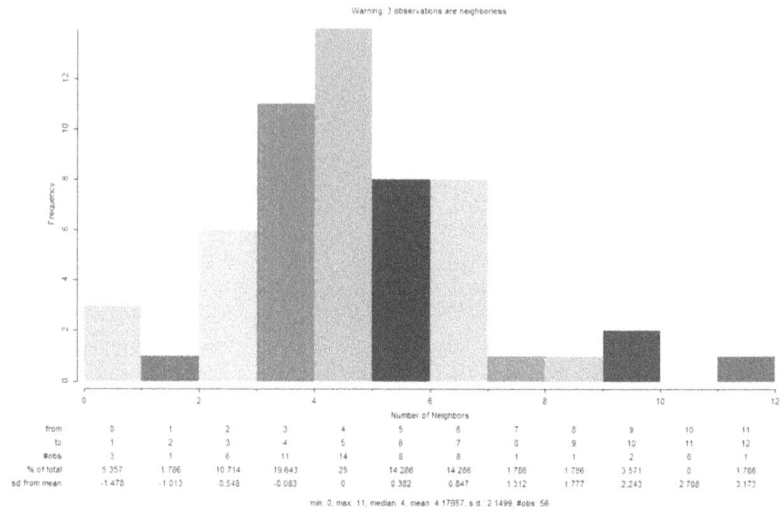

Fig. A9.2. Histogram of Scottish regions by adjacency

Now let's choose a connectivity chorochromatic map for analysis, which will show us an adjacency graph of each region's connections with other regions (Fig. A9.3). If we go back to the histogram and select an interval of our choosing, then only the result we are looking for will come up in the chorochromatic map. Right-clicking on the chorochromatic map allows us to perform several important functions. For example, you can change the colour of neighbouring regions or hide the map entirely, leaving only the adjacency graph.

Fig. A9.3. Adjacency neighbourhood graph for Scotland

Neighbourhood by Distance

As we saw in the previous example, the adjacency graph did not include an analysis of the islands of Scotland, which are located near the coast but do not have common points with other regions. In such cases, it is better to use a different method for calculating the spatial weights of the neighbourhood by distance. Euclidean distance (on a plane) is calculated using the formula:

$$d_{ij} = \sqrt{(x_i - x_j)^2 + (y_i - y_j)^2}$$

where d_{ij} is the distance between i and j with the coordinates (x_i, y_i) and (x_j, y_j).

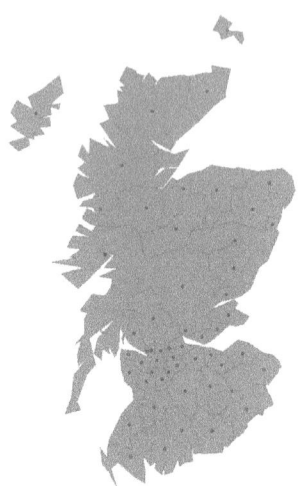

Fig. A9.4. Centroids of the regions of Scotland

However, there are problems with this calculation too. First, if we are studying regions, then what point should the distances be calculated from? By default, software packages calculate distance from the geometric centre (centroid) (Fig. A9.4). Second, if our projection is geographic, then the points will have latitude and longitude coordinates in degrees, and they cannot be measured this way.

When we have data in latitudes and longitudes, we use an *arc distance* – the shortest line between two points on a section of a curve. First, the latitude and longitude are transformed into radians:

$$Lat_r = (Lat_d - 90) * \frac{\pi}{180}; Lon_r = Lon_d * \frac{\pi}{180}$$

Further, if $\Delta Lon = Lon_{r(j)} - Lon_{r(i)}$, the distance will be calculated according to the following formula, where R is taken as the Earth's radius:

$$d_{ij} = R * arccos \frac{\left[cos(\Delta Lon) * sinsin\, Lat_{r(i)} * sinsin\, Lat_{r(j)} \right)}{+ cosLat_{r(i)} * cosLat_{r(j)}]}$$

Now that you know how to calculate the distance between centroids, you can create a spatial weights matrix by distance. To do this, you need to set a conditional range of distances that allow us to consider regions as neighbours. In other words, i and j are considered neighbours if j falls within a given range of distances δ from centroid i:

$$w_{ij} = 1, \text{ if } d_{ij} \leq \delta,$$

$$w_{ij} = 0, \text{ if } d_{ij} > \delta.$$

To properly calculate a spatial weights matrix by distance, at least one neighbour must fall into the given range of distances for each object. That is, we need to use the so-called maximum–minimum range selection principle: the maximum distance to the nearest region should be the extreme of the range. However, this figure is often too high because of remote enclaves or islands.

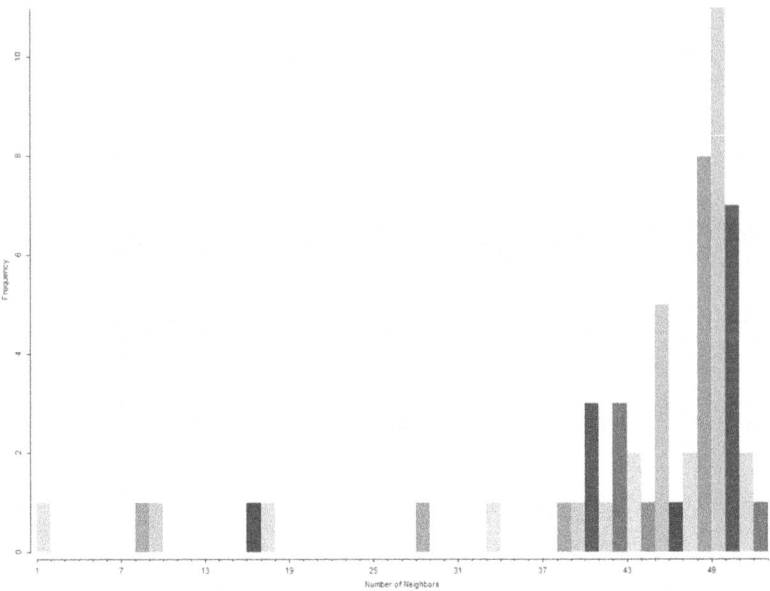

Fig. A9.5. Histogram of the distribution of the regions of Scotland by distance

To build spatial weights matrices by distance in GeoDa, use Tools → Weights Manager, create new weights, set a code or region number as a variable, select Distance Weight and determine whether they will be based on Euclidean distance or arc distance. The program will tell us the maximum–minimum criterion for the range of distances. Save the data and analyse the resulting histogram (Fig. A9.5). As we can see, our adjacency graph is not particularly suitable because of the remote Scottish islands – almost all on the mainland are neighbours with other regions (Fig. A9.6).

Fig. A9.6. Adjacency neighbourhood graph for Scotland by distance

Note here that although we have used the geographic coordinates of the centroids as reference points, we can set nonspatial weights for the calculation, with socioeconomic indicators serving as reference points. For example, we can calculate the distance in train fares, the number of plane transfers, or something else, instead of in kilometres.

Methodological Guide 345

The k-Nearest Neighbours Method

The methods described above assume a symmetrical neighbourhood principle. However, if we move away from this axiom, then it is possible to work with isolated regions. This is done using the k-nearest neighbours method. In its simplified form, the k-nearest neighbours algorithm involves assigning an object to the class that is most common among k-neighbours whose classes are already known.

In GeoDa, the function for creating spatial weights matrices using the k-nearest neighbours method is located in the same place as that for creating spatial weights matrices by distance – just select K-Nearest neighbors instead of distance. Selecting the number below lets you set the number of nearest neighbours needed for each element you are analysing. Compare, for example, the Scottish adjacency graph for $k = 2$, $k = 3$ $k = 4$ and $k = 5$ (Fig. A9.7).

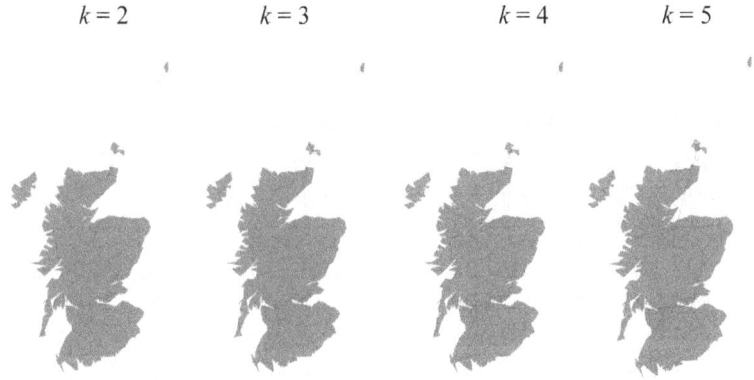

Fig. A9.7. Adjacency neighbourhood graph for Scotland using the k-Nearest neighbours method

Complex and Relative Neighbourhood Weights

If we are not satisfied with the weights created using just one of the methods for determining neighbourhood, we can create combined neighbourhood weights.

For example, say we want to create weights based on the principle that each object can have no more than five neighbours, and the distance between them cannot exceed the maximum–minimum criterion for the distance range (that is, the minimum for detecting at least one neighbour for all the objects being analysed). To do this, we need to go to the Weights Manager, select the weights we created for the data on Scotland by distance and using the *k*-nearest neighbours method with k=5, and select Intersection.

The program will create new neighbourhood weights, where the neighbours of the objects will be those polygons that meet the two criteria we have set – i.e. they are present in both previous spatial neighbourhood weight matrices.

Let's look at another task. Say we want to create weights where areas that are not strictly neighbours to the object (they do not share a border with it) are considered as such if there are fewer than three territories that are truly contiguous with it. To do this, go to Weights Manager, select the adjacency weights for Scotland created according to the Queen neighbourhood rule and the k-nearest neighbours method for k=3 and choose Union.

The program will create new neighbourhood weights in which the neighbours of the objects will be those polygons that meet at least one of the two criteria we have set – i.e. they are present in at least one of the two previous spatial weights matrices.

Before that, we created proximity weights based on geographic proximity between objects. But we can also model in a relative neighbourhood system in which proximity reflects similarity between objects with respect to a given parameter. This can be useful, for example, when we need to know whether similarities are more pronounced between geographical neighbours or between objects grouped according to a specific attribute. We can take the political map of the world and check if there is more similarity between geographical neighbours or between countries belonging to different political and economic regional blocs or civilizational groups. To do this, we need to build a spatial weights matrix based on non-geographic neighbourhood – one in which neighbours are

Methodological Guide

objects that are similar (or identical) with respect to a given parameter or group of parameters.

The first way to do this is to use the Block Weights function found in the Contiguity Weight tab in Geoda's Weights Manager (Fig.9.8). Select the AFF parameter, a measurement of industrial development, and then build weights based on it.

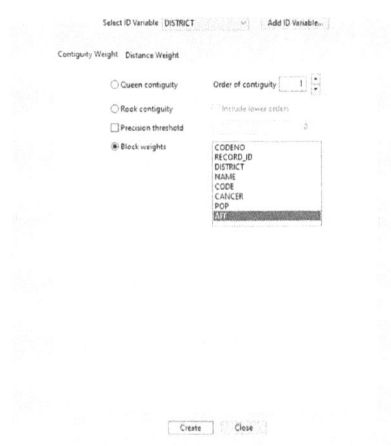

Fig. A9.8. Creating relative weights matrices. Method 1

Take a look at the resulting adjacency graph (Fig. A9.9). Note that the program has grouped the regions of Scotland into blocks with matching values of the AFF parameter. In other words, counties with similar AFF values are now considered neighbours.

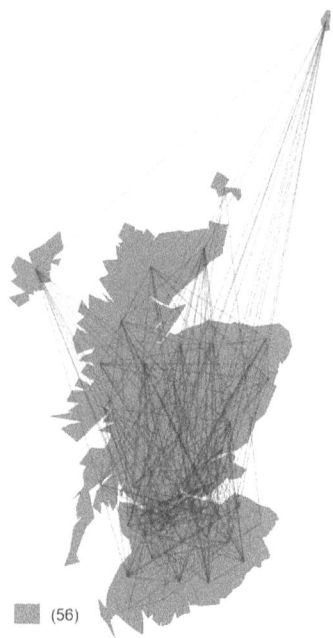

Fig. A9.9. Adjacency graph for relative weights matrix

If your parameter is expressed as text (which is the case with names of blocks or subregions), then you can display it on a chorochromatic map of unique values and save the resulting clusters in a category by right-clicking on the map and selecting the appropriate item. Categories stored in numerical format can then be used to build a spatial weights matrix by parameter. As you can see, this method is well suited for qualitative parameters.

The second method works best with quantitative parameters. Suppose we have a parameter with a range of values and we want to select the three nearest statistical neighbours for each country. Go to the Weights Manager, select the Distance Weight tab, click on Variables, and set one or more parameters that we want to use to search for similar objects (Fig. A9.10). Next, we'll select the k-nearest neighbours method, as we did with the geographic neighbourhood weights, set k=3 and save the weights. You can also set a range of values to constrain the search for similar polygons by parameter (or standardized parameters) – this is similar to the neighbourhood by distance method.

Methodological Guide 349

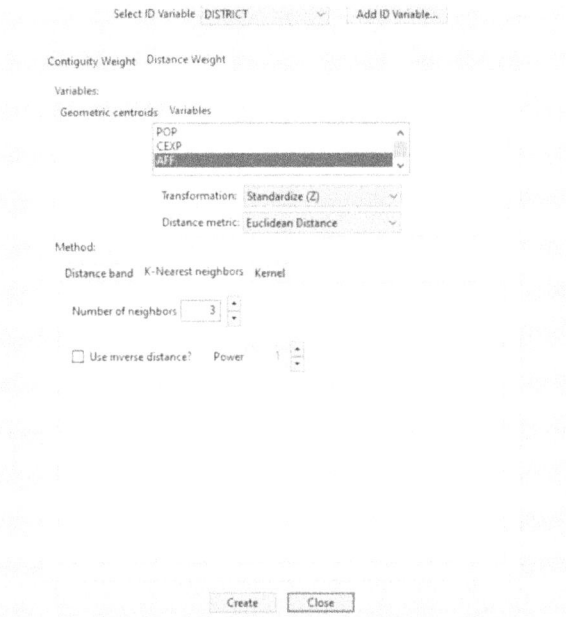

Fig. A9.10. Creating relative weights matrices. Method 2

Determine and set the optimal neighbourhood weights for clustering Scotland by POP (population). Now you will be able to assess the correlation coefficient – not between neighbouring polygons, but separately between densely populated (urban) and sparsely populated (rural) areas.

By checking the box next to Inverse, you can also apply the inverse distance weighting (IDW) method, when the neighbourhood weight for a given parameter is inversely proportional to the distance (geographic or statistical) between the parameters we are analysing (Fig. A9.11).

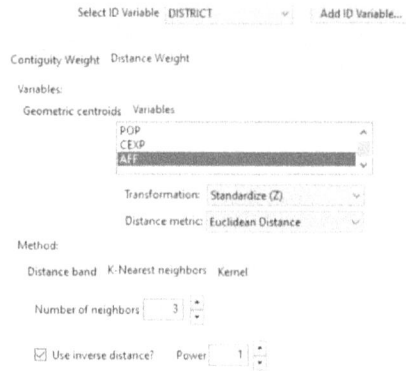

Fig. A9.11. Inverse distance weighting setting

Spatial Lag

When you know the spatial weights matrix, you can perform a multitude of operations to analyse the relationship between neighbouring objects. Let's start with the simplest operation – spatial lag. Spatial lag is the sum or average of the scores in neighbouring cells. The spatial average W_y for cell i is calculated using the following formula:

$$\lfloor W_y \rfloor_i = w_{j1} y_1 + w_{i2} y_2 + \ldots + w_{in} y_n$$

$$\lfloor W_y \rfloor_i = \sum_{j=1}^{n} w_{ij} y_j$$

where weights w_{ij} contains the indicators in row i of matrix W.

In other words, spatial lag is the sum of the values of the parameter in the cells adjacent to the cell we are studying, as $w_{ij} = 0$ for those that are not adjacent. For standardized series (where $\sum_j w_{ij} = 1$ is takes as the sum of the weights of the series), the spatial lag is the average value of the parameter in the cells adjacent to the cell we are studying.

To calculate spatial lag in GeoDa, select Table → Calculator → Spatial Lag. The first thing we need to do is to derive the spatial sum of adjacent parameters. To do this, select the weights we have saved (for example by metric k = 4, scotlip4), create a new variable LAG, parameter

CANCER (the number of cancer patients) and set the function LAG = scotlip4 * CANCER. Uncheck the Row-standardized weights and the Include diagonal of weights matrix boxes. Map the resulting LAG variable with ten equal intervals (Fig. A9.12). As we can see, this kind of spatial distribution has geographical dependency.

By checking the "Include diagonal of weights matrix" box, we will add the readings of the cell we are analysing to the sum, and if we also check the Row-standardized weights box, we will get the average, rather than the sum, of the distribution of the phenomenon in neighbouring cells.

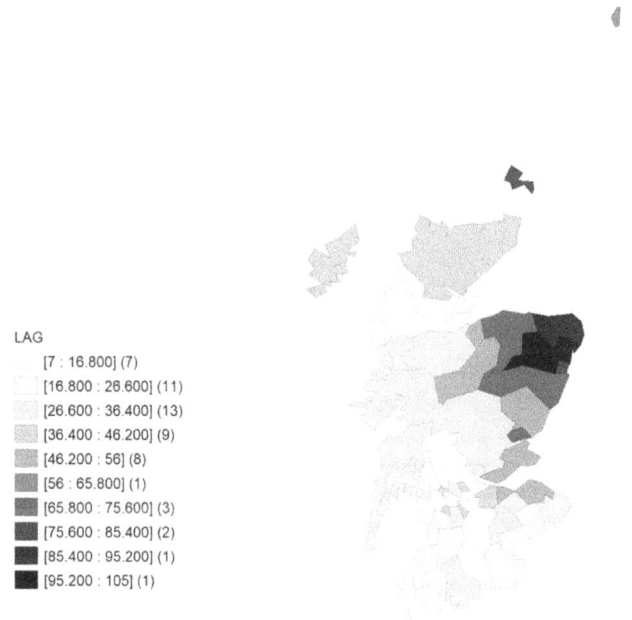

Fig. A9.12. Spatial lag in the distribution of cancer incidence in Scotland

Spatial Risk

In one of the previous chapters, we calculated the spatial distribution of relative values, but we did this without taking the neighbourhood into account. The number of infected was divided either by the population in a given region (i.e. we calculated the ratio between two parameters in a single region), or calculated in proportion to the population and the number of infected people throughout the country (i.e. the ratio of the parameter to the sum of the indicators from the sample). However, suppose we have two regions where the ratio of the population and the number of infected people is equal, but one is surrounded by areas where the epidemic is raging and the other is located on a remote island. Geographically speaking, the likelihood of infection in the first region will be higher. We can test these hypotheses by calculating the spatial risk based on an analysis of the ratio of a given indicator in neighbouring territories.

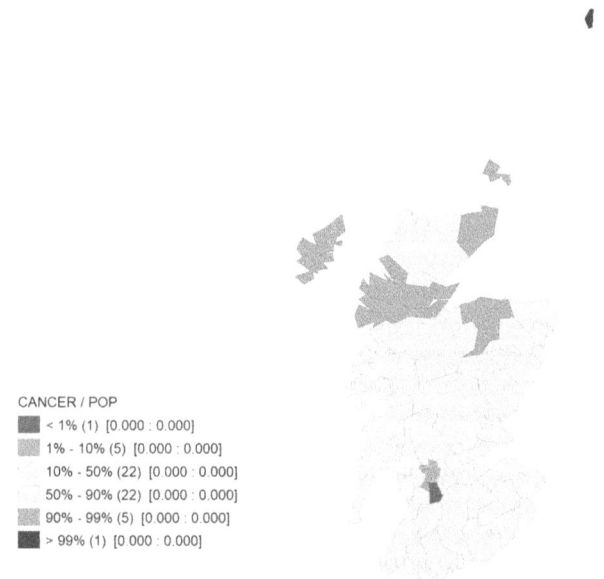

Fig. A9.13. Spatial risk distribution of cancer incidence in Scotland

In Geoda, you can calculate relative and probabilistic spatial risks (based on Bayes' theorem). To map relative spatial risk, select Map → Rates Calculated Map → Spatial Rate, and to map probabilistic spatial risk select Map → Rates Calculated Map → Spatial Empirical Bayes. Select the indicators to be analysed (CANCER and POP), the weights to be used and the type of mapping (Fig. A9.13).

Methodological Guide

Questions and Tasks

1. Compare the possible variants of spatial weights matrices for the regions of Scotland. Which is most suitable for the purposes of spatial analysis and why?
2. Calculate and map various spatial lags for cancer incidence in Scotland. Compare your results. What conclusions can you draw?
3. Calculate and map relative risk and Bayesian probability for cancer incidence in Scotland. Compare them with relative and probabilistic spatial risk indicators. What conclusions can you draw?
4. Calculate and map the relative and probabilistic spatial risk for the spread of the malaria epidemic in Colombia. Compare them with your results from the previous chapter. What conclusions can you draw?
5. What other indicators can be evaluated using spatial lag and spatial risk?

9. Spatial Autocorrelation Analysis

Objectives

- ✓ Build a correlogram
- ✓ Learn to distinguish between positive and negative spatial autocorrelation
- ✓ Calculate Moran's I spatial autocorrelation index
- ✓ Analyse a Moran scatter plot
- ✓ Calculate a bivariate Moran's I spatial autocorrelation index
- ✓ Map single- and dual-factor local indicators of spatial association

Building a Correlogram

One of the most basic tasks of spatial analysis is to understand how similar neighbouring phenomena are. This is the job of *spatial autocorrelation analysis*.

For set S of n geographical units, spatial autocorrelation is the ratio between the variable observed at each of the n units and the geographical proximity defined for all $n(n - 1)$ pairs of units from S.

Keep in mind that spatial autocorrelation can be either direct or inverse. In the first case, neighbouring cells are similar to each other, and in the second they are exact opposites. Zero spatial autocorrelation means that the spatial factor does not affect the distribution of the given phenomenon – in other words, the phenomenon is distributed randomly in space.

In this chapter, we will use a shapefile of the results of the 2012 and 2016 U.S. presidential elections is available on our Google Drive (https://drive.google.com/file/d/1HTZCkbh9ZDgLQMP4zEsTn5pPF7DxE4mo/view?usp=drive_link) and our website (https://mgimo.ru/upload/2023/05/spatial-autocorrelation-analysis.zip). Open the shapefile in GeoDa.

The easiest way to attempt to estimate spatial autocorrelation is to build a correlogram. Select Space → Spatial Correlogram and the desired parameter (for example, the share of votes received by the Republican candidate in the 2016 presidential election – pct_gop_16). Our correlogram depicts the relationship between all pairs of units depending on the distance between them (Fig. A10.1). This is not the most effective method as it does not specify a spatial weights matrix.

Now we will create a spatial weights matrix using the Queen neighbourhood rule. This will be of use to us later when we attempt more complex methods of estimating spatial autocorrelation. Let's analyse the resulting histogram of the distribution of an adjacency (Fig. A10.2).

Methodological Guide

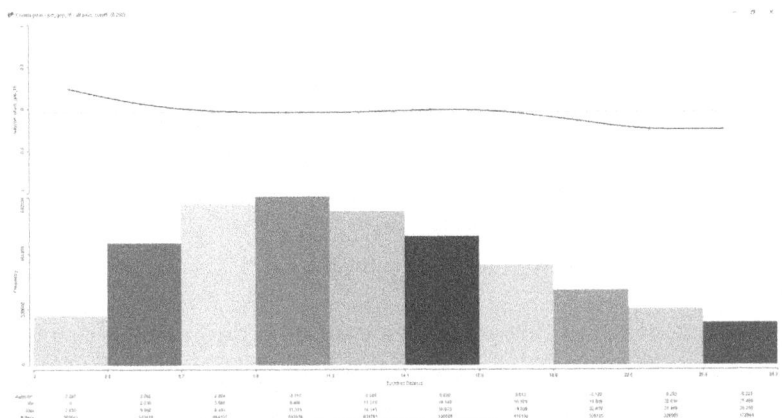

Fig. A10.1. Correlogram of the spatial autocorrelation of Republican votes in the 2016 U.S. Presidential Election

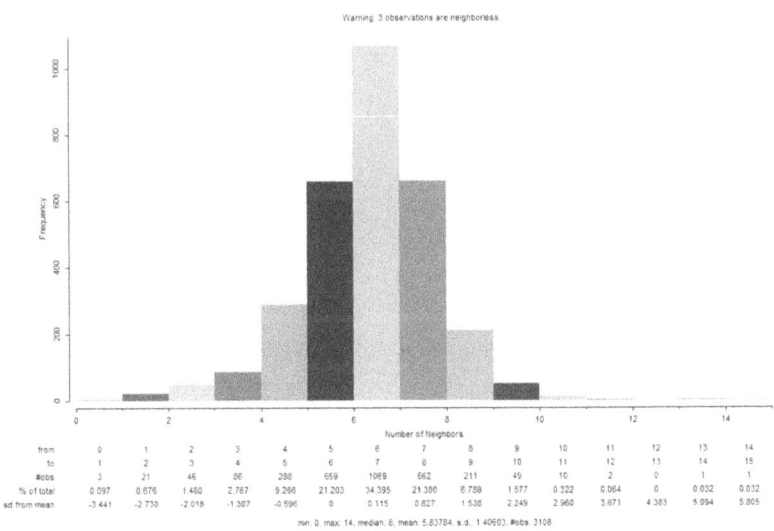

Fig. A10.2. Histogram of the distribution of U.S. Counties by contiguity

Moran's I Spatial Autocorrelation Index

To determine the degree of spatial autocorrelation, we need to calculate the Moran's I index, named after the Australian statistician who proposed the method. Moran's I is calculated using the following formula:

$$I = \frac{N}{W} \frac{\sum_i \sum_j w_{ij}(x_i - \bar{x})}{\sum_i (x_i - \bar{x})}$$

where i and j are units; x_i and x_j are values in the i-th and j-th units; $\bar{x}$ is the sample average of all the units; w_{ij} is the weight of the spatial relationship between the i-th and j-th units; N is the number of units; and W is the sum of the spatial weights.

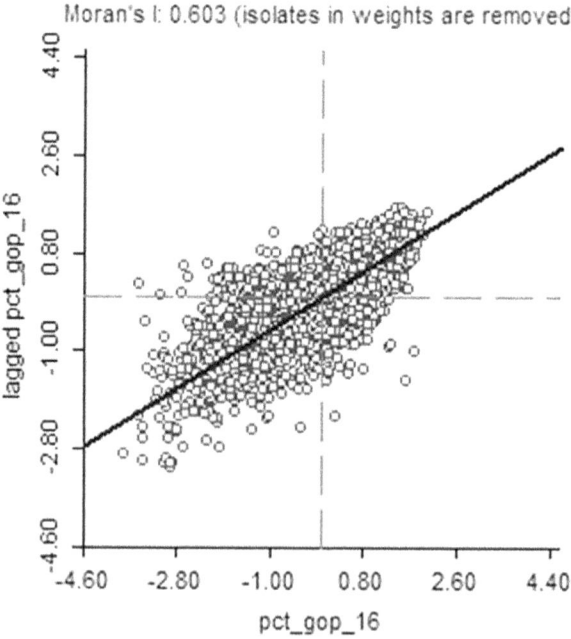

Fig. A10.3. Moran scatter plot of the spatial autocorrelation of Republican votes in 2016

As we can see, Moran's I is similar to the Pearson correlation coefficient. The difference is that it takes the neighbourhood effect into account.

Let's calculate spatial autocorrelation of Republican votes in the 2016 U.S. presidential election. To do this, select GeoDa Space → Univariate Moran's I, pct _ gop _ 16a as the variable, and the matrix we saved in accordance with the Queen neighbourhood rule as our spatial weights matrix. The result is a Moran scatterplot, where the value in each territorial unit is plotted along the x-axis, and its spatial lag – a weighted average of all neighbours – is plotted along the y-axis (Fig. A10.3). In a Moran scatterplot, the dotted lines represent the mean values along both axes, and the slant represents the linear regression of these values. It should be noted that, when calculating, the program automatically removes isolates (objects that do not have neighbours in a spatial weights matrix) from the analysis. Moran's I scatterplot also reports positive spatial autocorrelation (high-high and low-low) and negative spatial autocorrelation (high-low and low-high).

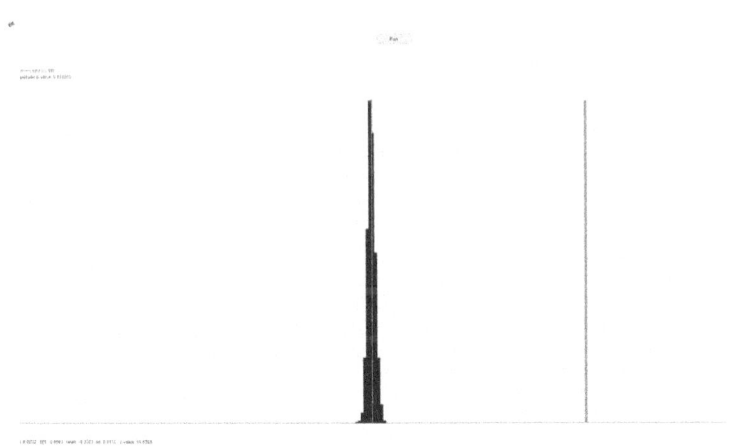

Fig. A10.4. Graph showing the distribution of permutation results

In our example, the Moran's index was 0.6. This means that there is a significant direct spatial autocorrelation, or, in other words, there was a spatial dependence in the voting for Republican candidates in the 2016 U.S. presidential election equal to 0.6 on the Moran's index.

The statistical significance of the Moran's I calculation depends on the number of permutations, which you can set by right clicking Randomization on the Moran scatterplot – 999 permutations are usually enough. When choosing the number of permutations, a graph showing the distribution of permutation results is displayed (Fig. A10.4), which, among other things, contains the p-value for the Moran's I we have calculated. In this case, a p-value of 0.001 with 999 with 999 permutations can be considered valid. At the same time, the significance level of the p-value is very high.

Now let's attempt to compare the spatial autocorrelation of Republican votes in Texas with the national average. To do this, select the Table → Selection Tool function and then use the attribute table to choose the data for the state of Texas (STATEFP = 48). Now right click on the Moran scatterplot and select View → Regimes Regression. This will produce three charts – the figures for Texas on the left, and for the entire country minus Texas on the right (Fig. A10.5). As we can see, the Moran's index for Texas was higher than for the rest of the country (0.65 compared to 0.59).

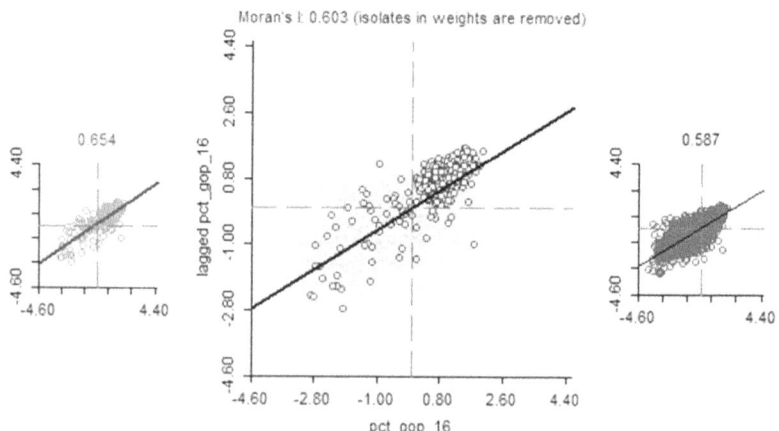

Fig. A10.5. Moran scatterplot of the spatial autocorrelation of Republican votes in Texas and rest of the country in the 2016 U.S. Presidential Election

Bivariate Moran's I Spatial Autocorrelation Index

Up until now, we have been calculating the Moran's index for a single indicator. But what if we want to test a hypothesis where the spatial distribution of one variable correlates with the spatial distribution of another variable? We may say, for example, that the spatial autocorrelation of Republican voters is related to the spatial autocorrelation of income level. Let's plot a Moran scatterplot for the shapefile we have been using that denotes income per person for the past 12 months (INC910213). As we can see, the Moran's index turned out to be even higher (Fig. A10.6).

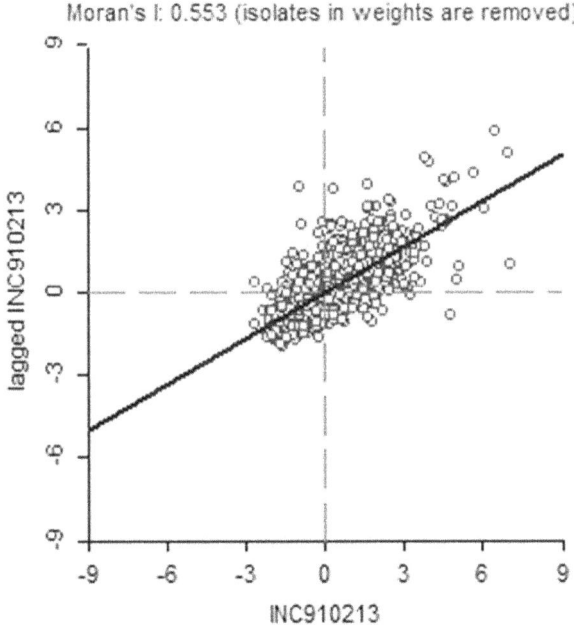

Fig. A10.6. Moran scatterplot of spatial autocorrelation by income level in the United States

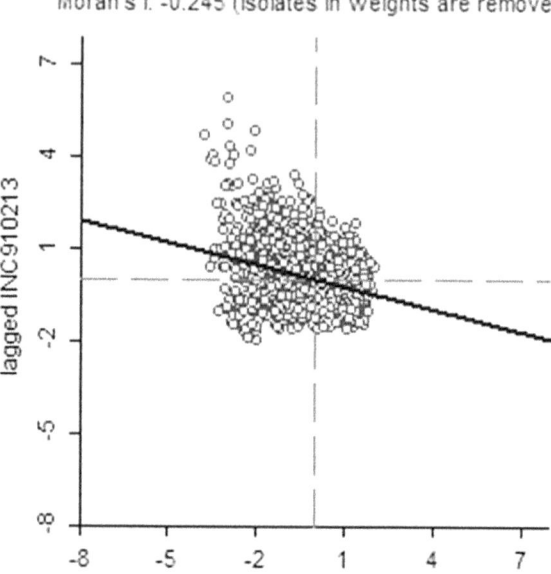

Fig. A10.7. Bivariate Moran's I scatterplot of spatial autocorrelation by income level in the United States and Republican votes

Now select Space → Bivariate Moran's I for the two parameters. The spatial autocorrelation index for them will be negative, at −0.24 (Fig. A10.7).

Finally, we'll repeat these calculations for the state of Texas (Fig. A10.8).

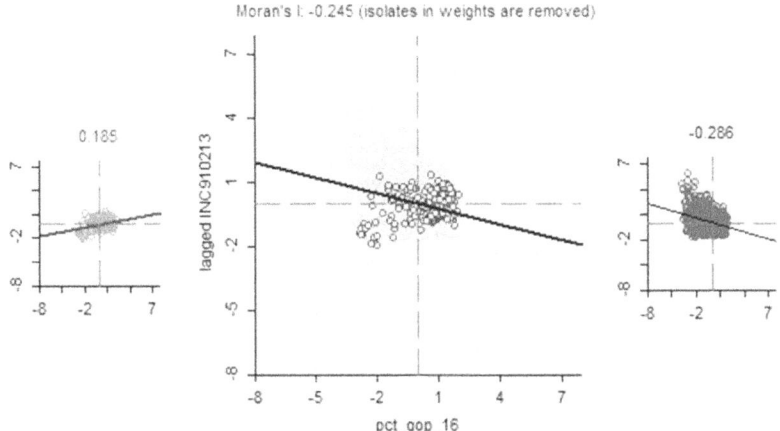

Fig. A10.8. Bivariate Moran's I scatterplot of the spatial autocorrelation of 2016 Republican votes and income levels in Texas and the remaining 49 States

Local Indicators of Spatial Autocorrelation

The Moran's index is used to estimate the spatial autocorrelation for the entire dataset. However, for research purposes, it may be important to weigh the spatial autocorrelation between neighbouring units. To do this, we need to calculate *Local Indicators of Spatial Autocorrelation* (*LISA*). This method allows us to identify four local clusters:

- *high-high* – a cluster of spatial autocorrelations with high indicators of a phenomenon
- *low-low* – a cluster of spatial autocorrelations with low indicators of a phenomenon
- *high-low* – cells in which there is a statistical expectation of a high level of spatial autocorrelation, but this is not observed in practice
- *low-high* – cells in which there is a statistical expectation of a low level of spatial autocorrelation, this is not observed in practice

The LISA formula looks like this:

$$L = \frac{N}{\sum_i \sum_j w_{ij}} \frac{\sum_j w_{ij}(z_{ij}-\underline{z})(z_j - \underline{z})}{\sum_i -(z_j - \underline{z})^2}$$

where N is the number of cells; z_i is the indicator for cell i we are calculating; w_{ij} is an estimate of the spatial weights that reflects whether i and j are neighbours such that if they are not, it is equal to zero, and if they are, it is equal to $\frac{1}{|\delta_i|}$, where δ_i is the number of neighbours in cell i.

To calculate LISA, select Space → Univariate Local Moran's I, set the same spatial weight matrix using the Queen neighbourhood rule, and choose the number of votes for Republican candidates in the 2016 U.S. presidential election (GOPvotes16) as the parameter for analysis.

Geoda generates two chorochromatic maps based on the results of the local indicators of spatial autocorrelation analysis: a significance chorochromatic map and a cluster chorochromatic map. One of them (Fig. A10.9) shows the p-value for various local indicators of spatial autocorrelation. For some states, the correlation with their neighbours is insignificant, while for some central states, for New England and for California it is extremely high (p-value = 0.001).

LISA map outcome also reports positive spatial autocorrelation (high-high and low-low) and negative spatial autocorrelation (high-low and low-high). On the other chorochromatic map (Fig. A10.10) we can see two types of spatial clusters: high-high (an area of high Republican support) and low-low (an area of low Republican support). We have also mapped states with significant deviation: high-low (where high numbers of votes for Republican candidates were statistically likely due to their proximity, but in reality, the number of votes was low), and low-high (where, vice versa, low numbers of votes for Republican candidates were statistically likely due to their proximity, but in reality, the number of votes was high).

Methodological Guide

Fig. A10.9. Significance chorochromatic map of local indicators of spatial autocorrelation

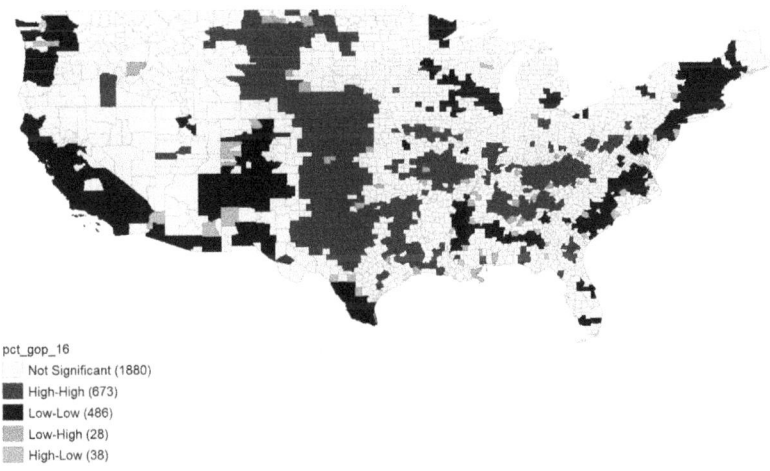

Fig. A10.10. Cluster chorochromatic map of local indicators of spatial autocorrelation

Now let's compare the results on a scatter map, where Republican votes are also divided into six equal zones (Fig. A10.11).

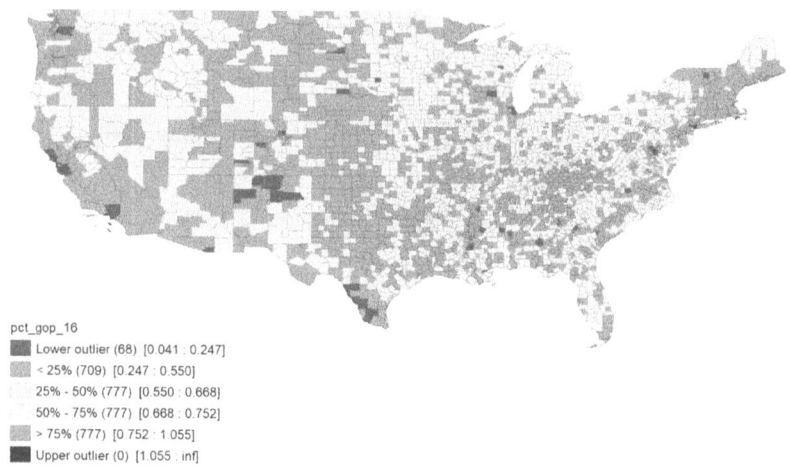

Fig. A10.11. Scatter map of Republican Support in the 2016 U.S. Presidential Elections

Bivariate Local Indicators of Spatial Autocorrelation

Now let's try to identify clusters of spatial autocorrelations between two parameters – votes for Republican candidates and income level. To start off, let's take a look at the scatter map of per capita income over the past 12 months (Fig. A10.12). It is obvious that the clusters here look nothing like the previous scatter map depicting support for the Republicans.

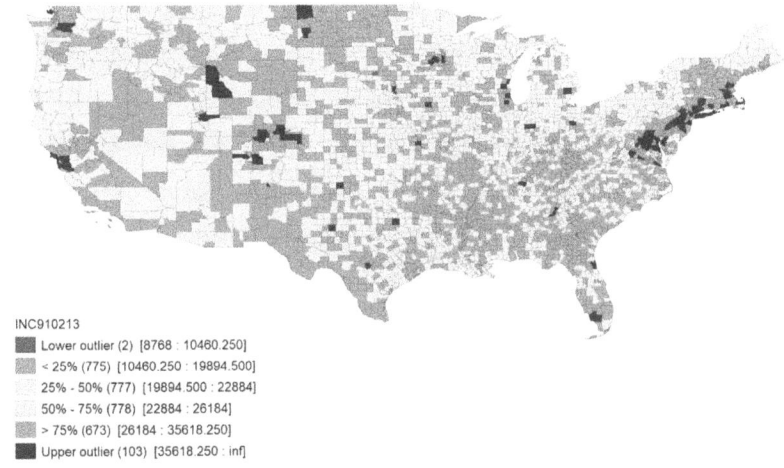

Fig. A10.12. Scatter map of Republican Support in the 2016 U.S. Presidential Election

Our first option in analysing the data is to create a conditional map (Fig. A10.13). To do this, right click Show As Conditional Map on the cluster chorochromatic map we generated in the previous step.

Fig. A10.13. Conditional map of spatial autocorrelation of Republican Support and per capita income

Another option is to select Space → Bivariate Local Moran's I, the two indicators that are of interest to us, and the spatial weights matrix. This will give us a p-value significance chorochromatic map (Fig. A10.14) and a cluster chorochromatic map (Fig. A10.15).

Methodological Guide 367

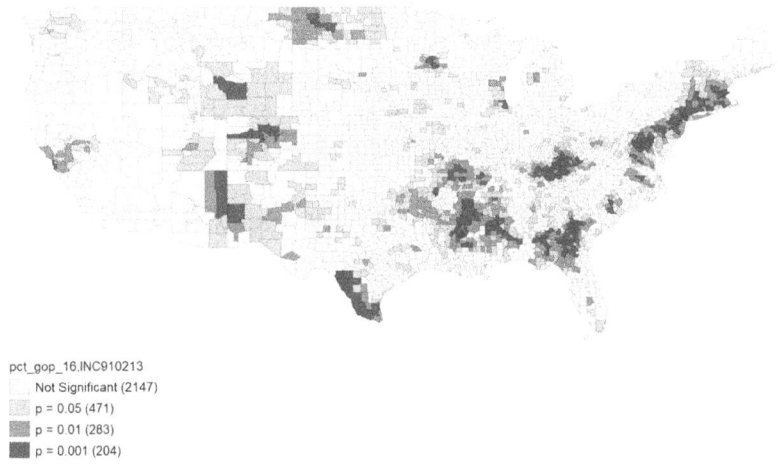

Fig. A10.14. Significance chorochromatic map of bivariate local indicators of spatial autocorrelation

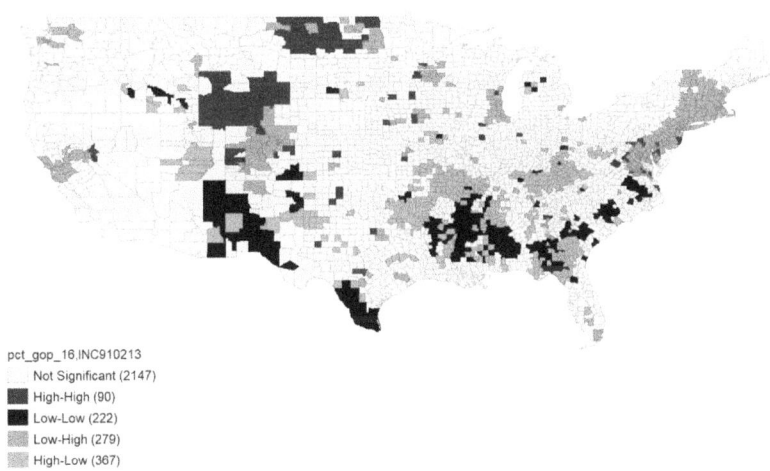

Fig. A10.15. Cluster chorochromatic map of bivariate local indicators of spatial autocorrelation

The cluster chorochromatic map gives us information about two clusters: high-high (high distributions of the phenomena) and low-low (low distributions of the phenomena). We have also mapped districts with significant deviation: high-low (where, due to their proximity, high distributions of the phenomena were statistically likely, but this was not observed in practice) and low-high (where, due to their proximity, low distributions of the phenomena were statistically likely, but this was not observed in practice).

Geary and Getis–Ord Indices of Spatial Autocorrelation

You can also recheck the results of the spatial autocorrelation using alternative global and local Geary and Getis–Ord indices.

Geary's C, as it is called, is similar to Moran's I, the difference is that it is more sensitive to differences in small regions. The value of Geary's C lies within the range of 0 to 2: a value of less than 1 indicates a positive (direct) spatial autocorrelation, and a value of greater than 1 indicates a negative (inverse) spatial autocorrelation. The following formula is used to calculate the global Geary's C index:

$$C = \frac{(n-1)}{2\sum_{i}^{n}\sum_{j}^{n}w_{ij}} \frac{\sum_{i}^{n}\sum_{j}^{n}w_{ij}(x_i - x_j)^2}{\sum_{i}^{n}(x_i - \underline{x})^2},$$

where n is the total number of spatial units,

x_i is the attribute value of the i-th object,

x_j is the attribute value of the j-th object,

$\underline{x}$ is the mean attribute value,

w_{ij} is a matrix of spatial weights between the i-th and j-th objects, and

$\sum_{i}^{n}\sum_{j}^{n}w_{ij}$ is the sum of all spatial weights.

To calculate the local Geary spatial autocorrelation index in GeoDa, select Space → Univariate Local Geary (Fig. A10.16).

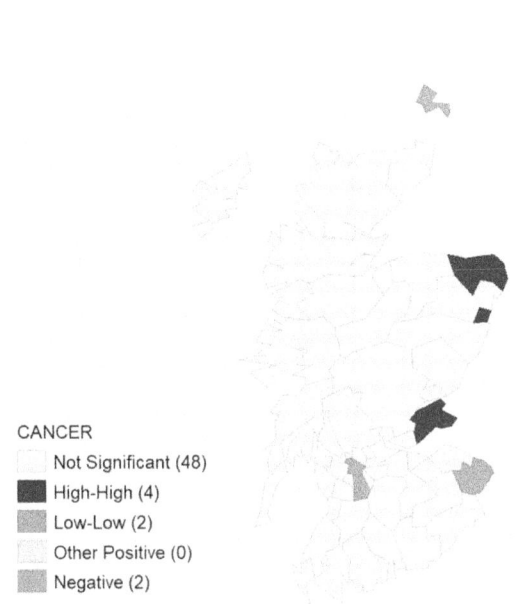

Fig. A10.16. Univariate Local Geary

To analyse cumulative spatial autocorrelation within a group of indicators, you can calculate and map the multivariate local Geary spatial autocorrelation index (using the Multivariate Local Geary function). The following formula is used for the calculation:

$$LG_i = \sum_j w_{ij}\left(x_i - x_j\right)^2.$$

In the resulting chorochromatic maps in Geoda, dark blue indicates clusters of positive autocorrelation (with similarly low statistics), light blue indicated negative spatial autocorrelation (with similarly high statistics), grey indicates clusters with a p-value above the threshold, and black indicates countries for which data is incomplete.

Fig. A10.17. Multivariate Local Geary

Another metric for measuring spatial autocorrelation is the Getis–Ord index, which is best for confirming the existence of local high–high clusters (also called clusters of high values or hot spots) and low–low clusters (clusters of low values or cold spots). In GeoDa, you can calculate the local Getis–Ord index either including or excluding the object of analysis (the country around which the spatial lag is calculated) from the formula. If you want to include the object of analysis in the calculation, select Space → Local G*. If you don't, select Space → Local G. The following formula is used:

$$G_i(d) = \frac{\sum_j w_{ij}(d) x_j}{\sum_{j=1}^{n} x_j}, j \neq i$$

Methodological Guide 371

where d is the estimated range of observed spatial autocorrelation,

$\sum_{j} w_{ij}(d)$ is the sum of weights for the pairs $j \uparrow i$ and distance d,

n is the total number of observations, and

x_j is the attribute value of the j-th object.

The $\forall j \neq i$ constraint does not apply for the G* index.

Build Geary and Getis–Ord (Figs. 10.18 and 10.19) chorochromatic maps of local indicators of spatial autocorrelation for the above example and compare the results with those obtained using Moran's *I*.

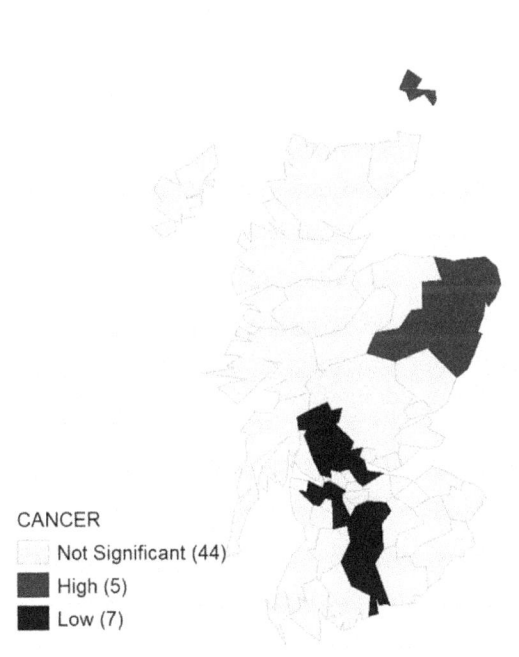

Fig. A10.18. Local G chorochromatic map

Fig. A10.19. Local G* chorochromatic map

Questions and Tasks

1. Explore the alternatives to spatial weights matrices. How will Moran's I and local indicators of spatial autocorrelation change?
2. Compare the spatial autocorrelation indices of support for the Republican Party in 2012 and 2016. Now compare them to the spatial autocorrelation indices of support for the Democratic party for the same years. What conclusions can you draw?
3. Carry out a bivariate spatial autocorrelation analysis of all the other parameters of the U.S. electoral statistics shapefile. Dependence between the spatial distribution of which variables yields the greatest autocorrelation?

Methodological Guide 373

4. Map bivariate local indicators of spatial autocorrelation of all the other parameters of the U.S. electoral statistics shapefile. Dependence between the spatial distribution of which variables gives us stable clusters?
5. What possible explanations are there for high-low and low-high spatial deviations?
6. Give examples of socio-political processes that have inverse spatial autocorrelation.

10. Spatial Autocorrelation Analysis in R

Objectives

✓ Build a spatial weights matrix in R
✓ Plot an adjacency graph
✓ Calculate Moran's I spatial autocorrelation index in R
✓ Calculate and map local indicators of spatial autocorrelation
✓ Perform segregation (multivariate autocorrelation) analysis

We'll be using the same U.S. electoral data from the "election" archive as in previous chapters to carry out spatial autocorrelation in R. Load the packages we need and load the shapefile.

```{r}
library(tidyverse)
library(sf)
library(sp)
library(spdep)
usa <- read_sf("County_election_2012_16.shp")
```

We'll use the following code to generate an adjacency matrix according to the Rook and Queen neighbourhood rules:

```{r}
usa_rook <- poly2nb(usa %>% as("Spatial"), queen = F)
```

```
usa_rook
```

```
Neighbour list object:
Number of regions: 3108
Number of nonzero links: 17472
Percentage nonzero weights: 0.1808759
Average number of links: 5.621622
3 regions with no links:
1268 2730 2764
```

````
```{r}
usa_queen <- poly2nb(usa %>% as("Spatial"), queen = T)

usa_queen
```
````

```
Neighbour list object:
Number of regions: 3108
Number of nonzero links: 18148
Percentage nonzero weights: 0.187874
Average number of links: 5.839125
3 regions with no links:
1268 2730 2764
```

Now we'll plot a matrix using the *k*-nearest neighbours algorithm with $k = 6$.

````
```{r}
usa_metrix <- usa %>% as("Spatial") %>% coordinates()
%>% knearneigh(6) %>% knn2nb()
usa_metrix
```
````

```
Neighbour list object:
Number of regions: 3108
Number of nonzero links: 18648
Percentage nonzero weights: 0.1930502
```

```
Average number of links: 6
Non-symmetric neighbours list
```

The next step is to calculate the spatial weights matrix, say, using the *k*-nearest neighbours method, with *k*=6.

```{r}
usa_metrix_weights <- nb2listw(usa_met-
rix, zero.
policy=TRUE)

usa_metrix_weights
```

```
Characteristics of weights list object:
Neighbour list object:
Number of regions: 3108
Number of nonzero links: 18648
Percentage nonzero weights: 0.1930502
Average number of links: 6
Non-symmetric neighbours list

Weights style: W
Weights constants summary:
```

To map the adjacency graph, we will perform the following operation, opening the shapefile with the `readOGR` function.

```{r}
library(rgdal)
usa.nb <- readOGR(dsn=".",layer="County_
election_2012_16")

plot(usa.nb,border=gray(.5))
plot(usa_metrix,coordinates(usa.
nb),col="blue",add=TRUE)
```

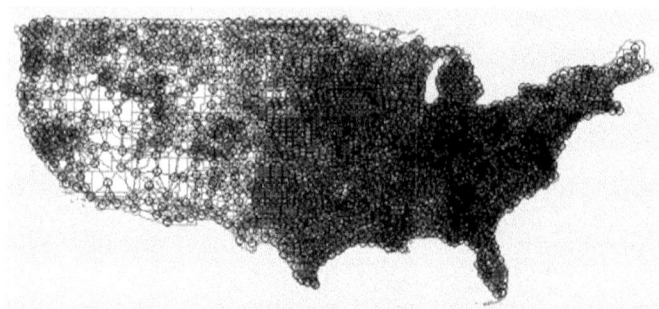

Now let's calculate Moran's I for the percentage of votes cast for the 2016 Democratic Party presidential candidate (pct_dem_16). The resulting data contains information about *Expectation* – the mathematical expectation of the index under the null hypothesis; *Variance* – the variability of the expected value under the null hypothesis; *Moran I statistic standard deviate* – the z-score of the Moran's index that has been calculated; and *p-value* – the p=value of the Moran's index that has been calculated. The Moran's Index (0.62) with an insignificant p-values indicates that there is a positive spatial autocorrelation for this indicator.

```
```{r}
usa %>% pull(pct_dem_16) %>% moran.test(usa_metrix_weights)
```

Moran I test under randomisation
data: .
weights: usa_metrix_weights

Moran I statistic standard deviate = 61.364, p-value < 2.2e-16
alternative hypothesis: greater
sample estimates:
Moran I statistic Expectation Variance
    0.6151143620 -0.0003218539 0.0001005858
```

Let's build a Moran scatterplot for the indicator.

```
```{r}
moran.plot(usa$pct_dem_16, usa_metrix_weights)
```
```

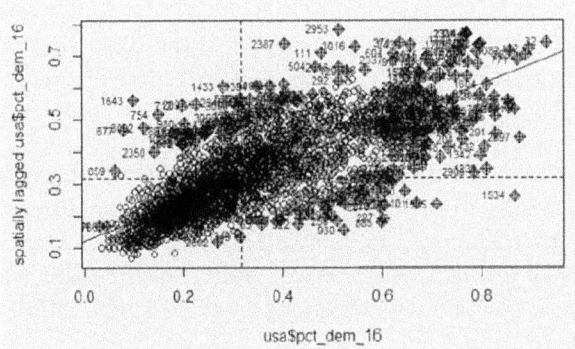

You can also calculate Geary's C using the `geary.test` command, for example:

`World$7ArLand %>% geary.test(weights)`

Now you can try to calculate and map local indicators of spatial autocorrelation for the same indicator using the `rgeoda` package. You can read more on geoda functions available in R here: https://geodacenter.github.io/rgeoda/. Local Getis-Ord's G statistics can also be calculated using this package (`local _ g()` command).

R also allows you to carry out a multivariate spatial autocorrelation analysis, that is, it allows you to find clusters with several parameters that stand out at the same time. For example, a ghetto will stand out in terms of housing prices and the intensity of ties with other areas. This kind of spatial autocorrelation analysis is called "segregation index," as it lets you estimate the level of segregation of certain social strata from others.

Spatial segregation analysis thus makes it possible to assess the degree to which the characteristics of one cell differ from those of an entire country or region. The segregation index is calculated using the following formula:

$$D = \sum_n \sum_i \frac{N_n}{2NI} |t_{ni} - t_i|$$

where N is the total population; i is the ethnic/social stratum we are studying; n are the cells adjacent to it; t_i is the percentage of the total population; t_{ni} is the standard presence of stratum i in the neighbourhood with n; and $I = \sum_i (t_i)(1-t_i)$.

For our segregation index, we will use data on the Brazilian state of Roraima available in the "RO_race" archive on original source: https://jonnyphillips.github.io/Intro_Spatial/Day_4/Data/RO_race.zip. We'll have to install the seg application and run the downloaded shapefile. As you can see, it contains data on the spatial distribution of the population in the state by race.

```{r}
library(seg)
Segregation <- read_sf("RO_race.shp")
Segregation %>% plot()
```

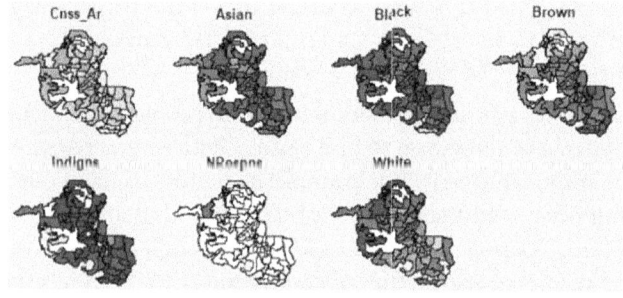

The next step is to calculate the segregation index, which is 0.0283.

```{r}
Segregation_data <- Segregation %>% 
  dplyr::select(-Cnss_Ar)   %>%   st_set_geometry(NULL)
Segregation_SP <- spseg(x = (Segregation %>% as("Spatial")), data = Segregation_data)
Segregation_SP
```

```
Reardon and O'Sullivan's spatial segregation
measures
Dissimilarity (D) : 0.0283
Relative diversity (R): 0.0029
Information theory (H): 0.0019
Exposure/Isolation (P):
Asian Black Brown Indigns NRespns White
Asian 0.01061310 0.07475775 0.6460878 0.007387445
0.01825601 0.2428979
Black 0.01045961 0.07508068 0.6450300 0.007473957
0.01855884 0.2433969
Brown 0.01043888 0.07448599 0.6481331 0.007672142
0.01925227 0.2400176
Indigns 0.01017173 0.07354746 0.6539233
0.008261193 0.02131349 0.2327828
NRespns 0.01005805 0.07307799 0.6567437
0.008532341 0.02226229 0.2293257
White 0.01051998 0.07534640 0.6434488 0.007325170
0.01803820 0.2453215
--
The exposure/isolation matrix should be read
horizontally.
Read 'help(spseg)' for more details.
```

Now let's map our result. We can see that the further south you move in the state, the higher the degree of segregation of racial groups.

```
```{r}
Segregation_map <- Segregation %>% gather(key =
"Race", value = "Pct_Race", -Cnss_Ar, -geometry)
Segregation_map['Pct_Race'] %>% plot()
```
```

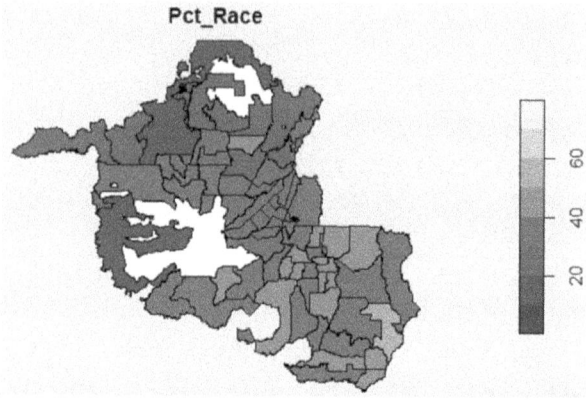

The algorithm for calculating the segregation index thus consists of three steps: opening the necessary packages and shapefile, calculating the segregation index, and mapping it.

```{r}
library(seg)

Segregation <- read_sf("RO_race.shp")
```

```{r}
Segregation_data <- Segregation %>%
dplyr::select(-Cnss_Ar)    %>%    st_set_geometry(NULL)
Segregation_SP <- spseg(x = (Segregation %>%
as("Spatial")), data = Segregation_data)
Segregation_SP
```

```{r}
Segregation_map <- Segregation %>% gather(key = "Race", value = "Pct_Race", -Cnss_Ar, -geometry)
Segregation_map['Pct_Race'] %>% plot()
```

Questions and Tasks

1. Create adjacency graphs for the Russia and France shapefiles we analysed earlier. What do you think your selection of spatial weights matrix will depend on?
2. Consider alternatives to spatial weights matrices. Calculate Moran's I and local indicators of spatial autocorrelation for each.
3. Compare the spatial autocorrelation indices of support for the Republican Party in 2012 and 2016. Now compare them to the spatial autocorrelation indices of support for the Democratic party for the same years. What conclusions can you draw?
4. What possible explanations are there for high-low and low-high spatial deviations?
5. Calculate the segregation index for the U.S. demographics data contained in the shapefile we have already analysed.

11. Spatial Cluster Analysis

Objectives

✓ Reduce the dimension of variables through principal component analysis and multidimensional scaling
✓ Estimate bivariate local indicators of spatial autocorrelation of principal components
✓ Create a spatial weights matrix based on multidimensional scaling
✓ Map statistical clusters calculated with account of geometric centroids
✓ Build a dendrogram with account of geometric centroids

The Principal Component Method

In this chapter, we need to make a distinction between the concepts of "spatial" and "statistical" clusters. A spatial cluster is a localized sample population, it shows up through the calculation of local indicators of spatial autocorrelation (a task we performed earlier). A statistical cluster is not necessarily a localized sample population. In statistical cluster analysis, the entire sample is divided into separate groups called statistical clusters.

The first step in cluster analysis is to identify the factors that will be used to divide the population into clusters. This can be done either logically or mathematically. The purpose of these methods is to reduce dimensionality, i.e. they are suitable for situations where there are many independent variables. We will look at two such methods: principal component analysis and multidimensional scaling

Principal component analysis (PCA) is the definition of new features, which are linear combinations of the variables being studied and which essentially "absorb" most of their variability and thus convey most of the information about them.

In this lab, we will use data on crime statistics in the southern U.S. states for the 1960s–1990s available in the "south" archive on our Google Drive (https://drive.google.com/file/d/1j5NVYi6UuBXEEm3jh XxyRAUelypn6aK9/view?usp=drive_link) and website (https://mgimo.ru/upload/2023/05/spatial-cluster-analysis.zip). Original source: https://geodacenter.github.io/data-and-lab//south/. Original source: https://geodacenter.github.io/data-and-lab//south/. Open the shapefile in GeoDa.

We'll calculate the principal components for the following indicators for1990 from the database: homicides per 100,000 people (HR90), population (PO90), unemployment rate (UE90), divorce rate (DV90), median age (MA90), number of African Americans (BLK90) and the Gini index (GI89). To do this, select Clusters → PCA, then choose the variables you need on the left and start the calculation. The results will be displayed in the field on the right. Save the indicators as separate columns in the attribute table.

```
---

PCA method: svd

Standard deviation:
1.522672 1.134305 1.066838 0.917720 0.820289
0.659754 0.553466
 Proportion of variance:
0.331218 0.183807 0.162592 0.120316 0.096125
0.062182 0.043761
Cumulative proportion:
```

```
0.331218 0.515025 0.677617 0.797933 0.894057
0.956239 1.000000
Kaiser criterion: 3.000000
95% threshold criterion: 5.000000

Eigenvalues:
 2.31853
 1.28665
 1.13814
 0.84221
0.672874
0.435275
0.306325

Variable Loadings:
 PC1 PC2 PC3 PC4 PC5 PC6 PC7
HR90 0.430269 0.396496 0.161889 -0.39595
0.0740999 -0.677622 -0.100061
PO90 0.00156157 0.688617 0.0398353
0.501666 0.492638 0.168569 0.0379591
UE90 0.457565 -0.338453 0.288758 0.40659
0.0747065 0.0278924 -0.648868
DV90 -  0.128919 0.279455 0.746311 0.0508258
-0.561927 0.158559 0.0694139
MA90 -0.340758 0.153847 0.467564 -0.465374
0.607478 0.168244 -0.166408
BLK90 0.462791 0.253881 -0.189503
-0.448526 -0.101938 0.670868 -0.154361
GI89 0.508517 -0.296358 0.278058 0.0768652
0.225952 0.0901333 0.714969

Squared correlations:
 PC1 PC2 PC3 PC4 PC5 PC6 PC7
HR90 0.429233 0.202272 0.0298283 0.132039
0.0036946 0.199866 0.003067
PO90 5.65413e-006 0.61012 0.00180608 0.211958
0.163301 0.0123686 000441381
```

```
UE90  0.485419  0.147385  0.0948995  0.13923
0.00375534  0.00033864  0.128972
DV90  0.0385342  0.100481  0.633922  0.00217565
0.212468  0.0109433  0.0014759
MA90  0.269218  0.0304534  0.248816  0.182399
0.248309  0.0123209  0.00848258
BLK90 0.496572  0.0829315  0.0408722  0.169432
0.00699205  0.195902  0.00729886
GI89  0.599548  0.113003  0.0879967  0.00497598
0.0343531  0.0035362  0.156587
```

The program has calculated seven components that describe our population. However, the principal components for the purpose of our study are those whose value (eigenvalues) exceeds 1 (the *Kaiser criterion*), since the sum of the values of the components will be equal to the sum of the values themselves, in our case seven. In our case, the first three components will be the principal ones, and the explanatory power of the first component (2.32) will be significantly higher than the power of the second and third components (1.29 and 1.14, respectively). The proportion of variance gives us an idea of the percentage of change each of our principal components accumulates – 33, 18 and 16, respectively, or 67 % in total (approximately two thirds). To understand how many criteria are needed to explain the maximum number of changes, we can use another criterion when we are deciding on the number of principal components; assuming the criterion of 95 % of explanations will give us five components.

Our task now is to try and understand exactly what our components denote. To do this, we need to look at the columns beneath the graph, which describe the correlation coefficient of each component with the original variables. As we can see, our second variable is closest to the "population" indicator, the third is closest to "divorce rate," and the first accumulates the remaining indicators. We'll plot the first two components on a scatter plot to make sure there is no correlation between them (Fig. A12.1).

Now let's create a chorochromatic map with four classes for the first principal component to visually analyse the territorial pattern of its distribution (Fig. A12.2).

Our next step after creating a spatial weights matrix (say, using the k-nearest neighbours method, with $k = 6$) is to calculate and map local indicators of spatial autocorrelation for the first principal component in order to identify those spatial clusters where it has a high explanatory power and those where it does not (Fig. A12.3).

To finish up, we will analyse bivariate local indicators of spatial autocorrelation of the first and second principal components in order to identify spatial clusters where they cumulatively have explanatory power (Fig. A12.4).

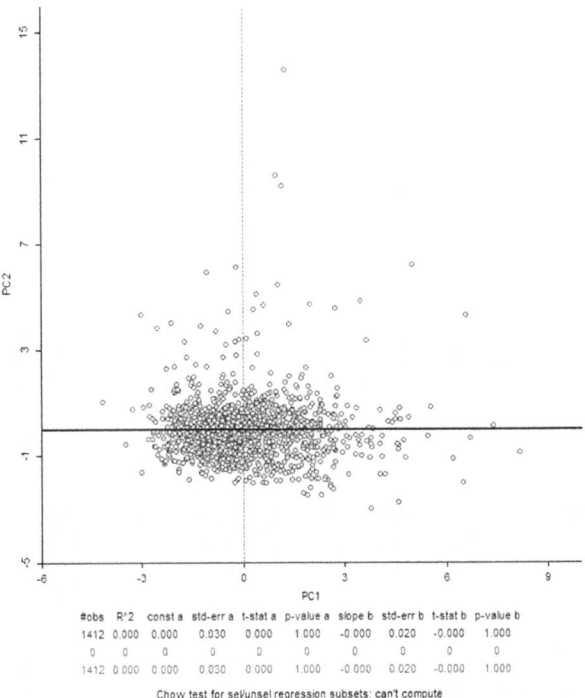

Fig. A12.1. Scatter plot of two principal components

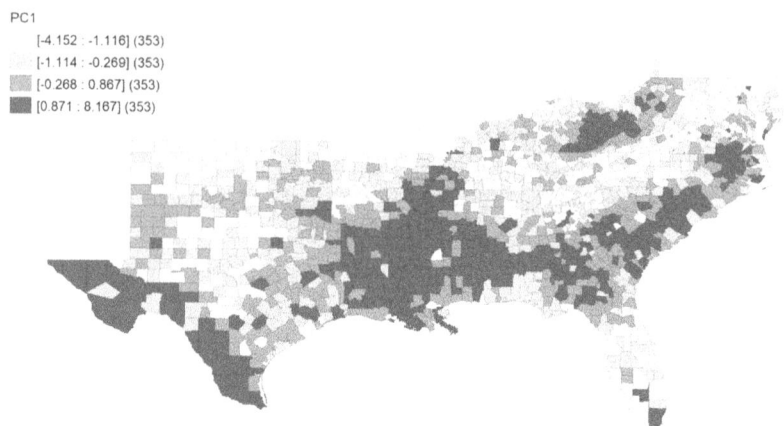

Fig. A12.2. Chorochromatic map of the first principal component

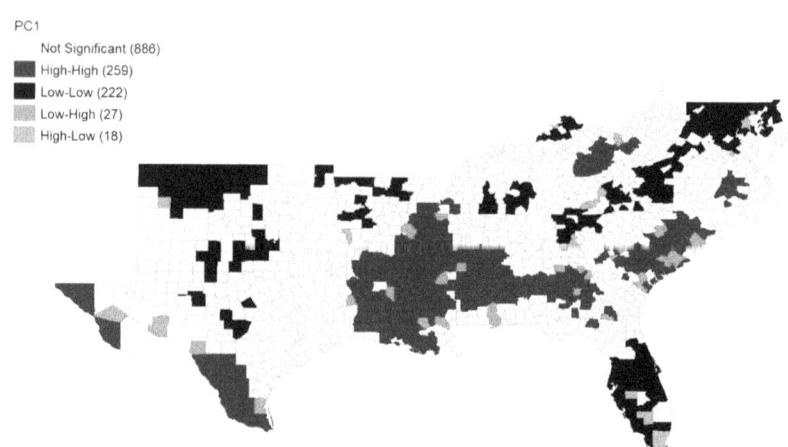

Fig. A12.3. Local indicators of spatial autocorrelation of the first principal component

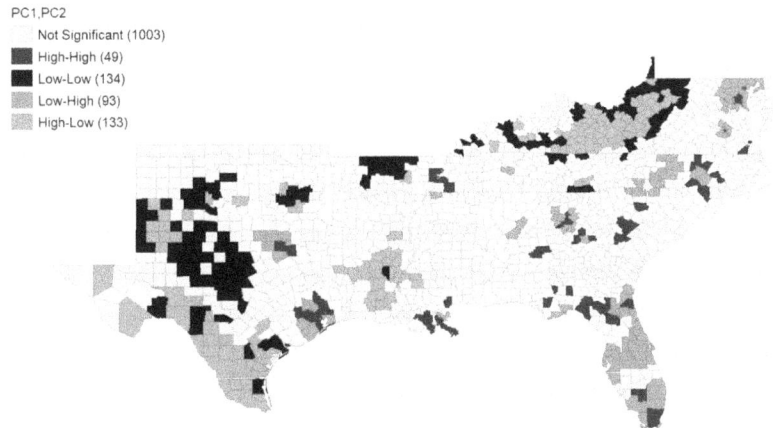

Fig. A12.4. Bivariate local indicators of spatial autocorrelation of the first and second principal components

Multidimensional scaling

Another possible dimensionality reduction technique is multidimensional scaling (MDS). This involves showing the differences between objects in the form of a two-dimensional graph. In other words, the closer two points are on a graph, the more similar they are.

In Euclidean (geographical) metric, the distance between observations x_i and x_j for p variables in p-dimensional space is defined as

$$d_{ij} = \| x_i - x_j \| = \sum_{k=1}^{p} \left(x_{ik} - x_{jk} \right)^2$$

In the Manhattan metric, where the distance between two points is equal to the sum of the modules of their coordinate differences, the distance is defined as

$$d_{ij} = \sum_{k=1}^{p} \left| x_{ik} - x_{jk} \right|$$

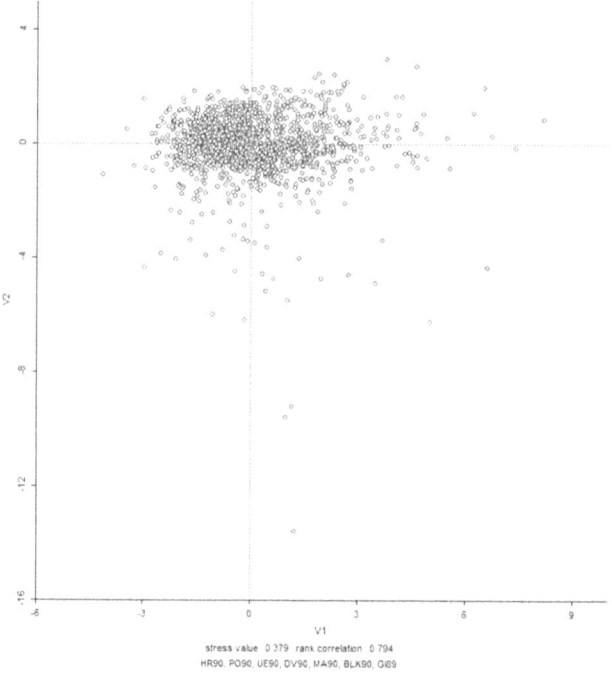

Fig. A12.5. Scatter plot of multidimensional scaling indicators

The task of multidimensional scaling in this case is to find points z_1, z_2, ..., z_n in two-dimensional space, which will correspond to the distance between them in multidimensional space as accurately as possible. This is done using the function

$$S(z) = \sum_i \sum_j \left(d_{ij} - \| z_i - z_j \| \right)^2$$

To carry out multidimensional scaling in GeoDa, select Clusters → MDS, choose the same parameters that we used in the principal components analysis, and save the two resulting variables. We'll create a scatter plot based on these two parameters (Fig. A12.5).

An interesting spatial application of multidimensional scaling is that it allows you to create a spatial weights matrix based not on physical proximity, but on analytic proximity – that is, it presents the data in terms of multidimensional scaling. To do this, right click on the scatter

Methodological Guide 389

plot we have just created, select create spatial weights matrix, then set the variable name and adjacency type in the window that pops up.

Figure A12.6 shows the Washington adjacency graph calculated on the basis of a spatial weights matrix using the Queen neighbourhood rule derived from multivariate scaling. We can see that the counties with crime rates that are most similar to those of Washington are not located closest to it geographically.

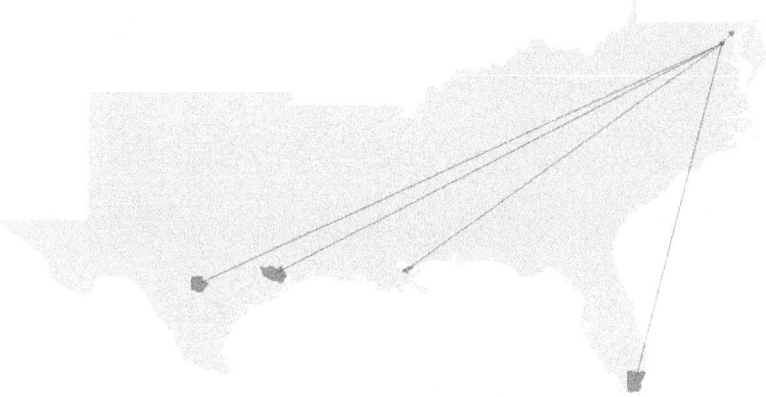

Fig. A12.6. Washington adjacency graph based on multivariate scaling

k-Means Clustering

k-means clustering divides the entire set of objects according to the given variables into the number of statistical clusters specified by the researcher so that the average values for the clusters differ as much as possible for each of the variables.

To perform cluster analysis in GeoDa, select Clusters → K-Means, set the same parameters from the statistics of the southern United States we analysed earlier, and instruct the program to divide the data into five clusters. This will give us a calculation of the indicators and a chorochromatic map of clusters (Fig. A12.7). We could also take the principal components or multivariate scaling variables we calculated earlier.

Method: KMeans

```
Number of clusters: 5
Initialization method: KMeans++
Initialization re-runs: 150
Maximum iterations: 1000
Transformation: Standardize (Z)
Distance function: Euclidean
Cluster centers:
|  |HR90    |PO90     |UE90     |MA90    |BLK90    |GI89     |
|--|--------|---------|---------|--------|---------|---------|
|C1|-0.355108|0.0164058|-0.655911|-0.374367|-0.270286|-0.807039 |
|C2|-0.400201|-0.207273|-0.173737|1.02858 |-0.621908|-0.0671793|
|C3|0.954409 |-0.130423|0.36325  |-0.499628|1.34748 |0.680908 |
|C4|-0.167246|-0.211003|1.58305  |-0.445108|-0.529936|1.08996  |
|C5|1.10682  |5.44676  |-0.419974|-0.345402|0.38528 |-0.455118 |

The total sum of squares: 8466
Within-cluster sum of squares:
|  |Within cluster S.S.|
|--|-------------------|
|C1|986.839 |
|C2|932.818 |
|C3|1281.37 |
|C4|583.495 |
|C5|484.772 |

The total within-cluster sum of squares: 4269.3
The between-cluster sum of squares: 4196.7
The ratio of between to total sum of squares: 0.495712
```

Methodological Guide

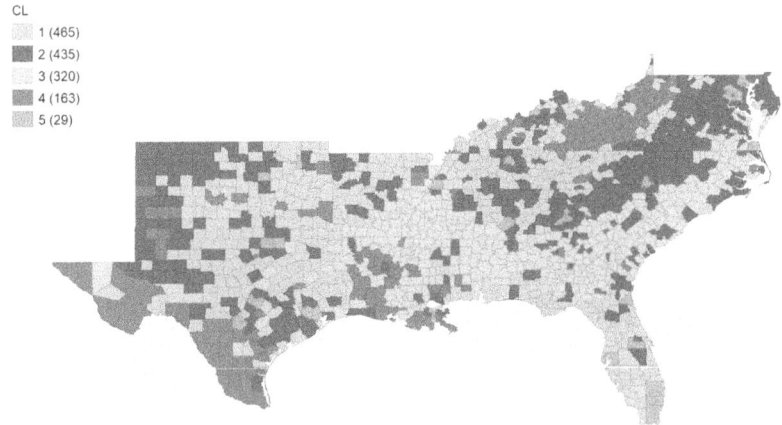

Fig. A12.7. *k*-Means cluster analysis

As the chart shows, the resulting clusters do not reflect the spatial concentration of the phenomenon. This can be avoided by checking the box next to Use geometric centroids. This is because *k*-means cluster analysis necessarily involves the calculation of centroids – an abstract entity representing a typical representative of the cluster – for input indicators. Adding geometric centroids with a weight of 1 to the calculation will mean that the cluster analysis will ignore all indicators that are not geometric centroids – we will be left with a chorochromatic map with clusters that are not related to the indicators we are analysing (Fig. A12.8).

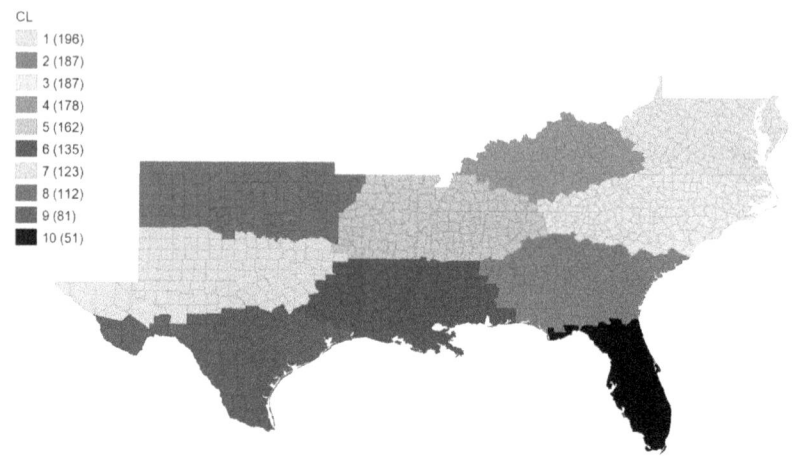

Fig. A12.8. *k*-Means cluster analysis based on geometric centroids

```
------
Method: KMeans
Number of clusters: 10
Initialization method: KMeans++
Initialization re-runs: 150
Maximum iterations: 1000
Transformation: Standardize (Z)
Distance function: Euclidean
Cluster centers:
|   |HR90     |PO90      |UE90     |MA90     |BLK90      |GI89      |
|---|---------|----------|---------|---------|-----------|----------|
|C1 |-0.265622|0.0593503 |-0.270023|0.23591  |0.00865551 |-0.723444 |
|C2 |0.514566 |-0.115185 |-0.104533|-0.518431|0.951052   |0.187771  |
|C3 |0.144451 |0.0148992 |-0.492501|0.0778701|0.13377    |-0.489476 |
|C4 |-0.361488|-0.158817 |0.460467 |0.0224999|-0.734039  |0.174035  |
```

```
|C5  |0.0853657|-0.128272 |0.210605
|0.0452145|0.253826 |0.136193 |
|C6  |0.382076  |-0.0403079|0.854282
|-0.538698|0.795113 |0.812601 |
|C7  |-0.280757|0.00159706|-0.302066|0.167406
|-0.602937 |0.113036 |
|C8  |-0.495594|-0.165156 |-0.257092|0.346711
|-0.753259 |-0.0888331|
|C9  |0.0592566|0.279199 |0.252628 |-0.268395|-
0.603422 |0.676409 |
|C10|0.254918 |1.12349 |-0.418705|1.1737
|-0.293255 |-0.456318 |
```

The total sum of squares: 8466
Within-cluster sum of squares:

	Within cluster S.S.
C1	1209.65
C2	733.233
C3	585.836
C4	647.946
C5	807.511
C6	645.611
C7	676.706
C8	317.271
C9	866.51
C10	498.434

The total within-cluster sum of squares: 6988.71
The between-cluster sum of squares: 1477.29
The ratio of between to total sum of squares: 0.174497

If, however, we set the weight of geometric centroids to, say, 0.5 in relation to the other data, then we will get a mixed picture, with the resulting clusters having both a spatial and a statistical variable (Fig. A12.9).

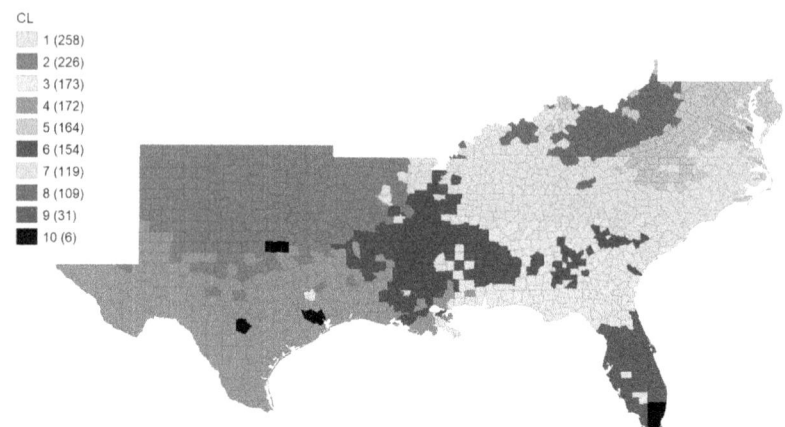

Fig. A12.9. *k*-Means cluster analysis taking geometric centroids into account

```
------
Method: KMeans
Number of clusters: 10
Initialization method: KMeans++
Initialization re-runs: 150
Maximum iterations: 1000
Transformation: Standardize (Z)
Distance function: Euclidean
Cluster centers:
|   |HR90      |PO90      |UE90     |MA90     |BLK90   |GI89      |
|---|----------|----------|---------|---------|--------|----------|
|C1 |-0.395368 |-0.133923 |-0.275666|0.240403 |-0.608086 |-0.434628 |
|C2 |-0.470121 |-0.197107 |-0.28802 |0.588709 |-0.67312  |-0.0627084|
|C3 |0.450911  |-0.0645901|-0.221208|-0.503458|0.548568  |-0.0322181|
|C4 |-0.0321908|-0.0554961|0.273049 |-0.435826|-0.368107 |0.520465  |
|C5 |-0.516523 |0.0902608 |-0.917959|0.157386
```

```
|-0.274637 |-1.22211 |
|C6 |0.949778 |-0.197046 |1.17729
|-0.703209|1.83766 |1.39534 |
|C7 |0.804247 |0.317997 |-0.288767|-
0.158945|1.26065 |-0.120192 |
|C8 |-0.24302 |-0.244336 |1.69052 |0.0675769|-
0.825423 |0.797434 |
|C9 |-0.0975169|1.08562 |-0.493967|1.93964
|-0.417262 |-0.555081 |
|C10|1.35168 |10.6934 |-0.154748|-0.507819|-
0.0529265|-0.0129437|

The total sum of squares: 8466
Within-cluster sum of squares:
| |Within cluster S.S.|
|---|-------------------|
|C1 |450.71 |
|C2 |730.378 |
|C3 |479.95 |
|C4 |746.948 |
|C5 |447.014 |
|C6 |555.566 |
|C7 |532.452 |
|C8 |298.5 |
|C9 |187.924 |
|C10|97.7154 |

The total within-cluster sum of squares: 4527.16
The between-cluster sum of squares: 3938.84
The ratio of between to total sum of squares: 0.465254
```

Hierarchical Clustering Method

k-means clustering is limited inasmuch as the number of clusters is determined by the researcher. We can overcome this drawback by using the hierarchical clustering method, which allows you to explore the structure of the differences between objects and choose the optimal number of clusters. With hierarchical clustering, all the possible differences between objects are calculated and the two most similar ones are combined into a cluster. Then, all the differences between the remaining objects and the centroid of the newly formed cluster are calculated again, and so on, until the entire set is divided into clusters. Using this method, we can create a dendrogram, which you can then use to determine the number of clusters you need. Hierarchical clustering also allows you to include data on the geometric location of centroids in the analysis.

To perform hierarchical clustering in GeoDa, select Clusters → Hierarchical. The results of our analysis of the same population using the hierarchical clustering method are given below.

```
------
Number of clusters: 5
Transformation: Standardize (Z)
Method: Ward's-linkage
Distance function: Euclidean
Cluster centers:
|  |HR90    |PO90      |UE90      |DV90      |MA90    |BLK90    |GI89    |
|--|--------|----------|----------|----------|--------|---------|--------|
|C1|-0.418412|-0.0212725|-0.542825|0.121594
|0.417875  |-0.507175|-0.571796|
|C2|-0.325759|-0.192452 |0.853611 |-0.344841|-
0.462496  |-0.489107|0.56984  |
|C3|0.812949 |-0.158392 |-0.178157|0.371982
|-0.0865803|0.576899 |0.0500174|
|C4|1.05044  |5.62763  |-0.320484|1.03297
|-0.398421 |0.375365 |-0.210469|
|C5|0.834766 |-0.164982 |0.845145 |-0.518017|-
0.583734  |1.85592  |1.10465  |

The total sum of squares: 9877
Within-cluster sum of squares:
|  |Within cluster S.S.|
```

```
|--|------------------|
|C1|2240.68  |
|C2|1478.32  |
|C3|1260.13  |
|C4|406.793  |
|C5|781.94   |

The total within-cluster sum of squares: 6167.86
The between-cluster sum of squares: 3709.14
The ratio of between to total sum of squares:
0.375533
```

Spatial Cluster Analysis

The goal of conventional statistical cluster analysis is to divide a set of observations into statistical clusters based on their similarity. Spatial cluster analysis aims to divide a set of observations into statistical clusters based on their similarity while at the same time grouping them into spatially continuous regions. In this sense, spatial cluster analysis is a mathematical expression of the traditional geographical regionalization method.

This means we are dealing with zoning – the combining of multi-dimensional data into statistical clusters while respecting spatial constraints. The resulting clusters (regions) should: (1) be as different from each other as possible; (2) contain elements that are as similar to each other as possible; and (3) include objects that are geographically located next to each other. Spatially bounded clustering can be soft (when geographic coordinates are introduced into the feature set or, as shown above, the weight of geographic centroids in the feature set changes), or hard (when it is impossible to create a cluster that combines observations of entities that are not neighbours). For the latter, the program needs to know which neighbourhood principle to use in the analysis, so you need to specify the spatial neighbourhood weights you want.

GeoDa includes a number of spatial cluster analysis methods:

1. Density-Based Spatial Clustering of Applications with Noise (DBSCAN)

2. Hierarchical Density-Based Spatial Clustering of Applications with Noise (HDBSCAN)
3. Spatially Constrained Hierarchical Clustering (SCHC)
4. Spatial 'K'luster Analysis by Tree Edge Removal (SKATER)
5. Regionalization with Dynamically Constrained Agglomerative Clustering and Partitioning (REDCAP)
6. Automatic Zoning Procedure (AZP)
7. Max-p Regions Model (max-p).

The first two are used exclusively to analyse point objects (or centroids of polygons) based on the density of their placement and the identification of noise points (outliers). Methods 3–5 are used to analyse all objects, and the algorithm assumes their hierarchical clustering with the spatial restraint that only neighbouring observations can fall into a single cluster. The REDCAP model is preferable, as it combines the procedures contained in the SCHS and SKATER models in its algorithm. Finally, methods 6 and 7 are not based on hierarchical clustering, so instead of looking for similarities between observations (with spatial constraints), they carry out zoning based on differences between regions. Note that the max-p model is the only one where you do not need to specify the desired number of clusters, only the maximum number of observations in a given cluster, which apparently makes it the most progressive for research in the social sciences.

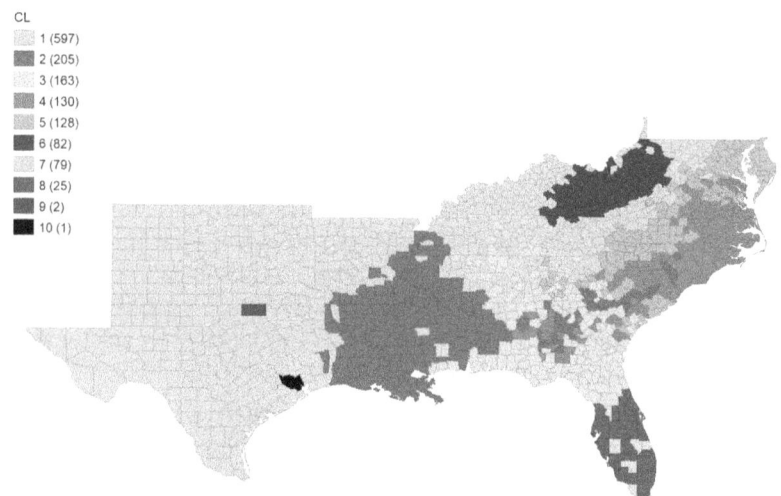

Fig. A12.10. REDCAP analysis

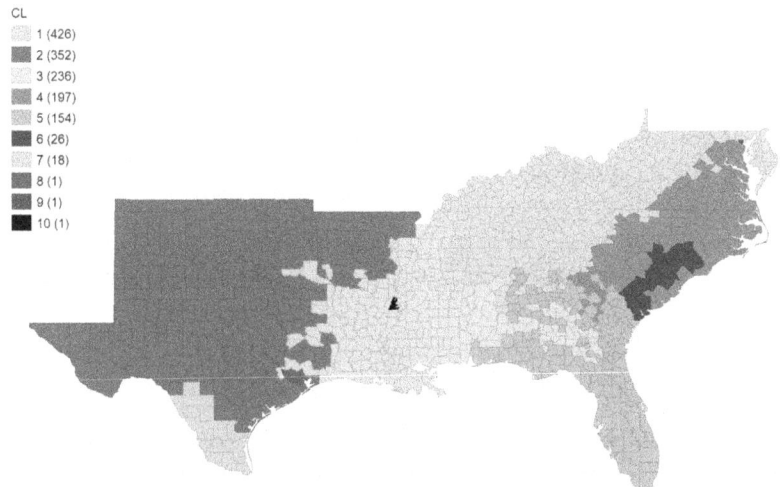

Fig. A12.11. AZP analysis

Perform clustering based on the parameters we used earlier, but with the spatial constraint that regions should be continuous. Create the optimal neighbourhood matrix for this data. Compare the resulting regions with the statistical clusters we obtained previously (Figs. 12.10 and 12.11).

Questions and Tasks

1. Select a different set of indicators from the shapefile we have been using, calculate the principal components and perform multivariate scaling and spatial cluster analysis.
2. Map bivariate local indicators of spatial autocorrelation for principal component pairs we have not yet analysed and interpret the results.
3. Compare the analysis results in Euclidean and Manhattan metrics. What might the difference between them suggest?
4. Compare the results of k-means cluster spatial analysis with differently weighted geometric centroids.
5. Analyse the resulting dendrogram. What options for selecting the number of clusters are the most useful?

12. Spatial Regression

Tasks:

- Calculate a valid spatial regression model
- Use the Lagrange multiplier to determine the appropriate spatial regression model: spatial lag (SAR) or spatial error (SEM)
- Interpret the results of spatial regression analysis and learn how to compare the spatial lag regression model (SAR) and the spatial error regression model (SEM)
- Calculate geographic regression discontinuity in R
- Plot a geographic regression discontinuity graph

Regression is the statistical dependence of the mean value of a random variable on the values of another random variable (or several random variables). The formula for a simple bivariate linear regression model is as follows:

$$y = \alpha + \beta x + \varepsilon,$$

where y is the dependent variable,

x is the independent variable,

α is a constant term, the point of intersection with the Y-axis, i.e. the value of the dependent variable when the independent variable equals 0,

β determines the slope of the regression line, i.e. the expected increment in the dependent variable when the independent variable changes by 1, and

ε is the error term (residual, discrepancy) of the model, i.e. the random deviation from a deterministic function.

Multiple linear regression is calculated using the formula:

$$y_i = \beta_0 + \beta_1 x_{i1} + \beta_2 x_{i2} + \ldots + \beta_m x_{im} + \varepsilon_i$$

where $i = 1, \ldots, n$,

y_i is the i-th value of the independent variable,

n is the total number of observations,

Methodological Guide

m is the total number of independent variables,

x_{im} is the i-th observation of the m-th independent variable,

β_1, β_2, …, β_m are coefficients,

β_0 is the intersection point that determines the value of the equation when all the variables and coefficients equal zero, and

ε_i is the i-th error.

Spatial regression (GWR) is needed when we need to explain certain indicators using other indicators with due account of the spatial effect. In other words, GWR tells us the degree to which the spatial effect strengthens or weakens the regression. The goal of GWR, as well as of linear regression in general, is to find a model that reliably explains the dependent variable while taking the neighbourhood effect into account.

In general terms, the GWR formula looks like this:

$$y_i = \beta_{0i} + \beta_{1i} x_{1i} + \beta_{2i} x_{2i} + \ldots + \beta_{mi} x_{mi} + \varepsilon_i,$$

where $i = 1, \ldots, n$,

n is the total number of locations or observations,

y_i is the value of the dependent variable Y at the i-th location,

x_{mi} is the observed value of variable X_m at the i-th location,

β_{0i} is the intersection point in the model at the i-th location,

$\beta_{1i}, \beta_{2i}, \ldots, \beta_{mi}$ are coefficients at the i-th location,

m is the total number of independent variables, and

ε_i is the error at the i-th location.

The coefficient β_i at each location i in GWR is estimated using the following formula:

$$\beta_i = \left(X^T W_i X \right)^{-1} X^T W_i Y,$$

where Y is an n-dimensional vector in the value of the dependent variable,

X is a data matrix with n observations for m independent variables (together with a column of units denoting points of intersection),

W_i is an $n \times n$ neighbourhood matrix whose diagonal elements denote the spatial weights of the w_{ij} neighbourhood at location i.

The two main spatial regression models are the spatial lag model (SAR) and the spatial error model (SEM). The spatial lag model is used to test the hypothesis that the result (the dependent variable) is influenced not only by independent variables, but also by neighbouring values (spatial lag) of the dependent variable itself (the diagram on the right). In the spatial error model, it is the error (residual, discrepancy) of the model itself, rather than the dependent variable, which is considered to be autoregressive (the diagram on the left) (Fig. A13.1).

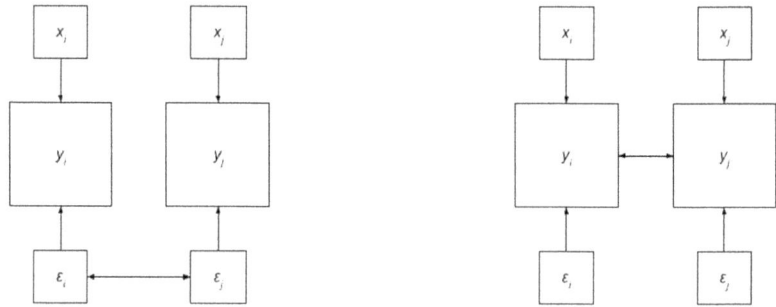

Fig. A13.1. Spatial regression models

As part of this practical task, we will learn how to select and evaluate an explanatory model – i.e. the set of independent variables that best explains the dependent variable. For this task we will use the "WorldR" archive available on our Google Drive (https://drive.google.com/file/d/1iY2ruAQH_41qPjdwZihXUiyy3k9LP8gF/view?usp=drive_link) and our website (https://mgimo.ru/upload/2023/05/worldr-geographically-weightedregression.rar). Open the WorldR .shp file in GeoDa, and load the World_W weights. Open the WorldR .shp file in GeoDa, and load the World _ W weights. The attribute table contains a set of seven variables: GDPPC – GDP per capita; CREDACCS – number of credit cards per 1000 people; CO2EM – CO_2 emissions (metric tons per capita); DEM – Democracy Index; ELECTRICIT – energy supply; RTA – participation in regional trade agreements; HITECH – the share of medium- and hi-tech industries in industrial production.

We need to find out whether the GDPPC variable can be explained by a model consisting of three explanatory variables – HITECH, ELECTRICIT and RTA. This choice of explanatory variables will allow us to determine the extent to which a country's GDP per capita determines this particular set of variables.

Select Regression in the control panel and select the following parameters in the dialog box (Fig. A13.2):

Dependent Variable: GDPPC
Covariates (independent variables): HITECH, RTA, ELECTRICIT
Weights File: World_W
Models: Classic
Once you have set the parameters, click Run to get the regression report.

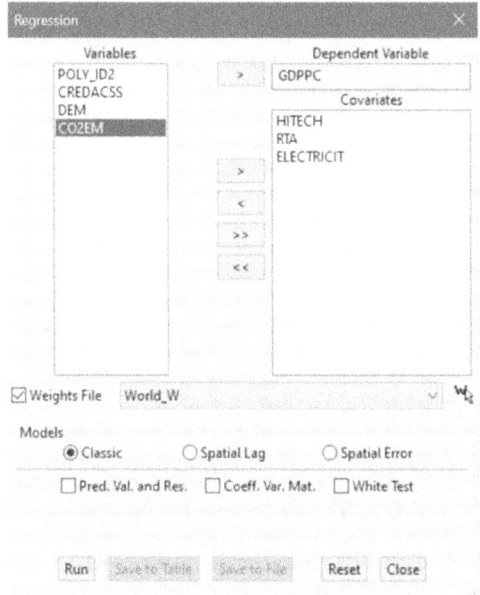

Fig. A13.2. Classic regression settings

When you have the regression report, you need to run a few tests to determine whether the model is statistically correct and can therefore be used.

1. p-value (probability) represents the level of statistical significance of the given value: the smaller the p, the higher the level of significance. The maximum allowable p-value in the social sciences has been agreed at p = 0.05. All p-values other than CONSTANT must satisfy this condition. The p-value in our case is below this threshold, so Test 1 has been passed successfully. The p-value of the diagnostics for heteroskedasticity (BP test or KB test) also has to be under 0.05,[8] while the p-value for the test on normality of errors (Jarque–Bera) should be higher than 0.1 (i.e. there should be no such normality), although this test is only valid for a large number of observations and can thus be disregarded for now.

```
РЕГРЕССИЯ
----------
SUMMARY OF OUTPUT: ORDINARY LEAST SQUARES ESTIMATION
Data set            :   v22_2608
Dependent Variable  :        GDPPC   Number of Observations:   144
Mean dependent var  :      23593,6   Number of Variables   :     4
S.D. dependent var  :      22068,2   Degrees of Freedom    :   140

R-squared           :      0,654078  F-statistic           :       88,2385
Adjusted R-squared  :      0,646665  Prob(F-statistic)     :  4,12296e-032
Sum squared residual:2,42592e+010    Log likelihood        :      -1568,17
Sigma-square        : 1,7328e+008    Akaike info criterion :       3144,34
S.E. of regression  :      13163,6   Schwarz criterion     :       3156,22
Sigma-square ML     :1,68467e+008
S.E of regression ML:     12979,5

----------------------------------------------------------------------
       Variable     Coefficient     Std.Error     t-Statistic   Probability
----------------------------------------------------------------------
       CONSTANT         1045,6       1919,74       0,544658      0,58685
         HITECH        425,992       81,4701       5,22881       0,00000
            RTA        339,21        86,935        3,90188       0,00015
      ELECTRICIT       1,54489       0,18615       8,29918       0,00000
----------------------------------------------------------------------
```

2. R squared (the coefficient of determination) is the proportion of the variation explained by the model. It is generally accepted that a model with an R squared (R^2) of less than 0.5 cannot be used, and that above 0.8 is best. In this case, the R^2 is 0.65, which makes the model usable. This means that the set of variables explains the value of GDP per capita by 65 %.

---

[8] Heteroskedastic behaviour is when a single error has equal variance and constant standard deviation.

```
РЕГРЕССИЯ
----------
SUMMARY OF OUTPUT: ORDINARY LEAST SQUARES ESTIMATION
Data set            :   v22_2608
Dependent Variable  :      GDPPC  Number of Observations:    144
Mean dependent var  :    23593,6  Number of Variables   :      4
S.D. dependent var  :    22068,2  Degrees of Freedom    :    140

R-squared           :   0,654078  F-statistic           :      88,2385
Adjusted R-squared  :   0,646665  Prob(F-statistic)     :4,12296e-032
Sum squared residual:2,42592e+010 Log likelihood        :    -1568,17
Sigma-square        : 1,7328e+008 Akaike info criterion :     3144,34
S.E. of regression  :    13163,6  Schwarz criterion     :     3156,22
Sigma-square ML     :1,68467e+008
S.E of regression ML:    12979,5

---------------------------------------------------------------------
      Variable     Coefficient     Std.Error    t-Statistic  Probability
---------------------------------------------------------------------
      CONSTANT        1045,6        1919,74      0,544658      0,58685
        HITECH       425,992        81,4701      5,22881       0,00000
           RTA       339,21         86,935       3,90188       0,00015
     ELECTRICIT      1,54489        0,18615      8,29918       0,00000
---------------------------------------------------------------------
```

3. You also need to pay attention to the Akaike information criterion (AIC): the lower the AIC, the more preferable the model. While there is no upper limit to the AIC, it can nevertheless be used when comparing two linear regressions.

The SAR and SEM models are not always equally appropriate, and spatial tests are carried out to determine the acceptable model using the Lagrange Multiplier (LM), which estimates the level of spatial autocorrelation of the dependent variable (for the spatial lag model) or the residuals (for the spatial error model). Take a look at the last column in the section "Diagnostics for Spatial Dependence," titled PROB (p-value), and choose the model – Lagrange Multiplier (lag) or Lagrange Multiplier (error) – for which the p-value is less than or equal to 0.05. If both models pass this test, then do the same with Robust LM (lag) and Robust LM (error). If the models pass this test as well, then choose the model with the lowest p-value. In this case, the Lagrange Multiplier (error) is above 0.05, meaning that the Lagrange Multiplier (lag) is the only usable model here.

```
                    RTA        339,21          86,935       3,90188     0,00015
              ELECTRICIT       1,54489          0,18615      8,29918     0,00000

REGRESSION DIAGNOSTICS
MULTICOLLINEARITY CONDITION NUMBER      4,671902
TEST ON NORMALITY OF ERRORS
TEST                      DF           VALUE                PROB
Jarque-Bera                2           463,8198             0,00000

DIAGNOSTICS FOR HETEROSKEDASTICITY
RANDOM COEFFICIENTS
TEST                      DF           VALUE                PROB
Breusch-Pagan test         3           145,3096             0,00000
Koenker-Bassett test       3            28,3950             0,00000

DIAGNOSTICS FOR SPATIAL DEPENDENCE
FOR WEIGHT MATRIX : World_W
     (row-standardized weights)
TEST                           MI/DF         VALUE          PROB
Moran's I (error)              0,1287        2,2060         0,02738
Lagrange Multiplier (lag)          1        11,1768         0,00083
Robust LM (lag)                    1         9,8346         0,00171
Lagrange Multiplier (error)        1         3,1326         0,07674
Robust LM (error)                  1         1,7904         0,18088
Lagrange Multiplier (SARMA)        2        12,9672         0,00153
============================== END OF REPORT ==============================
```

Follow the same steps that you used with linear regression, but choose the Spatial Lag model, as it is the only valid model for this particular set of variables (Fig. A15.3).

Fig. A13.3. Spatial Lag regression settings

Once you have your regression report, look at the p-values of the variables, the Diagnostics for Spatial Dependence, and the Diagnostics for Heteroskedasticity (BP test). Note that a new variable has appeared (compared to linear regression) – W _ GDPPC, which is the spatial lag of the dependent variable we included in the regression model (in the spatial error model, this will be the LAMBDA variable, reflecting the spatial autocorrelation of the residuals). In each of these cases, the p-value should not be greater than 0.05. The next thing you need to do is look at the R^2 and the AIC, which we have already covered, as well as at the Schwarz information criterion.

```
РЕГРЕССИЯ
----------
SUMMARY OF OUTPUT: SPATIAL LAG MODEL - MAXIMUM LIKELIHOOD ESTIMATION
Data set            : v22_2608
Spatial Weight      : World_W
Dependent Variable  :         GDPPC  Number of Observations:    144
Mean dependent var  :       23593,6  Number of Variables   :      5
S.D. dependent var  :       22068,2  Degrees of Freedom    :    139
Lag coeff.   (Rho)  :      0,315804

R-squared           :      0,699586  Log likelihood        :    -1560,29
Sq. Correlation     :      -         Akaike info criterion :     3130,57
Sigma-square        :  1,46304e+008  Schwarz criterion     :     3145,42
S.E of regression   :       12095,6

----------------------------------------------------------------------
      Variable      Coefficient    Std.Error      z-value   Probability
----------------------------------------------------------------------
        W_GDPPC        0,315804     0,0070651      4,46271      0,00001
       CONSTANT        -1354,9       1880,41     -0,720538      0,47119
          HITECH       343,969       76,5332      4,49438      0,00001
             RTA       242,665       85,2539      2,84638      0,00442
       ELECTRICIT      1,19979      0,179967      6,66675      0,00000
----------------------------------------------------------------------

REGRESSION DIAGNOSTICS
DIAGNOSTICS FOR HETEROSKEDASTICITY
RANDOM COEFFICIENTS
TEST                                DF        VALUE          PROB
Breusch-Pagan test                   3      131,2247        0,00000

DIAGNOSTICS FOR SPATIAL DEPENDENCE
SPATIAL LAG DEPENDENCE FOR WEIGHT MATRIX : World_W
TEST                                DF        VALUE          PROB
Likelihood Ratio Test                1      15,7634         0,00007
============================ END OF REPORT =============================
```

Just like with the AIC, there is no upper limit for Schwarz, and the optimal model is the one that shows the lowest values for this criterion. As the GWR report shows, the p-value tests have been passed, and the R^2 is sufficiently high, meaning that the model can be used as an explanatory model.

There are a number of ways to compare regression models: the Akaike information criterion can be used to compare all kinds of models – for example, non-spatial and spatial models, non-spatial models with other non-spatial models, and spatial models with other spatial models (and, in turn, lags and errors, lags with other lags, and errors with other errors).

The Schwarz information criterion only allows you to compare spatial models with other spatial models. This makes it more accurate for comparing spatial models than the Akaike information criterion. The regression report for the lag model produces an R^2 of 0.69, which also allows this set of variables to be used as an explanatory variable.

To compare these two models, we need to use the Akaike information criterion, which shows that the GWR turned out to be slightly more accurate than the linear regression model, with an acceptable R^2 value in both.

As an additional task, you can select a different set of explanatory variables and test these models yourself.

In the models described above, we were dealing with the assessment of global (continuous) spatial heterogeneity that is characteristic of the entire sample. However, in practice, heterogeneity can only be characteristic of certain regions (say, a continent or a part of a city) – i.e. be discrete. Heterogeneity is identified using the spatial regimes method. This is when the regression model (spatial or nonspatial) is calculated separately for different regions (or groups of observations).

To do this, you will need to use the special GeoDaSpace software, which is available at https://geodacenter.github.io/GeoDaSpace/. After you have opened the program, load the shapefile and neighbourhood weights and specify the dependent (Y) and independent (X) variables. Then, in the Regimes section, select a variable that marks a certain group of observations numerically. For example, observations in Europe are marked 1, and observations in Asia are marked 2. Now you may calculate the regression (spatial or nonspatial). The computer will run the regression model separately for each spatial regime, which will allow you to compare regions with each other in terms of their R^2 value and their level of significance. This will, in turn, allow you to draw conclusions about the spatial heterogeneity of the regression model.

At the end of the sample, you will be presented with the Chow coefficient. A low Chow coefficient indicates that the phenomenon is distributed in a spatially heterogenous manner, and thus suggests the need to search for zones in which the regression model changes its predictive capacity.

GeoDa Space (available for download at https://geodacenter.github.io/GeoDaSpace/download.html) allows you to perform regression analysis within individual groups of objects in a population, for example, regions of the world.

To analyse an entire set of objects, from the start window (Fig. A15.4), select a .dbf file in the Data File field, the dependent variable (in this case, GDPPC) in the Y field, and explanatory variables (HITECH, RTA, ELECTRICT) in the X field. Select which regression model you need in the Model Type window. The resulting report will be identical to those discussed earlier.

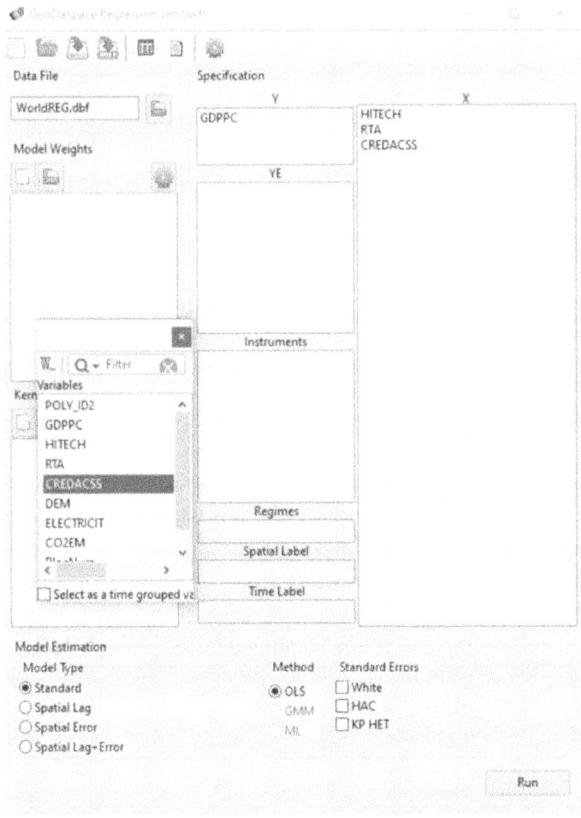

Fig. A13.4. Regression settings in GeoDa Space

```
REGRESSION
---------

SUMMARY OF OUTPUT: ORDINARY LEAST SQUARES ESTIMATION - REGIME 1
---------------------------------------------------------------
Data set            :       1.dbf
Weights matrix      :  File: 1.gal
Dependent Variable  :     1_Clergy        Number of Observations:        17
Mean dependent var  :      34.0588        Number of Variables    :         3
S.D. dependent var  :      19.0475        Degrees of Freedom     :        14
R-squared           :       0.1560
Adjusted R-squared  :       0.0354
Sum squared residual:     4899.649        F-statistic            :    1.2934
Sigma-square        :      349.975        Prob(F-statistic)      :    0.3052
S.E. of regression  :       18.708        Log likelihood         :  -72.263
Sigma-square ML     :      288.215        Akaike info criterion  :  150.527
S.E of regression ML:      16.9769        Schwarz criterion      :  153.027

------------------------------------------------------------------------------
        Variable     Coefficient       Std.Error     t-Statistic    Probability
------------------------------------------------------------------------------
      1_CONSTANT      54.0802479      17.5090850      3.0886964      0.0080102
        1_Infants     -0.0011902       0.0007401     -1.6082290      0.1300962
        1_Suicids     -0.0000024       0.0004748     -0.0051597      0.9959560
------------------------------------------------------------------------------

Regimes variable: CATEGORIES

REGRESSION DIAGNOSTICS
MULTICOLLINEARITY CONDITION NUMBER              7.985

TEST ON NORMALITY OF ERRORS
TEST                           DF        VALUE           PROB
Jarque-Bera                     2        1.723         0.4225

DIAGNOSTICS FOR HETEROSKEDASTICITY
RANDOM COEFFICIENTS
TEST                           DF        VALUE           PROB
Breusch-Pagan test              2        0.644         0.7248
Koenker-Bassett test            2        1.556         0.4593

DIAGNOSTICS FOR SPATIAL DEPENDENCE
TEST                        MI/DF        VALUE           PROB
Lagrange Multiplier (lag)       1        0.949         0.3300
```

The spatial regimes function allows you to analyse individual groups of objects. To do that, you need to select a variable in the Regime field that marks an observation as belonging to a group (in this case, we will select BlocNum – a numerical value that corresponds to one of the geopolitical blocs of the world) (Fig. A15.5). The resulting report will be presented both for the entire set of objects and for each group (Regime) individually.

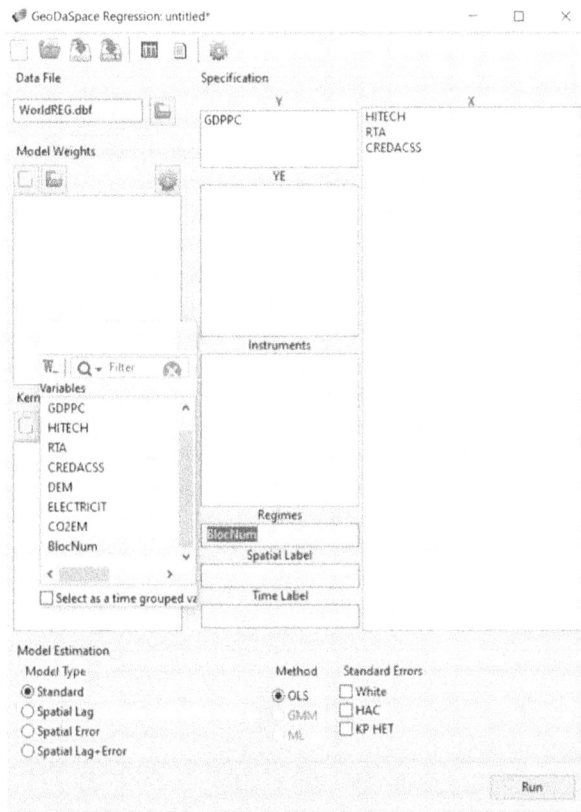

Fig. A13.5. Spatial regimes settings

```
REGRESSION
----------
SUMMARY OF OUTPUT: ORDINARY LEAST SQUARES
-----------------------------------------
Data set            :          1.dbf
Weights matrix      : File: 1.gal
Dependent Variable  :         Clergy      Number of Observations:          85
Mean dependent var  :        43.9294      Number of Variables   :           3
S.D. dependent var  :        24.7130      Degrees of Freedom    :          82
R-squared           :         0.1067
Adjusted R-squared  :         0.0849
Sum squared residual:      45829.901      F-statistic           :      4.8950
Sigma-square        :        558.901      Prob(F-statistic)     :    0.009812
S.E. of regression  :         23.641      Log likelihood        :    -387.937
Sigma-square ML     :        539.175      Akaike info criterion :     781.873
S.E of regression ML:         23.2201     Schwarz criterion     :     789.201

-----------------------------------------------------------------------------
       Variable      Coefficient     Std.Error     t-Statistic    Probability
-----------------------------------------------------------------------------
       CONSTANT       50.8084117     6.2691545       8.1045078      0.0000000
         Infants       0.0001476     0.0003044       0.4848826      0.6290525
         Suicids      -0.0002651     0.0000855      -3.0995030      0.0026560
-----------------------------------------------------------------------------

REGRESSION DIAGNOSTICS
MULTICOLLINEARITY CONDITION NUMBER             5.325

TEST ON NORMALITY OF ERRORS
TEST                              DF          VALUE           PROB
Jarque-Bera                        2          3.215           0.2004

DIAGNOSTICS FOR HETEROSKEDASTICITY
RANDOM COEFFICIENTS
TEST                              DF          VALUE           PROB
Breusch-Pagan test                 2          0.507           0.7760
Koenker-Bassett test               2          0.968           0.6164

DIAGNOSTICS FOR SPATIAL DEPENDENCE
TEST                           MI/DF          VALUE           PROB
Lagrange Multiplier (lag)          1         26.676           0.0000
Robust LM (lag)                    1          2.885           0.0894
Lagrange Multiplier (error)        1         23.878           0.0000
Robust LM (error)                  1          0.088           0.7673
Lagrange Multiplier (SARMA)        2         26.763           0.0000

============================ END OF REPORT =================================
```

Geographic Regression Discontinuity Analysis

Borders between countries and regions are not merely formal lines, they are also barriers that tear the natural fabric of social processes. For example, language areas should spread naturally, but they are often interrupted by state borders, since different regimes of language support operate on each side – it could be a state language on one side of the border and banned as a language of school education on the other side. Geographic regression discontinuity allows us to see how social phenomena transform as a result of the border effect. To illustrate, let's see Figs. 14.1 and 14.2 that represent right voting in countries bordering Russia in the electoral cycles A and B. Cycle A (Fig. A14.1) saw low right-wing support close to the Russian border, which then showed a rise followed by a steady decline as we moved deeper into the territory. In cycle B (Fig. A14.2) we could see the opposite trend, with low mid-territory rates of right voting getting consistently higher as we approached the border with Russia or moved deeper into the country.

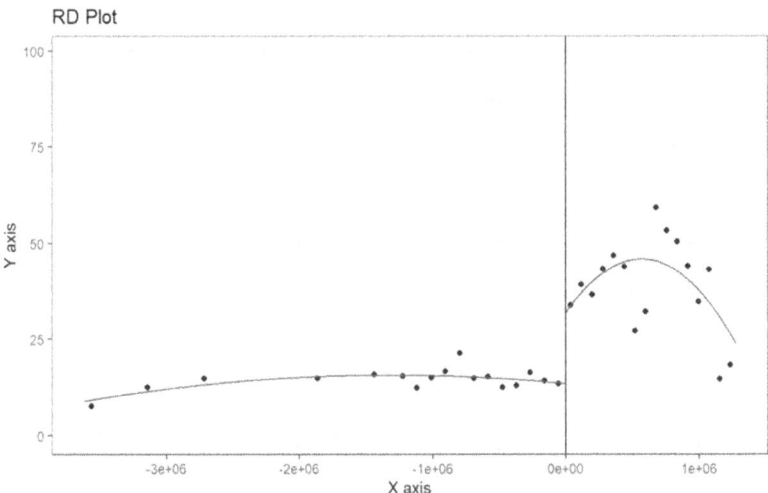

Fig. A14.1. Geographic regression discontinuity analysis of the right-wing voting in countries bordering Russia in cycle A

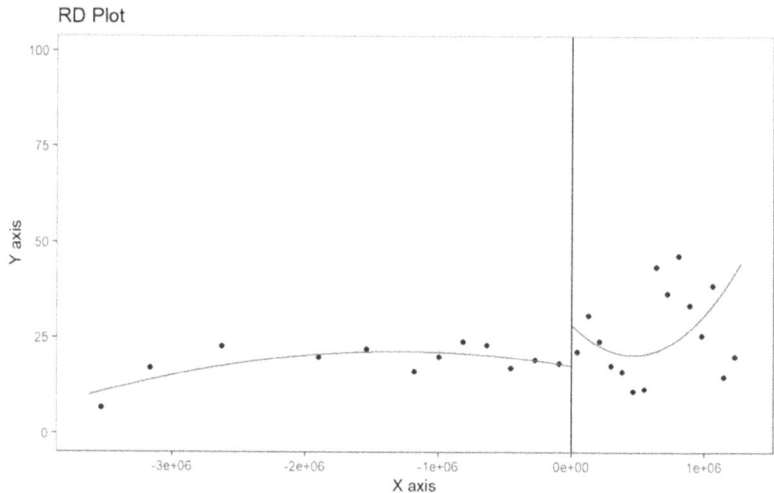

Fig. A14.2. Geographic regression discontinuity analysis of the right-wing voting in countries bordering Russia in cycle B

The following formula is used to calculate geographic regression discontinuity:

$$y_i = \alpha + f\begin{pmatrix} x + y + x^2 + y^2 + x^3 + y^3 + x^4 + y^4 + x*y*x^2*y^2 \\ + x^3*y^3 + x*y^2 + x*y^3 + x^2*y + x^3*y \end{pmatrix} + \varepsilon_i +$$

Or more generally:

$$y_i = \alpha + \beta_i + \gamma_i + \varepsilon_i$$

where x and y are the latitudinal and longitudinal coordinates of the border β, and γ is the distance to this border.

Let's calculate geographic regression discontinuity in R. To do this, we need information about the coordinates of the border and data calculated for some points on both sides of the border. We'll use shapefiles of the Brazilian border and data on the population of settlements along that border, which are available in the "Br_border_interior" and "Popn_at_point" archives on original sources: https://jonnyphillips.github.io/Intro_Spatial/Day_5/Data/Br_Border_interior.zip and https://jonnyphillips.github.io/Intro_Spatial/Day_5/Data/Popn_at_point.zip.

Install the required packages and load the files.

```{r}
library(sf)
library(tidyverse)
library(units)
library(rdrobust)

population <- read_sf("Popn_at_point.shp")
border <- read_sf("Br_border_Interior.shp")
```

Let's introduce a new variable that measures the distance between the points and the border. Set negative values for data outside of Brazil.

```{r}
population <- population %>% mutate(Dist_to_border=st_distance(.,border))

population <- population %>% mutate(Dist_to_border=ifelse(In_Brazil==1,   Dist_to_border, Dist_to_border*-1))
```

Now we'll calculate the regression and analyse the results.

```{r}
regression <- population %>% lm(lacpopd00 ~ In_Brazil + Dist_to_border, data =.)
summary(regression)
```

Call:
lm(formula = lacpopd00 ~ In_Brazil + Dist_to_border, data = .)
Residuals:
Min 1Q Median 3Q Max
 -6.68 -5.38 -3.62 -2.57 984.31

Coefficients:

```
            Estimate Std. Error t value Pr(>|t|)
(Intercept)  4.714e+00 5.223e-01 9.025 <2e-16 ***
In _ Brazil  1.566e+00 9.326e-01 1.679 0.0931 .
Dist _ to _ border 2.016e-05 4.139e-05 0.487 0.6263
---
Signif. codes: 0 '***' 0.001 '**' 0.01 '*' 0.05
'.' 0.1 ' ' 1

Residual standard error: 24.09 on 10067 degrees
of freedom
Multiple R-squared: 0.001669, Adjusted R-squared:
0.00147
F-statistic: 8.413 on 2 and 10067 DF, p-value:
0.0002236
```

Let's plot the resulting geographic regression discontinuity graph.

```{r}
rdplot(population$lacpopd00,population$Dist _ to _ 
border,p=1,y.lim = c(0,20))
```

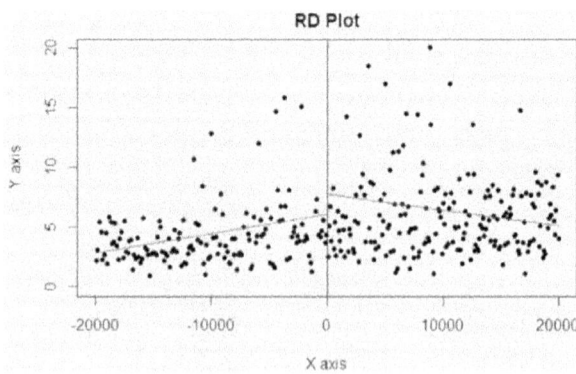

The final program code for calculating geographic regression discontinuity in R is as follows and consists of four steps:

```{r}
```

```
library(sf)
library(tidyverse)
library(units)
library(rdrobust)

population <- read_sf("Popn_at_point.shp")
border <- read_sf("Br_border_Interior.shp")
```

```{r}
population <- population %>% mutate(Dist_to_
border=st_distance(.,border))
population <- population %>% mutate(Dist_to_
border=ifelse(In_Brazil==1, Dist_to_border, Dist_
to_border*-1))
```

```{r}
regression <- population %>% lm(lacpopd00 ~ In_
Brazil + Dist_to_border, data =.)
```

```{r}
rdplot(population$lacpopd00,population$Dist_to_
border,p=1,y.lim = c(0,20))
```

The results of the analysis show that the population of settlements increases as you move closer to the border, and the figure is slightly higher on the Brazilian side. Geographic regression discontinuity, along with other spatial analysis techniques, allows us to effectively visualize data on maps and see geographical patterns that may not be immediately evident, and that can be used to solve a wide range of problems in the social sciences and humanities.

Questions and Tasks

1. What factors do you think affect the rental price of an apartment? Would rental prices in neighbouring apartment buildings be a factor? How can you test this hypothesis using spatial analysis methods?
2. Try to imagine a real-life situation in which a phenomenon is influenced by a spatial lag. What other factors would be at play here?
3. Select another dependent variable from the task and try to find a regression model for it that meets the criteria as has the highest R^2 value.
4. Is spatial regression applicable in this model?
5. Come up with a hypothesis about the spatial heterogeneity of a regression model. How can you check it?
6. Come up with research problems that will require you to calculate geographic regression discontinuity.
7. Map the geographic regression discontinuity using data from one of the files we used in previous chapters.

Bibliography

In Russian:

1. Aleskerov F.T., Borodin A.D., Kaspe' S.I., Marshakov V.A., Salmin A.M. Analiz e'lektoral'ny'x predpochtenij v Rossii v 1993–2003 gg.: dinamika indeksa polyarizovannosti // E'konomicheskij zhurnal Vy'sshej shkoly' e'konomiki. — 2005. — T. 9, № 2. — S. 173–184.
2. Aleskerov F.T., Orteshuk P. Vy'bory'. Golosovanie. Partii. — M.: Akademiya, 1995. — 208 s.
3. Aleskerov F.T., Platonov V.V. Indeksy' predstavitel'nosti parlamenta // Politiya. — 2003. — № 1 (28). — S. 193–200.
4. Aleskerov F.T., Xabina E'.L., Shvarcz D.A. Binarny'e otnosheniya, grafy' i kollektivny'e resheniya. — M.: FIZMATLIT, 2012. — 344 s.
5. Aksenov K.E'. Dvuxpartijnost' massovogo politicheskogo soznaniya i geografiya izbiratel'nogo povedeniya v Sankt-Peterburge // Regional'naya politika. — 1994. — № 6. — S. 51–71.
6. Aksenov K.E'., Zinov'ev A.S., Pleshhenko D.V. Krupny'j gorod — region — Rossiya: dinamika e'lektoral'nogo povedeniya na parlamentskix vy'borax // Polis. — 2005. — № 2. — S. 41–52.
7. Anisimov A.A. Vozmozhnost' modelirovaniya partijnoj sistemy' (na primere teorii S. Rokkana) // Vlast' i e'lity' sovremennoj Rossii / Pod red A.V. Duki. — SPb.: Sociologicheskoe obshhestvo im. M.M. Kovalevskogo, 2003. — S. 354–399.
8. Arbatskaya M.N. Organizaciya izbiratel'ny'x procedur: sovremennaya mirovaya praktika. — Irkutsk: Izbiratel'naya komissiya Irkutskoj oblasti, 2007. — 244 s.
9. Arbatskaya M.N. Struktura i geografiya informacii o xode golosovaniya // Zhurnal o vy'borax. — 2005. — № 6. — S. 61–64; 2006. — № 1. — S. 33–38.

10. Arbatskaya M.N. Territorial'ny'e politicheskie sistemy': e'lektoral'naya aktivnost' i ciklichnost' e'volyucii // Izvestiya RAN. Seriya geograficheskaya. — 2013. — № 1. — S. 31–41.
11. Arbatskaya M.N. Transformaciya publichnoj politiki i e'lektoral'noe prostranstvo. — Irkutsk: Izbiratel'naya komissiya Irkutskoj oblasti, 2006. — 218 s.
12. Arbatskaya M.N. E'lektoral'noe prostranstvo i upravlenie izbiratel'ny'mi pravami grazhdan: metodologicheskie i metodicheskie osnovy' analiza // Polite'ks. — 2006. — T. 2. № 1. — S. 62–80.
13. Arxipova G.G. Izbiratel'ny'e sistemy' v polie'tnicheskix gosudarstvax // Vestnik Yakutskogo gosudarstvennogo universiteta. — 2006. — T. 3. № 1. — S. 84–91.
14. Axremenko A.S. Kolichestvenny'j analiz rezul'tatov vy'borov: sovremenny'e metody' i problemy'. — M.: Izd-vo MGU, 2008. — 160 s.
15. Axremenko A.S. Prostranstvennoe modelirovanie e'lektoral'nogo vy'bora: razvitie, sovremenny'e problemy' i perspektivy' // Polis. — 2007. — № 1. — S. 153–167; № 2. — S. 165–179.
16. Axremenko A.S. Prostranstvenny'j e'lektoral'ny'j analiz: xarakteristika metoda, vozmozhnosti krossnacional'ny'x sravnitel'ny'x issledovanij // Politicheskaya nauka. — 2009. — № 1. — S. 32–59.
17. Axremenko A.S. Social'ny'e razmezhevaniya i struktury' e'lektoral'nogo prostranstva Rossii // Obshhestvenny'e nauki i sovremennost'. — 2007. — Vy'p. 4. — S. 80–92.
18. Axremenko A.S. Strukturirovanie e'lektoral'nogo prostranstva v rossijskix regionax: faktorny'j analiz parlamentskix vy'borov 1995–2003 gg. // Polis. — 2005. — № 2. — S. 26–60.
19. Axremenko A.S. Strukturny'e pozicii partij v e'lektoral'nom prostranstve i ix rezul'taty' na vy'borax: problema svyazi // Politiya. — 2006. — № 4 (43). — S. 45–67.
20. Axremenko A.S. Struktury' e'lektoral'nogo prostranstva. — M.: Social'no-politicheskaya my'sl', 2007. — 320 s.
21. Axremenko A.S. E'lektoral'noe uchastie i absenteizm v rossijskix regionax: zakonomernosti i tendencii // Vestnik Moskovskogo universiteta. Seriya 12. Politicheskie nauki. — 2005. — № 3. — S. 95–113.
22. Baburin V.L., Berezkin A.V., Sidorova E.V. Opy't geograficheskogo analiza nekotory'x rezul'tatov vy'borov na S''ezd narodny'x deputatov

SSSR // Vestnik Moskovskogo universiteta. Seriya 5. Geografiya. — 1990. — № 2. — S. 21–27.
23. Baranov A.V. Faktory' e'lektoral'nogo povedeniya v Rossii i Ukraine: sravnitel'ny'j analiz // PolitBook. — 2012. — № 1. — S. 42–52.
24. Baranov A.V. E'lektoral'naya geografiya Krasnodarskogo kraya // Chelovek. Soobshhestvo. Upravlenie. — 2000. — № 3–4. — S. 35–44.
25. Baranov A.V. E'lektoral'ny'e processy' v Rossii na federal'nom i regional'nom urovnyax: e'ffekty' izbiratel'noj sistemy' // Izvestiya Altajskogo gosudarstvennogo universiteta. — 2013. — № 4–2 (80). — S. 239–243.
26. Berezkin A.V., Kolosov V.A., Pavlovskaya M.E'., Petrov N.V., Smirnyagin L.V. Geografiya vy'borov narodny'x deputatov SSSR v 1989 g. (Pervy'e itogi) // Izvestiya AN SSSR. Seriya geograficheskaya. — 1989. — № 5. — S. 5–24.
27. Brumshtejn Yu.M., Grishin N.V. Issledovanie dinamiki e'lektoral'ny'x predpochtenij po mnogomerny'm traektoriyam // Informacionny'e texnologii modelirovaniya i upravleniya. — 2008. — № 7 (50). — S. 740–743.
28. Bryancev A.V. E'lektoral'naya geografiya Finlyandii po rezul'tatam parlamentskix vy'borov 2019 goda (golosovanie za partiyu «Zeleny'j soyuz») // Original'ny'e issledovaniya. — 2022. — T. 12. № 3. — S. 178–185.
29. Bunina A.A. Izbiratel'ny'e okruga v SShA: dzherrimendering kak aspekt partijnogo sopernichestva // Mirovaya e'konomika i mezhdunarodny'e otnosheniya. — 2020. — T. 64. № 10. — S. 52–63.
30. Buzin A.Yu. Vliyanie social'no-e'konomicheskogo razvitiya regionov Rossii na itogi vy'borov v Gosudarstvennuyu Dumu Federal'nogo Sobraniya RF vtorogo sozy'va // Polis. — 1996. — № 1. — S. 103–118.
31. Van'kov V.A. Poselencheskaya struktura v e'lektoral'nom povedenii po materialam parlamentskix vy'borov v Rossii // Polis. — 2003. — № 6. — S. 88–103.
32. Varyushin P.S. Geograficheskie zakonomernosti politicheskoj i e'lektoral'noj sistem SShA // Izvestiya Rossijskoj akademii nauk. Seriya geograficheskaya. — 2016. — № 4. — S. 34–42.

33. Varyushin P.S. Metody' issledovaniya geografii politicheskix predpochtenij naseleniya SShA // Vestnik Moskovskogo universiteta. Seriya 5: Geografiya. — 2014. — № 4. — S. 42–48.
34. Varyushin P.S., Tixoczkaya I.S. Territorial'ny'e faktory' e'lektoral'nogo povedeniya v Yaponii i SShA // Vestnik Moskovskogo universiteta. Seriya 5: Geografiya. — 2016. — № 2. — S. 102–111.
35. Vesna–89: geografiya i anatomiya parlamentskix vy'borov / Pod red. V.A. Kolosova, N.V. Petrova i L.V. Smirnyagina. — M.: Progress, 1990. — 382 s.
36. Vtoroj e'lektoral'ny'j cikl v Rossii (1999–2000) / Pod red. V.Ya. Gel'mana, G.V. Golosova, E.Yu. Meleshkinoj. — M.: Ves' Mir, 2002. — 216 s.
37. Gasparyan O.T. Primenenie metodov prostranstvennoj e'konometriki v prikladny'x politicheskix issledovaniyax // Politicheskaya konceptologiya. — 2017. — № 1. — S. 41–48.
38. Gel'man V.Ya. Izuchenie vy'borov v Rossii: issledovatel'skie napravleniya i metody' analiza // Politicheskaya nauka. — 2000. — № 3. — S. 16–51.
39. Gilev A.V., Semenov A.V., Shevczova I.K. «Politicheskie mashiny'» i ix «voditeli»: e'lektoral'noe administrirovanie na mestnom urovne // Politiya. — 2017. — № 3 (86). — S. 62–80.
40. Gimpel'son V.E., Chugrov S.V. Modeli e'lektoral'nogo povedeniya rossijskix regionov: opy't mnogomernogo statisticheskogo analiza itogov vy'borov 12 dekabrya 1993 g. // Mirovaya e'konomika i mezhdunarodny'e otnosheniya. — 1995. — № 4. — S. 22–32.
41. Golosov G.V. Partijny'e sistemy' stran mira: regional'noe i xronologicheskoe raspredelenie, modeli ustojchivosti // Politicheskaya nauka. — 2012. — № 3. — S. 71–104.
42. Golosov G.V. Sravnitel'naya politologiya. — SPb.: EUSPb, 2018. — 462 s.
43. Golosov G.V. E'lity', obshherossijskie partii, mestny'e izbiratel'ny'e sistemy' // Obshhestvenny'e nauki i sovremennost'. — 2000. — № 3. — S. 51–75.
44. Grekusis Dzh. Metody' i praktika prostranstvennogo analiza. Opisanie, issledovanie i ob"yasnenie s ispol'zovaniem GIS. — M.: DMK Press, 2021. — 500 s.

45. Grishin N.V. Bipolyarnaya model' strukturirovaniya geoe'lektoral'nogo prostranstva Rossii // Vlast'. — 2009. — № 4. — S. 86–90.
46. Grishin N.V. Dinamika e'lektoral'ny'x predpochtenij naseleniya yuga Rossii: sravnitel'noe issledovanie. — M.: MGU, 2008. — 181 s.
47. Grishin N.V. Kolichestvennoe izmerenie ustojchivosti territorial'ny'x razlichij politicheskix predpochtenij naseleniya // Yuzhno-rossijskij vestnik geologii, geografii i global'noj e'nergii. — 2006. — № 6 (19). — S. 27–30.
48. Grishin N.V. Kon"yunktura e'lektoral'nogo prostranstva politicheskix partij na parlamentskix vy'borax 2007 g. // Chelovek. Soobshhestvo. Upravlenie. — 2008. — № 1. — S. 23–30.
49. Grishin N.V. Social'ny'e faktory' e'lektoral'noj geografii // Uspexi sovremennogo estestvoznaniya. — 2006. — № 11. — S. 19–22.
50. Grishin N.V. Territorial'ny'e razlichiya e'lektoral'ny'x orientacij naseleniya na mikrourovne // Kaspijskij region: politika, e'konomika, kul'tura. — 2009. — № 4 (21). — S. 70–75.
51. Grishin N.V. Ustojchivost' territorial'ny'x razlichij e'lektoral'ny'x predpochtenij naseleniya (na primere yuga Rossii) // Polite'ks. — 2006. — T. 2. № 2. — S. 36–42.
52. Danilov A.V. Vliyanie cennostny'x raskolov na e'lektoral'noe povedenie: Dis... kand. polit. nauk: 23.00.02. — SPb.: SPbGU, 2006. — 158 s.
53. Dyuverzhe M. Politicheskie partii. — M.: Akademicheskij Proekt, 2007. — 544 s.
54. Evstifeev R.V. Melodii e'lektoral'ny'x prostranstv: politicheskij process i e'lektoral'ny'e predpochteniya izbiratelej (Vladimirskaya oblast', 1999–2009 gg.). — Vladimir: [b. i.], 2009. — 192 s.
55. Zheltov V.V., Zheltov M.V. E'lektoral'naya geografiya (k voprosu o nasledii A. Zigfrida) // Sociogumanitarny'j vestnik Kemerovskogo instituta (filiala) RGTE'U. — 2010. — № 2 (5). — S. 25–28.
56. Zhidkin A.P. Territorial'ny'e razlichiya v strukture i dinamike e'lektoral'ny'x predpochtenij naseleniya Rossii: Diss. ... kand. geogr. nauk: 25.00.24. — M.: MGU, 2002. — 240 s.

57. Zhirnova L.S. Regional'ny'e tendencii e'lektoral'noj podderzhki latvijskix partij: faktor sosedstva // Baltijskij region. — 2022. — T. 14. № 1. — S. 138–158.
58. Zhuravlev A.N. Vy'bory' v organy' zakonodatel'noj vlasti: predvaritel'ny'e itogi i territorial'naya specifika // Vlast'. — 1998. — № 2. — S. 31–36.
59. Zhuravlev A.N. Regional'ny'e vy'bory' — 96: osnovny'e rezul'taty' i territorial'naya specifika // Politiya. — 1997. — № 1. — S. 65–71.
60. Zhuribeda K.O. E'lektoral'noe povedenie grazhdan Rossii na federal'ny'x vy'borax 1996–2018 gg. na territorii za predelami RF // Polite'ks. — 2021. — T. 17. № 2. — S. 163–183.
61. Zaxarova E.A. E'lektoral'ny'e processy' v fyul'ke Norvegii cherez prizmu prostranstvennogo analiza // Pskovskij regionologicheskij zhurnal. — 2021. — № 1 (45). — S. 110–125.
62. Zimoxa A.Yu. Sravnitel'ny'j analiz zakonomernostej e'lektoral'noj geografii v stranax raznogo tipa: Diss. ... kand. geogr. nauk: 25.00.24. — M.: MGU, 2006. — 177 s.
63. Zimoxa A.Yu. Celi i metody' vozdejstviya na territorial'nuyu kartinu rezul'tatov vy'borov // Regional'ny'e issledovaniya. — 2006. — № 2 (8). — S. 17–24.
64. Zinov'ev A.S. Geopoliticheskij raskol v e'lektoral'noj geografii Litvy' // Regional'ny'e issledovaniya. — 2015. — № 3 (49). — S. 48–56.
65. Ignatov A.V., Popov P.L., Cherenev A.A. E'lektoral'ny'e predpochteniya zhitelej Vostochnoj Sibiri // Sociologicheskie issledovaniya. — 2022. — № 8. — S. 147–152.
66. Il'in M.V. Voronka prichinnosti: ot e'mpiricheskoj modeli k formirovaniyu paradigmy' mnogoslojnoj prichinnosti // Metod. — 2015. — № 5. — S. 442–451.
67. Isaev B.A. Teoriya partij i partijny'x sistem. — M.: Yurajt, 2016. — 370 s.
68. Isaev B.A., Vlaskina S.V. Teoriya social'ny'x raskolov Lipseta–Rokkana i vozmozhnosti ee primeneniya dlya analiza pervoj, vtoroj i tret'ej partijny'x sistem Rossii // Politicheskaya e'kspertiza. — 2016. — T. 12. № 4. — S. 43–64.

69. Kapranov N.V. Biopolyarnaya model' rossijskogo e'lektoral'nogo prostranstva // Vestnik Moskovskogo universiteta. Seriya 18. Sociologiya i politologiya. — 1998. — № 4. — S. 56–61.
70. Kayunov O.N. Ob optimal'nom raspredelenii izbiratel'ny'x okrugov // Zhurnal o vy'borax. — 2002. — № 2. — S. 50–51.
71. Kireev A.L. Geografiya i statistika rezul'tatov prezidentskix vy'borov v SShA — klyuch k ponimaniyu ix dostovernosti // E'lektoral'naya politika. — 2021. — № 1 (5). — S. 6.
72. Klima R.E'., Xodzh Dzh.K. Matematika vy'borov. — M.: MCzNMO, 2007. — 224 s.
73. Kovalev V.A. Vy'bornoe razdvoenie. Regional'ny'e politicheskie nravy' kak otrazhenie rossijskix politicheskix tendencij // Polite'ks. — 2007. — T. 3. № 2. — S. 265–286.
74. Kozlov V.N. Problemy' predstavitel'stva regionov v novoj Gosudarstvennoj Dume // Zhurnal o vy'borax. — 2007. — № 6. — S. 38–45.
75. Kolosov V.A. Politicheskie orientacii rossijskix regionov: proizoshel li v dekabre 1995 «obval»? (Analiz golosovaniya po partijny'm spiskam) // Polis. — 1996. — № 1. — S. 91–102.
76. Kolosov V.A., Borodulina N.A. E'lektoral'ny'e predpochteniya izbiratelej krupny'x gorodov Rossii: tipy' i ustojchivost' // Polis. — 2004. — № 4. — S. 70–79.
77. Kolosov V.A., Vendina O.I. Politicheskie predpochteniya moskvichej v xode izbiratel'ny'x kampanij // Vestnik RAN. — 1997. — T. 67. № 8. — S. 675–680.
78. Kolosov V.A., Turovskij R.F. Vy'bory' v Gosudarstvennuyu Dumu 1995 goda: bor'ba v odnomandatny'x okrugax // Vlast'. — 1996. — № 5. — S. 25–35.
79. Kolosov V.A., Turovskij R.F. Itogi gubernatorskix vy'borov // Vlast'. — 1997. — № 3. — S. 49–56.
80. Kolosov V.A., Turovskij R.F. Osenne-zimnie vy'bory' glav ispolnitel'noj vlasti v regionax: scenarii peremen // Polis. — 1997. — № 1. — S. 97–108.
81. Kolosov V.A., Turovskij R.F. E'lektoral'naya karta sovremennoj Rossii: genezis, struktura i e'volyuciya // Polis. — 1996. — № 4. — S. 33–46.

82. Korgunyuk Yu.G. Koncepciya razmezhevanij i faktorny'j analiz // Politiya. — 2013. — № 3 (70). — S. 31–61.
83. Korgunyuk Yu.G. Novy'e instrumenty' izmereniya e'lektoral'ny'x razmezhevanij: ot makro- k mikrourovnyu // E'lektoral'naya politika. — 2019. — № 1. — S. 1.
84. Korgunyuk Yu.G. Regional'naya karta e'lektoral'ny'x razmezhevanij po itogam dumskix vy'borov 2011 goda // Politiya. — 2014. — № 3 (74). — S. 75–91.
85. Korgunyuk Yu.G. E'lektoral'ny'e razmezhevaniya i motivy' golosovaniya // Politiya. — 2011. — № 2 (61). — S. 85–117.
86. Korneeva E.M. Lokal'ny'j uroven' golosovaniya v Rossii: prostranstvenno-e'konometricheskij podxod // Politicheskaya nauka. — 2021. — № 3. — S. 229–250.
87. Kostenko Yu.V., Penicyn Yu.A., Skorobogatov V.V. E'lektoral'naya politologiya. — Krasnodar: KubGU, 2019. — 159 s.
88. Krishtal' M.I. Prostranstvenny'e osobennosti e'lektoral'nogo konformizma v Rossii v 2000-e gody' // Vestnik Volgogradskogo gosudarstvennogo universiteta. Seriya 4. Istoriya. Regionovedenie. Mezhdunarodny'e otnosheniya. — 2021. — T. 26. № 3. — S. 237–248.
89. Kuleczkaya L.E. Prostranstvenny'e modeli e'lektoral'nogo vy'bora: obzor teoreticheskix i e'mpiricheskix podxodov // Prostranstvennaya e'konomika. — 2021. — T. 17. № 2. — S. 127–164.
90. Ky'nev A.V. Vy'bory' regional'ny'x parlamentov v Rossii 2009–2013: Ot partizacii k personalizacii. — M.: Centr «Panorama», 2014. — 728 s.
91. Ky'nev A.V. Metamorfozy' e'lektoral'noj geografii Rossii v 2007–2008 gg. i ix prichiny' // Rossijskoe e'lektoral'noe obozrenie. — 2008. — № 1. — S. 4–22.
92. Ky'nev A.V., Lyubarev A.E. Partii i vy'bory' v sovremennoj Rossii: E'volyuciya i devolyuciya. — M.: Fond «Liberal'naya missiya», 2011. — 792 s.
93. Lapkin V.V. Opy't kolichestvennogo opisaniya transformacij e'lektoral'nogo prostranstva Rossii v e'lektoral'nom cikle 1999–2000 gg. // Vestnik MGU. Seriya 12. Politicheskie nauki. — 2000. — № 6. — S. 51–67.

94. Lapkin V.V., Pantin V.I., Solov'ev A.I., Il'in M.V., Meleshkina E.Yu., Novinskaya M.I., Avdonin V.S., Yargomskaya N.B., Axremenko A.S., Malakanova O.A., Kuz'min A.S., Baranov S.D., Kovalev V.A., Diligenskij G.G., Petrov N.V. Struktura i dinamika rossijskogo e'lektoral'nogo prostranstva: krugly'j stol // Polis. — 2000. — № 2. — S. 80–110.
95. Lebedev V.A., Kandalov P.M., Nerovnaya N.N. Partii na vy'borax: opy't, problemy', perspektivy'. — M.: Izd-vo MGU, 2006. — 285 s.
96. Levanskij V.A., Lyubutov A.S. Politicheskij spektr Rossijskoj Federacii: strukturno-taksonometricheskij analiz: Partii, frakcii, vy'bory' v 1993–1996 gg. // Gosudarstvo i pravo. — 1997. — № 9. — S. 87–94.
97. Lixtenshtejn A.V. Federalizm i partiya vlasti: geografiya raspredeleniya e'lektoral'noj podderzhki // Politicheskaya nauka. — 2005. — № 2. — S. 40–67.
98. Lyubarev A. Zanimatel'naya e'lektoral'naya statistika. — M.: Golos konsalting, 2021. — 304 s.
99. Lyubarev A.E. Izbiratel'ny'e sistemy': rossijskij i mirovoj opy't. — M.: NLO, 2016. — 632 s.
100. Lyubarev A.E. Ocenka territorial'noj odnorodnosti itogov golosovaniya v Rossijskoj Federacii i rossijskix regionax // E'lektoral'naya politika. — 2021. — № 2 (6). — S. 1.
101. Lyubarev A.E., Buzin A.Yu., Ky'nev A.V. Mertvy'e dushi. Metody' fal'sifikacii itogov golosovaniya i bor'ba s nimi. — M.: CzPK «Nikkolo M», 2007. — 192 s.
102. Manakov A.G., Kapkina I.V. E'lektoral'naya geografiya Rossii i Pskovskoj oblasti. — Pskov: Centr «Vozrozhdenie», 1998. — 48 s.
103. Manujlov M.A., Shklyar A.B. Predstavlenie ob e'lektoral'noj geografii Sankt-Peterburga na osnovanii nekotory'x itogov vy'borov 12.12.93 g. // Regional'naya politika. — 1994. — № 1. — S. 132–142.
104. Meleshkina E.Yu. Voronka prichinnosti v e'lektoral'ny'x issledovaniyax // Polis. — 2002. — № 5. — S. 47–53.
105. Meleshkina E.Yu. Issledovaniya e'lektoral'nogo povedeniya: teoreticheskie modeli i problemy' ix primeneniya // Politicheskaya nauka. — 2001. — № 2. — S. 187–212.

106. Meleshkina E.Yu. Koncepciya social'no-politicheskix razmezhevanij: problema universal'nosti // Politicheskaya nauka. — 2004. — № 4. — S. 11–29.
107. Mel'vil' A.Yu. Metodologiya «voronki prichinnosti» kak promezhutochny'j sintez «struktury' i agenta» v analize demokraticheskix tranzitov // Polis. — 2002. — № 5. — S. 54–59.
108. Mirkin B.G. Problema gruppovogo vy'bora. — M.: Nauka, 1974. — 256 c.
109. Morozova O.S. Kriterii ocenki kachestva predstavitel'nosti izbiratel'ny'x sistem // Kaspijskij region: politika, e'konomika, kul'tura. — 2013. — № 2 (35). — S. 67–72.
110. Morozova O.S. Municipal'ny'e vy'bory': teoriya i praktika. — Ryazan': Ryazanskij institut biznesa i upravleniya, 2013. — 180 s.
111. Morozova O.S. Formirovanie izbiratel'ny'x okrugov kak metod e'lektoral'nogo targetirovaniya // Kaspijskij region: politika, e'konomika, kul'tura. — 2013. — № 1 (34). — S. 106–112.
112. Muxametov R.S. E'lektoral'ny'e predpochteniya izbiratelej: e'ffekt «druzej i sosedej» imeet znachenie v Rossii? // Politicheskaya nauka. — 2022. — № 4. — S. 165–184.
113. Myagkov M., Sitnikov A., Shakin D. E'lektoral'ny'j landshaft Rossii: Analiticheskij doklad. — M.: Institut otkry'toj e'konomiki, 2004. — 41 s.
114. Ob"edinenie regionov Rossijskoj Federacii: Sociologicheskie danny'e, glubinny'e interv'yu, sravnitel'ny'j analiz / I.Yu. Okunev, P.V. Oskolkov, M.I. Tislenko, E'.S. Bibina, R.S. Shilovskij. — M.: Aspekt Ppecc, 2020. — 208 s.
115. Ovchinnikov B.V. E'lektoral'naya e'volyuciya: prostranstvo regionov i prostranstvo partij v 1995 i 1999 godax // Polis. — 2000. — № 2. — S. 68–79.
116. Oreshkin D.B. Geografiya e'lektoral'noj kul'tury' i cel'nost' Rossii // Polis. — 2001. — № 1. — S. 73–93.
117. Oreshkin D.B. E'lektoral'naya demokratiya i celostnost' politicheskogo prostranstva Rossii // Zhurnal o vy'borax. — 2001. — № 2. — S. 28–33.
118. Oreshkin D.B., Kozlov V.N. Geografiya prezidentskix vy'borov 2004 goda // Zhurnal o vy'borax. — 2004. — № 2. — S. 6–9.

119. Okunev I.Yu. Geograficheskij favoritizm partij i izbiratel'ny'x sistem // Politicheskaya nauka. — 2022. — № 4. — S. 90–106.
120. Okunev I.Yu. Osnovy' prostranstvennogo analiza. — M.: Aspekt Press, 2020. — 255 s.
121. Okunev I.Yu. Politicheskaya geografiya. — M.: Aspekt Press, 2019. — 512 s.
122. Okunev I.Yu. Ciklichnost' idejno-politicheskix razmezhevanij v e'lektoral'nom prostranstve: k novomu prochteniyu koncepcii Lipseta–Rokkana // Vestnik Permskogo universiteta. Politologiya. — 2022. — T. 16. № 3. — S. 52–62.
123. Oskolkov P.V. Ocherki po e'tnopolitologii. — M.: Aspekt Press, 2021. — 176 s.
124. Oskolkov P.V. Partijnaya sistema E'stonii na sovremennom e'tape: e'lektoral'naya turbulentnost' i smena e'tnoregional'ny'x patternov // Baltijskij region. — 2020. — T. 12. № 1. — S. 4–15.
125. Panov P.V. Izmenenie e'lektoral'ny'x institutov v Rossii (krossregional'ny'j sravnitel'ny'j analiz) // Polis. — 2004. — № 6. — S. 16–28.
126. Panov P.V. Prostranstvennaya lokalizaciya e'tnicheskix grupp kak faktor golosovaniya na vy'borax v nacional'ny'x respublikax Rossijskoj Federacii // Vestnik Permskogo federal'nogo issledovatel'skogo centra. — 2019. — № 2. — S. 53–62.
127. Panov P.V. Reforma regional'ny'x izbiratel'ny'x sistem i razvitie politicheskix partij v regionax Rossii: Krossregional'ny'j sravnitel'ny'j analiz // Polis. — 2005. — № 5. — S. 102–117.
128. Partii i partijny'e sistemy': sovremenny'e tendencii razvitiya / B.I. Makarenko i dr. — M.: Politicheskaya e'nciklopediya, 2015. — 303 s.
129. Pervy'j e'lektoral'ny'j cikl v Rossii. 1993–1996 gg. / Obshh. red. V.Ya. Gel'man, G.V. Golosov, E.Yu. Meleshkina. — M.: Ves' Mir, 2000. — 248 s.
130. Petrenko E.S. Regional'ny'e vy'bory': predvaritel'ny'e rezul'taty' i vy'vody' // Vlast'. — 1997. — № 1. — S. 27–32.
131. Petrov N.V. Demografiya i vy'bory' // Naselenie i obshhestvo. — 1995. — № 4. — S. 1–3.
132. Podkolzina E.A., Demidova O.A., Kuleczkaya L.E. Prostranstvennoe modelirovanie e'lektoral'ny'x predpochtenij v Rossijskoj Federacii // Prostranstvennaya e'konomika. — 2020. — T. 16. № 2. — S. 70–100.

133. Politicheskaya geografiya i sovremennost'. Regional'ny'e i prikladny'e aspekty' / Otv. red. S.B. Lavrov. — SPb.: SPbGU, 1991. — 196 s.
134. Ponomarenko E.V., Isaev V.A. E'konomika i finansy' obshhestvennogo sektora (osnovy' teorii e'ffektivnogo gosudarstva). — M.: INFRA-M, 2009. — 427 s.
135. Popov P.L., Saraev V.G., Cherenev A.A. E'lektoral'ny'e predpochteniya na regional'nom i makroregional'nom urovnyax: 2016–2018 goda // Grazhdanin. Vy'bory'. Vlast'. — 2017. — № 4. — S. 59–68.
136. Popov P.L., Cherenev A.A., Gales D.A., Saraev V.G. Kartograficheskoe otobrazhenie e'lektoral'ny'x yavlenij (po materialam vy'borov v Gosudarstvennuyu dumu Rossijskoj Federacii 2016 g.) // Geodeziya i kartografiya. — 2018. — T. 79. № 5. — S. 37–44.
137. Popov P.L., Cherenev A.A., Saraev V.G., Gales D.A. E'lektoral'naya podderzhka partij v Vostochnoj Sibiri: makroregional'ny'j i regional'ny'j aspekty' // Geografiya i prirodny'e resursy'. — 2019. — № 3. — S. 146–153.
138. Ralko A.N. Vozniknovenie i razvitie e'lektoral'noj geografii v SShA i Zapadnoj Evrope // Ucheny'e zapiski Tavricheskogo nacional'nogo universiteta imeni V.I. Vernadskogo. Seriya: Geografiya. — 2012. — T. 25 (64), № 2. — S. 147–152.
139. Razmaxnina Yu.S. Vliyanie e'tnicheskogo faktora na itogi izbiratel'ny'x kampanij v Vostochnoj Sibiri // Geopolitika i e'kogeodinamika regionov. — 2020. — T. 6. № 3. — S. 251–259.
140. Razmaxnina Yu.S. E'tnicheskij faktor v e'lektoral'no-geograficheskix processax (na primere Irkutskoj oblasti) // Geograficheskij vestnik. — 2020. — № 2 (53). — S. 48–62.
141. Rachev P.A. Dinamika e'lektoral'ny'x predpochtenij zhitelej aglomeracij SShA v 2000–2016 gg. // Regional'ny'e issledovaniya. — 2020. — № 4 (70). — S. 58–71.
142. Regional'ny'e vy'bory' v Rossii: Sb. st. / Pod red. Z.M. Zotovoj, A.I. Kovlera. — M.: RCzOIT, 1996. — 85 s.
143. Rozanova V.V. E'lektoral'naya karta mestny'x vy'borov v Ukraine: faktor politicheskoj regionalizacii ili politicheskogo mnogoobraziya? // Polite'ks. — 2007. — T. 3. № 2. — S. 217–226.
144. Rossiya na vy'borax: uroki i perspektivy'. Politgeograficheskij analiz / Pod red. V.A. Kolosova, R.F. Turovskogo. — M.: CzPT, ROPCz, 1995. — 218 s.

145. Savenkov R.V. E'lektoral'ny'j potencial oppozicionny'x kandidatov po itogam vy'borov v regional'ny'e parlamenty' v Central'no-Chernozemnom regione (2010–2018) // Politicheskaya nauka. — 2019. — № 2. — S. 74–94.
146. Savoskul M.S. Sociologiya: teorii, metody' i vozmozhnosti ix primeneniya v geografii. — M.: Geograficheskij fakul'tet MGU, 2012. — 202 s.
147. Salmin A.M. Izbiratel'ny'e sistemy' i partii: vy'bor vy'borov // Politiya. — 2004. — № 1 (32). — S. 12–24.
148. Skachkov V.S. Problemy' issledovaniyami e'lektoral'noj geografii meksiki v XXI v. // Geograficheskaya sreda i zhivy'e sistemy'. — 2020. — № 2. — S. 44–51.
149. Sy'ropyatova M.V. Social'ny'e faktory' e'lektoral'nogo povedeniya na mestny'x vy'borax v Rossii // Politicheskaya nauka. — 2017. — № 1. — S. 301–310.
150. Tarasov I.N., Krishtal' M.I., Urazbaev E.E. Vliyatel'nost' e'tnoterritorial'ny'x faktorov na e'lektoral'ny'e praktiki v stranax Baltii // Politicheskaya nauka. — 2022. — № 4. — S. 207–239.
151. Tarasov I.N., Fidrya E.S. Geografiya e'lektoral'noj volatil'nosti v Varmin'sko-Mazurskom voevodstve Pol'shi // Baltijskij region. — 2016. — T. 8. № 4. — S. 78–89.
152. Titkov A.S. Istoricheskaya geografiya vy'borov kak formiruyushheesya napravlenie // Voprosy' geografii. Vy'p. 136: Voprosy' istoricheskoj geografii / Pod red. V.M. Kotlyakova, V.N. Streleczkogo. — M.: Kodeks, 2014. — S. 113–129.
153. Titkov A.S. Metody' e'lektoral'noj geografii 1990-x godov: rabotayut li oni v 2000-e gody' // Politicheskaya sociologiya / Otv. red. V.L. Rimskij. — M.: ROSSPE'N, 2008. — S. 91–101.
154. Tregubov N.A. Faktory' golosovaniya: voprosy' klassifikacii i analiza // Polis. — 2017. — № 3. — S. 119–134.
155. Tretij e'lektoral'ny'j cikl v Rossii / Pod red. V.Ya. Gel'mana. — SPb.: Izd-vo Evropejskogo universiteta v Sankt-Peterburge, 2007. — 294 s.
156. Turovskij R.F. Geograficheskie zakonomernosti e'lektoral'nogo tranzita v postkommunisticheskix stranax // Politiya. — 2004. — № 4 (35). — S. 110–149.

157. Turovskij R.F. Konceptual'naya e'lektoral'naya karta postsovetskoj Rossii // Politiya. — 2005. — № 4 (39). — S. 161–202.
158. Turovskij R.F. Nacionalizaciya i regionalizaciya partijny'x sistem: podxody' k issledovaniyu // Politiya. — 2016. — № 1 (80). — S. 162–180.
159. Turovskij R.F. Politicheskaya geografiya. — Smolensk: SGU, 1999. — 381 s.
160. Turovskij R.F. Regional'noe izmerenie e'lektoral'nogo processa // Obshhestvenny'e nauki i sovremennost'. — 2006. — № 5. — C. 5–19.
161. Turovskij R.F. Regional'ny'e osobennosti prezidentskix vy'borov 2000 g. // Vestnik MGU. Seriya 12. Politicheskie nauki. — 2000. — № 4. — S. 38–54.
162. Turovskij R.F. E'lektoral'noe prostranstvo Rossii: ot navyazannoj nacionalizacii k novoj regionalizacii? // Politiya. — 2012. — № 3 (66). — S. 100–120.
163. Turovskij R.F. E'lektoral'ny'e geostruktury' v zapadny'x demokratiyax: popy'tka sistemnogo sravnitel'nogo analiza // Politiya. — 2004. — № 1 (32). — S. 198–232; № 2 (33). — S. 200–217.
164. Uollerstajn M. Izbiratel'ny'e sistemy', partii i politicheskaya stabil'nost' // Polis. — 1992. — № 6. — S. 156–162.
165. Farty'shev A.N. Kolichestvenny'e metody' v rossijskix geopoliticheskix issledovaniyax // Politicheskaya nauka. — 2022. — № 4. — S. 18–40.
166. Frolov Yu.S. Kolichestvennaya xarakteristika formy' geograficheskix yavlenij // Izvestiya Vsesoyuznogo geograficheskogo obshhestva. — 1974. — № 4. — S. 281–291.
167. Cherenev A.A. Politiko-geograficheskie metody' i faktory' v formirovanii izbiratel'ny'x okrugov // Izvestiya Irkutskogo gosudarstvennogo universiteta. Seriya: Politologiya. Religiovedenie. — 2017. — T. 21. — S. 126–133.
168. Cherenev A.A. Social'no-e'konomicheskoe razvitie i e'lektoral'noe prostranstvo municipal'ny'x obrazovanij // Izvestiya Irkutskogo gosudarstvennogo universiteta. Seriya: Politologiya. Religiovedenie. — 2012. — № 2 (9), ch. 1. — S. 18–24.

169. Cherenev A.A. Usloviya formirovaniya e'lektoral'ny'x predpochtenij v sel'skoj mestnosti (na primere Irkutskoj oblasti) // Geografiya i prirodny'e resursy'. — 2009. — № 4. — S. 134–139.
170. Cherkashin K.V. Vliyanie xaraktera zhilishhnoj zastrojki na itogi golosovanij po izbiratel'ny'm uchastkam // Informacionny'e vojny'. — 2015. — № 2 (34). — S. 48–57.
171. Cherkashin K.V. Proporcional'nost' e'lektoral'noj podderzhki mezhdu territoriyami i vosstanovlenie itogov golosovanij // Vestnik Moskovskogo universiteta. Seriya 12. Politicheskie nauki. — 2021. — № 1. — S. 70–88.
172. Cherkashin K.V. E'lektoral'naya geografiya postperevorotnoj Ukrainy' i novaya rasstanovka sil v strane // Vestnik Moskovskogo universiteta. Seriya 12. Politicheskie nauki. — 2015. — № 5. — S. 103–115.
173. Chirkin V.E. Partii i vy'bory' (rossijskij i zarubezhny'j opy't) // Zhurnal o vy'borax. — 2001. — № 1. — S. 10–14.
174. Chuvilina N.B. Vy'bory' kak faktor razvitiya politiko-vlastny'x processov v stranax SNG: teoretiko-metodologicheskij aspekt // Polite'ks. — 2008. — T. 4. № 4. — S. 213–228.
175. Chugrov S.V. E'lektoral'noe povedenie rossijskix regionov: Statisticheskij analiz vy'borov dekabrya 1995 g. // Mirovaya e'konomika i mezhdunarodny'e otnosheniya. — 1996. — № 6. — S. 27–39.
176. Shestakova M.N. E'lektoral'ny'e issledovaniya prigranichny'x territorij v otechestvennoj nauke: sostoyanie i perspektivy' // Kaspijskij region: politika, e'konomika, kul'tura. — 2020. — № 1 (62). — S. 43–54.
177. Shmatkova L.P., Domanov A.O. Opy't sravnitel'nogo prostranstvennogo analiza e'lektoral'nogo povedeniya v regionax gosudarstv — sosedej Rossii // Politicheskaya nauka. — 2022. — № 4. — S. 145–164.
178. Shumilov A.V. E'lektoral'naya geografiya: regional'noe prostranstvo. — Cheboksary': ChGPU, 2009. — 159 s.
179. E'volyuciya e'lektoral'nogo landshafta / Pod red. A.A. Sidorenko. — M.: KomKniga, 2005. — 68 s.
180. E'lektoral'naya politologiya: Teoriya i opy't Rossii / Pod red. L.V. Smorgunova. — SPb.: Izd-vo SPbGU, 1998. — 202 s.

181. E'lektoral'naya Rossiya 2015. Ezhegodny'j sbornik statej o rossijskix vy'borax / Agentstvo strategicheskix kommunikacij «Nikkolo M». — M.: Izd-vo «Pero», 2016. — 272 s.
182. E'lektoral'naya Rossiya — 2016: bitva za Gosdumu. Ezhegodny'j sbornik statej o rossijskix vy'borax / Pod red. I. Mintusova. — M.: Grifon, 2017. — 272 s.
183. E'lektoral'naya sociologiya / Pod red. S.B. Suvorova. — Saratov: Saratovskaya gosudarstvennaya akademiya prava, 2009. — 216 s.

In other languages:

184. *Agnew J.* From Political Methodology to Geographical Social Theory? A Critical Review of Electoral Geography, 1960–1987 / Johnson R.J., Shelley F.M., Taylor P.J. (eds.) Developments in Electoral Geography. — London: Routledge, 1990. — P. 15–21.
185. *Agnew J.* Models of Spatial Variation in Political Expression: The Case of the Scottish National Party // International Political Science Review. — 1985. — Vol. 6. No. 2. — P. 171–196.
186. *Agnew J.* Mapping Politics: How Context Counts in Electoral Geography // Political Geography. — 1996. — Vol. 15. No. 2. — P. 129–146.
187. *Agnew J.* Place and Politics. The Geographical Mediation of State and Society. — London: Routledge, 1987. — 286 p.
188. *Alam M.S., Sivaramakrishnan K.C.* Fixing Electoral Boundaries in India: Laws, Processes, Outcomes, and Implications for Political Representation. — New Delhi: Oxford University Press, 2015. — 279 p.
189. *Almond G., Verba S.* The Civic Culture: Political Attitudes and Democracy in Five Nations. — Princeton: Princeton University Press, 1963. — 574 p.
190. *Almond G., Verba S.* (eds.) The Civic Culture Revisited: An Analytic Study. — Boston: Little, Brown, 1980. — 421 p.
191. *Alonso S.* Challenging the State: Devolution and the Battle for Partisan Credibility: A Comparison of Belgium, Italy, Spain and the United Kingdom. — Oxford: Oxford University Press, 2012. — 278 p.
192. *Amani K.Z.* Elections in Haryana, India: A Study in Electoral Geography // The Geographers. — 1970. — Vol. 17. — P. 27–40.

193. *Anselin L.* A Local Indicator of Multivariate Spatial Association, Extending Geary's c // Geographical Analysis. — 2019. — Vol. 51. No. 2. — P. 133–150.

194. *Anselin L.* Local Indicators of Spatial Association — LISA. // Geographical Analysis. — 1995. — Vol. 27. No. 2. — P. 93–115.

195. *Anselin L.* Spatial Econometrics: Methods and Models. — Boston: Kluwer, 1988. — 300 p.

196. *Anselin L., Xun L.* Tobler's Law in a Multivariate World // Geographical Analysis. — 2020. — Vol. 52. No. 4. — P. 494–510.

197. *Ansolabehere S., Maxwell P.* A Two-Hundred Year Statistical History of the Gerrymander // Ohio State Law Journal. — 2016. — Vol. 77. No. 4. — P. 741–762.

198. *Archer J.C.* The Geography of an Interminable Election: Bush v. Gore, 2000 // Political Geography. — 2002. — Vol. 21. No. 1. — P. 71–77.

199. *Archer C., Shelley F.* American Electoral Mosaics. — Washington: Association of American Geographers, 1986. — 97 p.

200. *Arminen I.* Social Functions of Location in Mobile Telephony // Personal and Ubiquitous Computing. — 2006. — Vol. 10. No. 5. — P. 319–323.

201. *Arrow K.* Social Choice and Individual Values. — N.Y.: J. Wiley and Sons, 1951. — 99 p.

202. *Bailey M.A., Strezhnev A., Voeten E.* Estimating Dynamic State Preferences from United Nations Voting Data // Journal of Conflict Resolution. — 2017. — Vol. 61. No. 2. — P. 430–456.

203. *Balinski M.L., Young P.H.* Fair Representation: Meeting the Ideal of One Man, One Vote. — Washington: Brookings Institution Press, 2001. — 195 p.

204. *Ball M.* Bloc Voting in the General Assembly // International Organization. — 1951. — Vol. 5. No. 1. — P. 3–31.

205. *Banzhaf J.F.* Weighted Voting Doesn't Work: A Mathematical Analysis // Rutgers Law Revue. — 1965. — Vol. 19. No. 2. — P. 317–343.

206. *Barnes R., Solomon J.* Gerrymandering and Compactness: Implementation Flexibility and Abuse // Political Analysis. — 2021. — Vol. 29. No. 4. — P. 448–466.

207. *Baur N., Hering L., Raschke A.L., Thierbach C.* Theory and Methods in Spatial Analysis. Towards Integrating Qualitative, Quantitative

and Cartographic Approaches in the Social Sciences and Humanities // Historical Social Research. — 2014. — Vol. 39. No. 2. — P. 7–50.
208. *Bawn K.* Voter Responses to Electoral Complexity: Ticket Splitting, Rational Voters and Representation in the Federal Republic of Germany // British Journal of Political Science. — 1999. — Vol. 29. No. 3. — P. 487–505.
209. *Beaujeu-Garnier J.* Essai de géographie électorale guinéenne // Cahiers d'Outre-Mer. — 1958. — Vol. 11. No. 44. — P. 309–333.
210. *Bekaroğlu E.A., Osmanbaşoğlu G.K.* (eds.) Turkey's Electoral Geography: Trends, Behaviors, and Identities. — L.: Routledge, 2021. — 204 p.
211. *Bickerstaff, S.* Election Systems and Gerrymandering Worldwide Studies in Choice and Welfare. — NY.: Springer, 2020. — 259 p.
212. *Billet J.* L'Expression Politique en Grésivaudan et son Interprétation Géographique // Revue de Géographie Alpine — 1958. — Vol. 46. Part 1. — P. 97–128.
213. *Black D.* On the Rationale of Group Decision-Making // Journal of Political Economy. — 1948. — Vol. 56. No. 1. — P. 23–34.
214. *Blake D.E.* The Measurement of Regionalism in Canadian Voting Patterns // Canadian Journal of Political Science. — 1967. — Vol. 5. No. 1. — P. 55–81.
215. *Boamfa I.* Atlasul electoral al României: (1990–2009). — Iaşi: Editura Universităţii «Alexandru Ioan Cuza», 2013. — 754 p.
216. *Boamfa I.* Geografie electorală. — Iaşi: Editura Universităţii «Alexandru Ioan Cuza», 2013. — 462 p.
217. *Bochsler D.* Measuring Party Nationalization: A New Ginibased Indicator That Corrects for the Number of Units // Electoral Studies. — 2010. — Vol. 29. No. 1. — P. 155–168.
218. *Bon F., Cheylan J.-P.* La France qui vote. — Paris: Hachette, 1988. — 464 p.
219. *Brimo A.* Méthode de la géo-sociologie électorale. — Paris: Pédone, 1968. — 142 p.
220. *Brown M., Knopp L., Morill R.* The Culture Wars and Urban Electoral Politics: Sexuality, Race, and Class in Tacoma, Washington // Political Geography. — 2005. — Vol. 24. No. 3. — P. 267–291.

221. *Brunn S.D.* Geography and Politics in America. — NY: Joanna Cotler Books, 1974. — 443 p.
222. *Brusa C.* Elezioni, territorio, società. — Milano: Unicopli, 1986. — 182 p.
223. *Brusa C.* Geografia elettorale nell'Italia del dopoguerra. Edizione aggiornata alle elezioni politiche. — Parma: Unicopli, 1984. — 180 p.
224. *Bunge W.* Gerrymandering, Geography and Grouping // Geographical Review. — 1966. — Vol. 56. No. 2. — P. 256–263.
225. *Bunge W.* Theoretical Geography. Lund Studies in Geography, Series C: General and Mathematical Geography. — Lund: Gleerup, 1962. — 210 p.
226. *Burden B.* Deterministic and Probabilistic Voting Models // American Journal of Political Science. — 1977. — Vol. 41. No. 4. — P. 1150–1169.
227. *Burghardt A.F.* The Bases of Support for Political Parties in Burgenland // Annals of the Association of American Geographers. — 1964. — Vol. 54. No. 3. — P. 372–390.
228. *Bürklin W.* Was leistet die Wahlgeographie? // Geographische Rundschau. — 1980. — Vol. 32. No. 9. — P. 396–403.
229. *Busteed M.A.* Geography and Voting Behavior. — London: Oxford University Press, 1975. — 60 p.
230. *Caldas de Castro M., Singer B.H.* Controlling the False Discovery Rate: A New Application to Account for Multiple and Dependent Tests in Local Statistics of Spatial Association // Geographical Analysis. — 2006 — Vol. 38. No. 2. — P. 180–208.
231. *Calvert R.* Robustness of the Multidimensional Voting Model: Candidate Motivations, Uncertainty, and Convergence // American Journal of Political Science. — 1985. — Vol. 29. No. 1. — P. 69–95.
232. *Cambell A., Converse P., Miller W., Stokes D.* The American Voter. — NY.: Wiley, 1960. — 576 p.
233. *Campbell R., Cowley P., Vivyan N., Wagner M.* Why Friends and Neighbors? Explaining the Electoral Appeal of Local Roots // The Journal of Politics. — 2019. — Vol. 81. No. 3. — P. 937–951.
234. *Caramani D.* The Nationalization of Politics: The Formation of National Electorates and Party Systems in Western Europe. — Cambridge: Cambridge University Press, 2004. — 347 p.

235. *Carnap R.* Der Raum: Ein Beitrag zur Wissenschaftslehre // Kant-Studien Ergänzungshefte. — 1922. — No. 56. — P. 21–208.
236. *Carstairs A.M.* Short History of Electoral Systems in Western Europe. — Abingdon: Routledge, 2010. — 256 p.
237. *Chalothorn S.* Greater Bangkok: An Analysis in Electoral Geography, 1957–1976. — Bangkok: The Social Science Association of Thailand, 1982. — 77 p.
238. *Chamberlin W.* The North Atlantic Bloc in the UN General Assembly // Orbis. — 1958. — Vol. 1, No. 4. — P. 459–473.
239. *Chand R.* Geography of Electoral Participation in Himachal Pradesh (1982–1990): A Spatial Perspective // Transactions, Institute of Indian Geographers. — 1996. — No. 18. — P. 31–44.
240. *Chevalier M.* Problèmes de la géographie électorale française // Revue Géographique de l'Est. — 1985. — Vol. 25. No. 1. — P. 93–118.
241. *Chiba T.* Research Trends in the Study of Electoral Geography // Geographical Review of Japan. — 1978. — Vol. 51. No. 3. — P. 235–244.
242. *Cliff A.D., Ord J.K.* Spatial Autocorrelation. — London: Pion, 1973. — 178 p.
243. *Cliff A.D., Ord J.K.* Spatial Processes, Models and Applications. — London: Pion, 1981. — 266 p.
244. *Cliff A.D. Ord J.K.* The Problem of Spatial Autocorrelation / Scott A.J. (ed.) London Papers in Regional Science. — London: Pion, 1969. — P. 25–55.
245. *Coates B.E., Rawstron E.M.* Regional Variations in Britain. — London: Batsford, 1971. — 300 p.
246. *Colín A.I.* (comp.) Elecciones y geografía electoral. — México: Instituto Electoral del Distrito Federal, 2007. — 406 p.
247. *Condorcet M. de.* Essai sur l'application de l'analyse à la probabilité des décisions rendues à la pluralité des voix. — Paris: De l'Imprimerie royale, 1785. — 304 p.
248. *Coughlin P.* Pareto Optimality of Policy Proposals with Probabilistic Voting // Public Choice. — 1982. — Vol. 39. No. 3. — P. 427–433.
249. *Coughlin P.* Probabilistic Voting Theory. — Cambridge: Cambridge University Press, 1992. — 268 p.

250. *Cover B.P.* Quantifying Partisan Gerrymandering: An Evaluation of the Efficiency Gap Proposal // Stanford Law Review. — 2018. — Vol. 70. No. 4. — P. 1131–1233.
251. *Cox K.R.* Conflict, Power and Politics in the City: A Geographical View. — NY: McGraw-Hill, 1973. — 154 p.
252. *Cox K.R.* The Spatial Components of Urban Voting Response Surfaces // Economic Geography. — 1971. — Vol. 47. No. 1. — P. 27–35.
253. *Cox K.R.* The Voting Decision in a Spatial Context // Progress in Geography. — 1969. — Vol. 1. No. 1. — P. 81–117.
254. *Cruz-Coke R.* Geografía electoral de Chile. — Santiago: Editorial del Pacífico, 1952. — 139 p.
255. *David Q., Hamme van G.* Pillars and Electoral Behavior in Belgium: The Neighborhood Effect Revisited // Political Geography. — 2011. — Vol. 30. No. 5. — P. 250–262.
256. *Dean V.K.* Geographic Aspects of the Newfoundland Referendum // Annals of the Association of American Geographers. — 1949. — Vol. 39. No. 2. — P. 159–176.
257. *Dekker A.* The Eurovision Song Contest as a 'Friendship' Network // Connections. — 2007. — Vol. 27. No. 3. — P. 53–58.
258. *Detterbeck K., Hepburn E.* (eds.) Handbook of Territorial Politics. — Cheltenham: Edward Elgar Publishing, 2018 — 432 p.
259. *Diamanti I.* La Lega: geografia, storia e sociologia di un nuovo soggetto politico. — Rome: Donzelli, 1993. — 127 p.
260. *Dikshit S.K.* Electoral Geography of India: With Special Reference to Sixth & Seventh Lok Sabha Elections. — Varanasi: Vishwavidyalaya Prakashan, 1993. — 141 p.
261. *Dikshit R.D.* On the Place of Electoral Studies in Political Geography // Transactions, Institute of Indian Geographers. — 1980. — No. 18. — P. 23–28.
262. *Dikshit R.D., Sharma V.* Voting Preferences in State vis a vis National Elections under a Federal System: A Case Study of Haryana (India) // Transactions, Institute of Indian Geographers. — 1993. — No. 15. — P. 51–70.

263. *Dikshit S.K., Giri H.H.* Concept and Purpose of Electoral Geography // Transactions, Institute of Indian Geographers. — 1984. — Vol. 6. No. 12. — P. 85–88.

264. *Dinas E., Gemenis K.* Measuring Parties' Ideological Positions with Manifesto Data: A Critical Evaluation of the Competing Methods // Party Politics. — 2010. — Vol. 16. No. 4. — P. 427–450.

265. *Dogan M., Rokkan S.* (eds.) Quantitative Ecological Analysis in the Social Sciences. — Cambridge: MIT Press, 1969. — 607 p.

266. *Downs A.* An Economic Theory of Democracy. — NY: Harper and Row, 1957. — 320 p.

267. *Dunham P.* Space, Place and Pollbooks: Incorporating a Neglected Electoral Geography // Area. — 1997. — Vol. 29. No. 2. — P. 141–150.

268. *Easton D.* System Analysis of Political Life. — NY: John Wiley & Sons, 1965. — 507 p.

269. *Ebeid M., Rodden J.* Economic Geography and Economic Voting: Evidence from the US States // British Journal of Political Science. — 2006. — Vol. 36. No. 3. — P. 527–547.

270. *Emmerich G.E., Ayala A.A.* Votos y mapas: estudios de geografía electoral en México. — Toluca: UAEM, 1993. — 340 p.

271. *Enelow J.M., Hinich M.J.* Advances in the Spatial Theory of Voting. — NY: Cambridge University Press, 1990. — 256 p.

272. *Enelow J.M., Hinich M.J.* Probabilistic Vote and the Importance of Centrist Ideologies in Democratic Elections // The Journal of Politics. — 1984. — Vol. 46. No. 2. — P. 459–478.

273. *Enelow J.M., Hinich M.J.* The Spatial Theory of Voting: An Introduction. — NY: Cambridge University Press, 1984. — 238 p.

274. *Erikson R.S.* Malapportionment, Gerrymandering and Party Fortunes in Congressional Elections // American Political Science Review. — 1972. — Vol. 66. No. 4. — P. 1234–1245.

275. *Ethington P.J., McDaniel J.A.* Political Places and Institutional Spaces: The Intersection of Political Science and Political Geography // Annual Review of Political Science. — 2007. — Vol. 10. — P. 127–142.

276. *Falter J.W., Winkler J.R.* Wahlgeographie und politische Ökologie // Falter J.W., Schoen H. (eds.). Handbuch Wahlforschung. — Wiesbaden: Springer, 2014. — P. 135–167.
277. *Felsenthal D.S., Machover M.* The Measurement of Voting Power. — Cheltenham: Edward Elgar, 1998. — 322 p.
278. *Field W.H.* Regional Dynamics: The Basis of Electoral Support in Britain. — London: Frank Cass, 1997. — 210 p.
279. *Fiorina M.* Formal Models in Political Science // American Journal of Political Science. — 1975. — Vol. 19. No. 1. — P. 133–159.
280. *Fiorina M.* Retrospective Voting in American National Elections. — New Haven: Yale University Press, 1981. — 288 p.
281. *Fischer M.M., Getis A.* (eds.) Handbook of Applied Spatial Analysis. — Berlin: Springer, 2010. — 811 p.
282. *Fischer M.M., Getis A.* (eds.) Recent Developments in Spatial Analysis: Spatial Statistics, Behavioural Modelling, and Computational Intelligence. — Berlin: Springer, 1997. — 443 p.
283. *Fitton M.* Neighbourhood and Voting: A Sociometric Explanation // British Journal of Political Science. — 1973. — Vol. 3. No. 4. — P. 445–472.
284. *Flaherty M.S., Crumplin W.W.* Compactness and Electoral Boundary Adjustment: An Assessment of Alternative Measures // The Canadian Geographer. — 1992. — Vol. 36. No. 2. — P. 159–171.
285. *Flora P. Stein Rokkan.* Staat, Nation und Demokratie in Europa. Die Theorie Stein Rokkans aus seinen gesammelten Werken rekonstruiert und eingeleitet von Peter Flora. — Frankfurt am Main: Suhrkamp, 2000. — 499 p.
286. *Foladare I.S.* The Effect of Neighborhood on Voting Behavior // Political Science Quarterly. — 1968. — Vol. 83. No. 4. — P. 526–529.
287. *Forest B.* Regionalism in Election District Jurisprudence // Urban Geography. — 1996. — Vol. 17. No. 7. — P. 572–578.
288. *Forrest J., Johnston R.J.* Spatial Aspects of Voting in the Dunedin City Council Elections of 1971 // New Zealand Geographer. — 1973. — Vol. 29. No. 2. — P. 166–181.
289. *Fotheringham A.S., Brunsdon C., Charlton M.* Quantitative Geography — Perspectives on Spatial Data Analysis. — London: SAGE, 2000. — 288 p.

290. *Fotheringham S.A., Rogerson P.A.* GIS and Spatial Analytical Problems // International Journal of Geographical Information Systems. — 1993. — Vol. 7. No. 1. — P. 3–19.
291. *Fotheringham A.S., Rogerson P.A.* (eds.) The SAGE Handbook of Spatial Analysis. — London: SAGE, 2009. — 528 p.
292. *Franklin M.* The Decline of Cleavage Politics // Franklin M. (ed.) Electoral Change: Responses to Evolving Social and Attitudinal Structures in Western Countries. — NY: Cambridge University Press, 1992. — P. 383–405.
293. *Gabel M., Huber J.* Putting Parties in Their Place: Inferring Party Left-Right Positions from Party Manifestos Data // American Journal of Political Science. — 2000. — Vol. 44. No. 1. — P. 94–103.
294. *Gachon L.* La géographie électorale. Ses hypothèses possibles de travail // Revue de Géographie de Lyon. — 1955. — Vol. 30. No. 4. — P. 353–356.
295. *Gallagher M.* Comparing Proportional Representation Electoral Systems: Quotas, Thresholds, Paradoxes and Majorities // British Journal of Political Science. — 1992. — Vol. 22. No. 4. — P. 469–496.
296. *Gallagher M.* Proportionality, Disproportionality and Electoral Systems // Electoral Studies. — 1991. — Vol. 10. No. 1. — P. 33–51.
297. *Gallagher M., Mitchell P.* The Politics of Electoral Systems. — Oxford: Oxford University Press, 2008. — 237 p.
298. *Geary R.* The Contiguity Ratio and Statistical Mapping // The Incorporated Statistician. — 1954. — Vol. 5. No. 3. — P. 115–145.
299. *Getis A.* A History of the Concept of Spatial Autocorrelation: A Geographer's Perspective // Geographical Analysis. — 2008. — Vol. 40. No. 3. — P. 297–309.
300. *Getis A., Aldstadt J.* Constructing the Spatial Weights Matrix Using a Local Statistic // Geographical Analysis. — 2004. — Vol. 36. No. 2. — P. 90–104.
301. *Getis A., Ord J.K.* The Analysis of Spatial Association By Use of Distance Statistics // Geographical Analysis — 1992. — Vol. 24. No. 3. — P. 189–206.
302. *Gimpel J.G., Karnes K.A., McTague J., Pearson-Merkowitz S.* Distance-decay in the Political Geography of Friends-and-neighbors

Voting // Political Geography. — 2008. — Vol. 27. No. 2. — P. 231–252.

303. *Goguel F.* Géographie des élections françaises de 1870 à 1951. — Paris: Armand Colin, 1951. — 144 p.

304. *Goguel F.* Initiation aux recherches de géographie électorale. — Paris: Centre de Documentation Universitaire, Centre d'Etudes Supérieures de Sociologie, 1947. — 95 p.

305. *Golosov G.V.* Party Nationalization and the Translation of Votes into Seats under Single-Member Plurality Electoral Rules // Party Politics. — 2018. — Vol. 24. No. 2. — P. 118–128.

306. *Goodey B.R.* The Geography of Elections: An Introductory Bibliography. Grand Forks: University of North Dakota, 1968. — 64 p.

307. *Goodchild M.F., Anselin L., Appelbaum R.P., Harthorn B.H.* Towards a Spatially Integrated Social Science // International Regional Science Review. — 2000. — Vol. 23. No. 2. — P. 139–159.

308. *Granberg D., Brown T.* The Perception of Ideological Distance // The Western Political Quarterly. — 1992. — Vol. 45. No. 3. — P. 727–750.

309. *Grofman B., Owen G., Noviello N., Glazer A.* Stability and Certainty of Legislative Choice in the Spatial Context // American Political Science Review. — 1987. — Vol. 81. No. 2. — P. 539–553.

310. *Gudgin G., Taylor P.* Seats, Votes, and the Spatial Organization of Elections. — London: Pion, 1979. — 240 p.

311. *Guillorel H.* 70 ans de géographie électorale // Politix. — 1989. — Vol. 2, No. 5. — P. 57–68.

312. *Häge F.M.* Choice or Circumstance? Adjusting Measures of Foreign Policy Similarity for Chance Agreement // Political Analysis. — 2011. — Vol. 19. No. 3. — P. 287–305.

313. *Hamelin J., Letarte J., Hamelin M.* Les élections provinciales dans le Québec // Cahiers de géographie du Québec. — 1959. — Vol. 4. No. 7. — P. 5.

314. *Harmel R., Janda K.* Parties and Their Environments: Limits to Reform? — NY: Longman, 1982. — 176 p.

315. *Harris C.* A Scientific Method of Districting // Behavioral Science. — 1964. — Vol. 3. No. 9. — P. 219–225.

316. *Heberle R.* Landbevölkerung und Nationalsozialismus. Eine soziologische Untersuchung der politischen Willensbildung in Schleswig-Holstein 1918 bis 1932. — Stuttgart: Deutsche Verlagsanstalt, 1963. — 171 p.

317. *Hinich M.J., Khmelko V., Ordeshook P.C.* Ukraine's 1998 Parliamentary Elections: A Spatial Analysis // Post-Soviet Affairs. — 1999. — Vol. 15. No. 2. — P. 149–185.

318. *Hinich M.J., Pollard W.* A New Approach to the Spatial Theory of Electoral Competition // American Journal of Political Science. — 1981. — Vol. 25. No. 2. — P. 323–341.

319. *Hodge D.C., Staeheli L.A.* Social Relations and Geographic Patterns of Urban Electoral Behavior // Urban Geography. — 1992. — Vol. 13. No. 4. — P. 307–333.

320. *Hoffman D.* A Model for Strategic Voting // SIAM Journal of Applied Mathematics. — 1982. — Vol. 42. No. 4. — P. 751–761.

321. *Holbrook T.M.* Did the Whistle-Stop Campaign Matter? // Political Science & Politics. — 2002. — Vol. 35. No. 1. — P. 59–66.

322. *Hopkins D.A.* Red Fighting Blue: How Geography and Electoral Rules Polarize American Politics. — Cambridge: Cambridge University Press, 2017. — 244 p.

323. *Huber J., Inglehart R.* Expert Interpretations of Party Space and Party Locations in 42 Societies // Party Politics. — 1995. — Vol. 1. No. 1. — P. 73–111.

324. *Huckfeldt R., Sprague J.* Citizens, Politics and Social Communication: Information and Influence in an Election Campaign. — Cambridge: Cambridge University Press, 1995. — 305 p.

325. *Hugonnier S.* Tempéraments Politiques et Géographie Électorale de Deux Grands Vallées Intraalpines des Alpes du Nord: Maurienne et Tarentaise // Revue de Géographie Alpine. — 1954. — Vol. 42. Part 1. — P. 45–80.

326. *Ikeda K., Huckfeldt R.* Political Communication and Disagreement among Citizens in Japan and the United States // Political Behavior. — 2001. — No. 23 (1). — P. 23–51.

327. *Inglehart R.* Culture Shift in Advanced Industrial Society. — Princeton: Princeton University Press, 1990. — 504 p.

328. *Inglehart R.* The Silent Revolution: Changing Values and Political Styles Among Western Publics. — Princeton: Princeton University Press, 1977. — 496 p.
329. *Jalan S.* Electoral Geography. — Jaipur: Rawat Publications, 2015. — 254 p.
330. *Jenks G.F. Caspall F.C.* Error on Choroplethic Maps: Definition, Measurement, Reduction // Annals of American Geographers. — 1971. — Vol. 61. No. 2. — P. 217–244.
331. *Johnston R.J.* Multivariate Statistical Analysis in Geography, A Primer on the General Linear Model. — London and NY: Longman, 1978. — 280 p.
332. *Johnston R.J.* Political, Electoral, and Spatial Systems: An Essay in Political Geography. — Oxford: Clarendon Press, 1979. — 221 p.
333. *Johnston R.J.* Spatial Elements in Voting Patterns at the 1968 Christchurch City Council Election // Political Science. — 1972. — Vol. 24. No. 1. — P. 49–61.
334. *Johnston R.J.* Spatial Structure, Plurality Systems and Electoral Bias // Canadian Geographer. — 1976. — Vol. 20. No. 3. — P. 310–328.
335. *Johnston R.J.* The Electoral Geography of an Election Campaign // Scottish Geographical Magazine. — 1977. — Vol. 93. No. 2. — P. 98–108.
336. *Johnston R.J.* The Neighbourhood Effect Revisited: Spatial Science or Political Regionalism? // Environment and Planning D: Society and Space. — 1986. — Vol. 4. No. 1. — P. 41–55.
337. *Johnston R.J., Manley D., Pattie Ch., Jones K.* Geographies of Brexit and its Aftermath: Voting in England at the 2016 Referendum and the 2017 General Election // Space and Polity. — 2018. — Vol. 22. No. 2. — P. 162–187.
338. *Johnson R.J., Pattie Ch.* Kevin Cox and Electoral Geography / Jonas A., Wood A. (eds.) Territory, the State and Urban Politics: A Critical Appreciation of the Selected Writings of Kevin R. Cox. — Burlington: Ashgate, 2012. — P. 23–44.
339. *Johnson R.J., Shelley F.M., Taylor P.J.* (eds.) Developments in Electoral Geography. — London: Routledge, 1990. — 292 p.
340. *Jones K., Johnson R.J., Pattie Ch.* People, Places and Regions: Exploring the Case of Multi-Level Modelling in the Analysis of Electoral

Data // British Journal of Political Science. — 1992. — Vol. 22. No. 3. — P. 343–380.

341. *Jones M.P., Mainwaring S.* The Nationalization of Parties and Party Systems: An Empirical Measure and an Application to the Americas // Party Politics. — 2003. — Vol. 9. No. 2. — P. 139–166.

342. *Jusko K.L.* Who Speaks for the Poor? Electoral Geography, Party Entry, and Representation. — Cambridge: Cambridge University Press, 2017. — 218 p.

343. *Kasperson R.E., Minghi J.V.* The Structure of Political Geography. — Chicago: Aldine, 1969. — 527 p.

344. *Kavianirad M.* Explanation of Relationship between Geography and Elections (Electoral Geography) // Geopolitics Quarterly. — 2015. — Vol. 10. No. 4. — P. 93–108.

345. *Kerekes D.* The City as an Onion? Case Studies of Electoral Geography in Prague and Warsaw // Sociológia. — 2020. — Vol. 52. No. 3. — P. 245–272.

346. *Kevický D.* Themes, Approaches, and Methods in the Geographical Analysis of Czech and Slovak Parliamentary Elections: A Systematic Review // AUC Geographica. — 2021. — Vol. 56. No. 2. — P. 248–261.

347. *Key V.O. Jr.* A Theory of Critical Elections // Journal of Politics. — 1955. — Vol. 17, No. 1. — Pp. 3–18.

348. *Key V.O. Jr.* Southern Politics in State and Nation. — NY: Alfred A. Knopf Inc., 1949. — 675 p.

349. *Key V.O. Jr.* The Responsible Electorate: Rationality in Presidential Voting, 1936–1960. — Cambridge: Belnap Press, 1966. — 158 p.

350. *Khan N., Rahman M.* Electoral Geography: Spatial Analysis of Voting Patterns in 2011 Assembly Election in West Bengal // International Journal of Informative and Futuristic Research. — 2015. — Vol. 3. No. 4. — P. 1179–1187.

351. *King G.* A Solution to the Ecological Problem: Reconstructing Individual Behavior from Aggregate Data. — Princeton: Princeton University Press, 1997. — 342 p.

352. *King G.* Why Context Should Not Count // Political Geography. — 1996. — Vol. 15. No. 2. — P. 159–164.

353. *Kinnear M.* The British Voter: An Atlas and Survey Since 1885. — London: Batsford, 1968. — 176 p.
354. *Kish G.* Some Aspects of the Regional Political Geography of Italy // Annals of the Association of American Geographers. — 1953. — Vol. 43. — P. 178.
355. *Kovács Z., Vida G.* Geography of the New Electoral System and Changing Voting Patterns in Hungary // Acta Geobalcanica. — 2015. — Vol. 1. No. 2. — P. 55–64.
356. *Koziełło T., Szczepański D.* Geografia wyborcza Polski: interpretacje postaw i zachowań obywateli. — Rzeszów: Wydawnictwo Uniwersytetu Rzeszowskiego, 2018. — 339 p.
357. *Kowalski M.* Geografia wyborcza Polski — przestrzenne zróżnicowanie zachowań wyborczych Polaków w latach 1989–1998. — Warszawa: Instytut Geografii i Przestrzennego Zagospodarowania PAN, 2000. — 137 p.
358. *Kowalski M., Śleszyński P.* (red.) Atlas wyborczy Polski. — Warszawa: Instytut Geografii i Przestrzennego Zagospodarowania PAN, 2018. — 369 p.
359. *Krehbiel E.* Geographic Influences in British Elections // Geographical Review. — 1916. — Vol. 2. No. 6. — P. 419–432.
360. *Laakso M., Taagepera R.* Effective Number of Parties: A Measure with Application to Western Europe // Comparative Political Studies. — 1979. — Vol. 12. No. 1. — P. 3–27.
361. *Lancelot M.-T., Lancelot A.* Atlas des élections françaises de 1968 et 1969 // Revue française de science politique. — 1970. — Vol. 20. No. 2. — P. 312–328.
362. *Lazarsfeld P., Berelson B., Gaudet H.* The People's How Voter Makes Up His Mind in a Presidential Campaign. — NY.: Columbia University Press, 1968. — 178 p.
363. *Laserwitz B.* Suburban Voting Trends: 1948–1956 // Social Forces. — 1960. — Vol. 39. No. 1. — P. 29–36.
364. *Laux H.D., Simms A.* Parliamentary Elections in West Germany: The Geography of Electoral Choice // Area. — 1973. — Vol. 5. No. 3. — P. 161–171.
365. *Leib J., Quinton, N.* On the Shores of the "Moribund Backwater"? Trends in Electoral Geography Research since 1980 // Warf B., Leib

J. (eds.) Revitalizing Electoral Geography. — Farnham: Ashgate, 2011. — P. 9–27.

366. *Lee S.I.* Developing a Bivariate Spatial Association Measure: An Integration of Pearson's r and Moran's I // Journal of Geographical Systems. — 2001. — No. 3(4). — P. 369–385.

367. *Lehmann S.H.* The German Elections in the 1870s: Why Germany Turned from Liberalism to Protectionism // The Journal of Economic History. — 2010. — Vol. 70. No. 1. — P. 146–178.

368. *Leleu C.* Géographie des élections françaises depuis 1936. — Paris: Presses universitaires de France, 1971. — 353 p.

369. *Lijphart A.* Electoral Systems and Party Systems: A Study of Twenty-Seven Democracies, 1945–1990. — Oxford: Oxford University Press, 1994. — 209 p.

370. *Lijphart A.* Patterns of Democracy: Government Forms and Performance in Thirty-Six Countries. — New Haven: Yale University Press, 1999. — 351 p.

371. *Lijphart A.* The Analysis of Bloc Voting in the General Assembly: A Critique and a Proposal // The American Political Science Review. — 1963. — Vol. 57. No. 4. — P. 902–917.

372. *Linke A.M., O'Loughlin J.* Spatial Analysis in Political Geography / Agnew J., Mamadouh V.A., Secor A., Sharp J. (eds.) Companion to Political Geography. — Oxford: Blackwell/Wiley, 2016. — P. 189–205.

373. *Lipset S.M. Rokkan S.* Cleavage Structures, Party Systems and Voter Alignment. Party Systems and Voter Alignments. — NY: Free Press, 1967. — 64 p.

374. *Listhaug O., MacDonald S., Rabinowitz G.* A Comparative Spatial Analysis of European Party Systems // Scandinavian Political Studies. — 1990. — Vol. 13. No. 3. — P. 227–254.

375. *Liu X., Liu Y.* The Trend and Progress in Electoral Geography // Human Geography. — 2019. — Vol. 34. No.1. — P. 37–45. [in Chinese].

376. *Macallister I., Johnston R.J., Pattie C.J., Tunstall H., Dorling D.F.L, Rossiter D.J.* Class Dealignment and the Neighbourhood Effect: Miller Revisited // British Journal of Political Science. — 2001. — Vol. 31. No. 1. — P. 41–59.

377. *Marzagao T.* A dimensão geográfica das eleições brasileiras // Opinião Pública. — 2013. — Vol. 19. No. 2. — P. 270–290.
378. *Massicotte L., Blais A.* Mixed Electoral Systems: A Conceptual and Empirical Survey // Electoral Studies. — 1999. — Vol. 18. No. 3. — P. 341–366.
379. *Mattila M.* Roll Call Analysis of Voting in the European Union Council of Ministers after the 2004 Enlargement // European Journal of Political Research. — 2009. — Vol. 48. No.6. — P. 840–857.
380. *McGee T.G.* The Malayan Elections of 1959: A Study in Electoral Geography. — Malaya: University of Malaya, 1962. — 99 p.
381. *McGhee E.* Partisan Gerrymandering and Political Science // Annual Review of Political Science. — 2020. — Vol. 23. No. 1. — P. 171–185.
382. *McGing C.* Towards a Feminist Electoral Geography // Political Geography. — 2015. — Vol. 47. — P. 86–87.
383. *McPhail I.R.* Recent Trends in Electoral Geography / Proceeding of the Sixth New Zealand Geography Conference. — Christchurch, 1971. — P. 7–12.
384. *Mehta S., Sekhon J.S.* Patterns of Voting Participation in Himachal Pradesh: A Spatial Perspective // Indian Journal of Political Science. — 1980. — Vol. 14. No. 20. — P. 79–90.
385. *Merrill S. III* A Probabilistic Model for the Spatial Distribution of Party Support in Multiparty Electorates // Journal of the American Statistical Association. — 1994. — Vol. 89. No. 428. — P. 1190–1197.
386. *Merrill S. III* Discriminating between the Directional and Proximity Spatial Models of Electoral Competition // Electoral Studies. — 1995. — Vol. 14. No. 3. — P. 273–287.
387. *Merrill S. III* Voting Behavior Under the Directional Spatial Model of Electoral Competition // Public Choice. — 1993. — Vol. 77. No. 4. — P. 739–756.
388. *Middleton A.* The Effectiveness of Leader Visits during the 2010 British General Election Campaign // The British Journal of Politics and International Relations. — 2014. — Vol. 17. No. 2. — P. 244–259.
389. *Miller W.L.* Electoral Dynamics in Britain since 1918. — London: Macmillan, 1977. — 242 p.

390. *Mintz E.* Election Campaign Tours in Canada // Political Geography Quarterly. — 1985. — Vol. 4. No. 1. — P. 47–54.

391. *Mirahmadi F.S., Kavyani Rad M.* Explanation of Dominant Approaches in Electoral Geography // Political Spatial Planning. — 2019. — Vol. 1. No. 2. — P. 105–116. [in Persian].

392. *Molinar J.* Counting the Number of Parties: An Alternative Index // American Political Science Review. — 1991. — Vol. 85. No. 4. — P. 1383–1391.

393. *Monmonier M.* Bushmanders and Bullwinkles: How Politicians Manipulate Electronic Maps and Census Data to Win Elections. — Chicago: University of Chicago Press, 2001. — 216 p.

394. *Monroe B.L.* Disproportionality and Malapportionment: Measuring Electoral Inequity // Electoral Studies. — 1994. — Vol. 13. No. 2. — P. 132–149.

395. *Moran P.A.P.* Notes on Continuous Stochastic Phenomena // Biometrika. — 1950. — Vol. 37. No.1–2. — P. 17–23.

396. *Morgenstern S., Swindle S.M., Castagnola A.* Party Nationalization and Institutions // Journal of Politics. — 2009. — Vol. 71. No. 4. — P. 1322–1341.

397. *Morrill E.L.* Redistricting Revisited // Annals of the Association of American Geographers. — 1976. — Vol. 66. No. 4. — P. 548–556.

398. *Morrill R., Knopp L, Brown M.* Anomalies in Red and Blue: Exceptionalism in American Electoral Geography // Political Geography. — 2007. — Vol. 26. No. 5. — P. 525–553.

399. *Muir R., Paddison R.* Politics, Geography and Behavior. — London and NY: Methuen, 1981. — 230 p.

400. *Myagkov M., Ordeshook P.* The Spatial Character of Russia's New Democracy // Public Choice. — 1998. — Vol. 97. No. 3. — P. 491–523.

401. *Nagle J.F.* How Competitive Should a Fair Single Member Districting Plan Be? // Election Law Journal: Rules, Politics, and Policy. — 2017. — Vol. 16. No. 1. — P. 196–209.

402. *Nagle J.F.* Measures of Partisan Bias for Legislating Fair Elections // Election Law Journal: Rules, Politics, and Policy. — 2015. — Vol. 14. No. 4. — P. 346–360.

403. *Neto O.A., Cox G.W.* Electoral Institutions, Cleavage Structures and the Number of Parties // American Journal of Political Science. — 1997. — Vol. 41. No. 1. — P. 149–174.
404. *Niemi R.G., Grofman B., Carlucci C., Hofeller Th.* Measuring Compactness and the Role of a Compactness Standard in a Test for Partisan and Racial Gerrymandering // The Journal of Politics. — 1990. — Vol. 52. No. 4. — P. 1155–1181.
405. *Nohlen D.* Elections and Electoral Systems. — Bonn: Forschungsinstitut der Friedrich-Ebert-Stiftung, 1984. — 105 p.
406. *Noragon J.L.* Redistricting, Political Outcomes, and the Gerrymandering of the 1960s // Annals of the New York Academy of Science. — 1973. — No. 219. — P. 314–333.
407. *Norris P.* Why Electoral Integrity Matters? — NY: Cambridge University Press, 2014. — 312 p.
408. *Nosek V., Netrdova P.* Measuring Spatial Aspecks of Variability. Comparing Spatial Autocorrelation with Regional Decomposition in International Unemployment Research // Historical Social Research. — 2014. — Vol. 39. No. 2. — P.292–314.
409. *Nurmi H.* Comparing Voting Systems. — Dodrecht: D. Reidel Publishing Company, 1987. — 209 p.
410. *Okunev I.* Political Geography / Translated from Russian by M. Ananyeva, N. Panich and N. Simakov. — Bruxelles: Peter Lang, 2021. — 474 p.
411. *Okunev I., Oskolkov P., Tislenko M.* Transforming the Matryoshka: Merger of Russian Regions // Regions and Cohesion. — 2019. — Vol. 9. No. 3. — P. 29–57.
412. *Okuniew I.* Geografia Polityczna / Tłumaczenie z języka angielskiego i rosyjskiego A. Kochanecki. — Krakow: Polskie Towarzystwo Geopolityczne, 2021. — 446 p.
413. *O'Loughlin J., Flint C., Anselin, L.* The Geography of the Nazi Vote: Context, Confession, and Class in the Reichstag Election of 1930 // Annals of the Association of American Geographers. 1994. — Vol. 84. No. 3. — P. 351–380.
414. *O'Loughlin J., Shin M., Talbot P.* Political Geographies and Cleavages in the Russian parliamentary elections // Post-Soviet Geography and Economics. — 1996. — Vol. 37. No. 6. — P. 355–385.

415. *O'Loughlin J.* Spatial Analysis in Political Geography // Agnew J., Mitchell K., Toal G. (eds.) A Companion to Political Geography. — Oxford: Blackwell, 2003. — P. 30–46.

416. *Openspaw S.* The Modifiable Areal Unit Problem / Concepts and Techniques in Modern Geography. Vol. 38. — Norwich: Geo Book, 1984. — 41 p.

417. *Ord K.* Estimation Methods for Models of Spatial Interaction // Journal of the American Statistical Association. — 1975. — Vol. 70. No. 349. — P. 120–126.

418. *Ord K., Getis A.* Local Spatial Autocorrelation Statistics: Distributional Issues and an Application // Geographical Analysis. — 1995. — Vol. 27. No. 4. — P. 286–306.

419. *Orr D.* The Persistence of Gerrymandering in North Carolina Redistricting // South Eastern Geographer. — 1969. — Vol. 9. — P. 39–54.

420. *Pareto V.* Cours d'Economie Politique. — Lausanne: Rouge, 1896. — 430 p.

421. *Pattie Ch., R . Johnston R.* Electoral Geography / Kitchin R., Thrift N. (eds.) International Encyclopedia of Human Geography. — Amsterdam: Elsevier, 2009. — P. 405–422.

422. *Pelling H.* The Social Geography of British Elections 1885–1910. — London: Macmillan, 1967. — 486 p.

423. *Perepechko A., Kolossov V., ZumBrunnen C.* Remeasuring and Rethinking Social Cleavages in Russia: Continuity and Changes in Electoral Geography 1917–1995 // Political Geography. — 2007. — Vol. 26. No. 2. — P. 179–208.

424. *Petrov N.* Dos necrológicas de Rusia: un estudio comparativo de la geografía electoral de 1917 y 1989 // Estudios geográficos. — 1991. — Vol. 52. No. 204. — P. 475–496.

425. *Pickel S.* Secularization of Electoral Behavior? The State-Church-Cleavage in Europe // Pickel G., Sammet K. (eds.) Transformations of Religiosity. Religion and Religiosity in Eastern Europe 1989–2010. — Wiesbaden: VS, 2011. — P. 111–134.

426. *Polsby D.D., Popper R.D.* The Third Criterion: Compactness as a Procedural Safeguard Against Partisan Gerrymandering // Yale Law & Policy Review. — 1991. — Vol. 9. No. 2. — P. 301–353.

427. *Prescott J.R.V.* The Function and Methods of Electoral Geography // Annals of the Association of American Geographers. — 1959. — Vol. 49. No. 3. — P. 296–304.
428. *Pulsipher A.G., Weatherby J.L., Jr.* Malapportionment, Party Competition, and the Functional Distribution of Government Expenditures // American Political Science Review. — 1962. — Vol. 62. No. 4. — P. 1207–1219.
429. *Put G.-J.* Is there a Friends-and-neighbors Effect for Party Leaders? // Electoral Studies. — 2021. — No. 71 (2). — Article No. 102338. — 32 p.
430. *Quinn K., Martin A.* An Integrated Computational of Multiparty Electoral Competition // Statistical Science. — 2002. — Vol. 17. No. 4. — P. 405–419.
431. *Rabinowitz G., MacDonald S.E.* A Directional Theory of Issue Voting // American Political Science Review. — 1989. — Vol. 83. No. 1. — P. 93–121.
432. *Rae D.W.* The Political Consequences of Electoral Laws. — New Haven: Yale University Press, 1971. — 209 p.
433. *Reményi P., Gekić H., Bidžan-Gekić A., Sümeghy D.* Electoral Geography of Bosnia and Herzegovina — is there Anything Beyond the Ethnic Rule? // East European Politics. — 2022. — Vol. 38. No. 2. — P. 227–253.
434. *Reock E.C.* Measuring Compactness as a Requirement of Legislative Apportionment // Midwest Journal of Political Science. — 1961. — Vol. 5. No. 1. — P. 70–74.
435. *Reynolds A., Reilly B., Ellis A.* (eds.) Electoral System Design: The New International IDEA Handbook. — Stockholm: IDEA, 2008. — 237 p.
436. *Reynolds D.R.* A Spatial Model for Analyzing Voting Behavior // Acta Sociologica. — 1969. — Vol. 12. No. 3. — P. 122–131.
437. *Riggs R.E., Hanson K.F., Heinz M., Hughes B.B.J. Volgy T.J.* Behavioralism in the Study of the United Nations // World Politics. — 1970. — Vol. 22. No. 2. — P. 197–236.
438. *Roberts M.C., Rumage K.W.* The Spatial Variations in Urban Left-wing Voting in England and Wales in 1951 // Annals of the Association of American Geographers. — 1965. — Vol. 55. No. 1. — P. 161–176.

439. *Rokkan S.* Citizens, Elections, Parties: Approaches to the Comparative Study of the Processes of Development. — Oslo: Universitetsforlaget, 1970. — 470 p.
440. *Romero Ballivián S.* Geografía electoral de Bolivia. — La Paz: Fundemos, 2003. — 510 p.
441. *Römmele A.* Cleavage Structure and Party Systems in East and Central Europe // Lawson K. et al. (eds.) Cleavages, Parties, and Voters: Studies from Bulgaria, the Czech Republic, Hungary. Poland, and Romania. — Westport: Praeger, 1999. — P. 3–17.
442. *Rose R., Urwin D.* Persistence and Change in Western Party Systems Since 1945 // Political Studies. — 1970. — Vol. 18. No. 3. — P. 287–319.
443. *Rose R., Urwin D.* Regional Differentiation and Political Unity in Western Nations. — London: SAGE, 1975. — 53 p.
444. *Rykiel Z.* Polish Electoral Geography and its Methods // Przestrzeń Społeczna (Social Space). — 2011. — Vol. 1. No. 1. — P. 17–48.
445. *Saari D.G.* Basic Geometry of Voting. — Berlin: Springer-Verlag, 1995. — 312 p.
446. *Samuels D., Snyder R.* The Value of a Vote: Malapportionment in Comparative Perspective // British Journal of Political Science. — 2001. — Vol. 31. No. 4. — P. 651–671.
447. *Sartori G.* Parties and Party Systems: Volume 1. A Framework for Analysis. — Cambridge: Cambridge University Press, 1976. — 383 p.
448. *Sauer C.O.* Geography and the Gerrymander // American Political Science Review. — 1918. — Vol. 12. No. 3. — P. 403–426.
449. *Schofield N.* The Spatial Model of Politics. — London, NY: Routledge, 2008. — 320 p.
450. *Schraff D., Vergioglou I., Demirci B.B.* The European NUTS-level Election Dataset: A Tool to Map the European Electoral Geography // Party Politics. — 2022. — Vol. 29. No. 1. — P. 1–10.
451. *Schwartzberg J.E.* Reapportionment, Gerrymanders, and the Notion of Compactness // Minnesota Law Review. — 1966. — Vol. 50. — P. 443–452.
452. *Secor A.* Feminizing Electoral Geography / *Staeheli L., Kofman E., Peake L.* (eds.) Mapping Women, Making Politics; Feminist Perspectives on Political Geography. — NY: Routledge, 2004. — P. 261–272.

453. *Schurr C.* Towards an Emotional Electoral Geography: The Performativity of Emotions in Electoral Campaigning in Ecuador // Geoforum. — 2013. — Vol. 49. — P. 114–126.
454. *Senftleben W.* Electoral Systems and Ethnic Pluralism in Developing Countries. Introductory Case-Studies in Comparative Geography // Geographic Research. — 1976. — No. S. — P. 105–145.
455. *Senftleben W.* Studies in Electoral Geography. — Taipei: National Taiwan Normal University, 1977. — 146 p.
456. *Shapley L.S., Shubik M.* A Method for Evaluating the Distribution of Power in a Committee System // American Political Science Revue. — 1954. — Vol. 48. No. 3. — P. 787–792.
457. *Shepsle K.* Models of Multiparty Electoral Competition. — London: Routledge, 1991. — 112 p.
458. *Shin M.* Democratizing Electoral Geography: Visualizing Votes and Political Neogeography // Political Geography. — 2009. — Vol. 28. No. 3. — P. 149–153.
459. *Shin M., Agnew J.* Spatial Regression for Electoral Studies: The Case of the Italian Lega Nord / Warf B., Leib J. (eds.) Revitalizing Electoral Geography. — Farnham: Ashgate, 2011. — P. 59–74.
460. *Siegfried A.* Géographie électorale de l'Ardèche sous la 3e République. — Paris: Armand Colin, 1949. — 139 p.
461. *Siegfried A.* Tableau politique de la France de l'ouest sous la 3e Republique. — Paris: Armand Colin, 1913. — 536 p.
462. *Signorino C.S., Ritter J.M.* Tau-b or Not Tau-b: Measuring the Similarity of Foreign Policy Positions // International Studies Quarterly. — 1999. — Vol. 43. No. 1. — P. 115–144.
463. *Singh P.* Geography of Voting: A Case Study in Electoral Geography // International Journal of Informative and Futuristic Research. — 2015. — No. 3 (1). — P. 10–18.
464. *Sinha M.* Electoral Geography of India. — New Delhi: Adhyayan, 2007. — 315 p.
465. *Smet de R., Evalenko, R.* Les élections belges. Explication de la répartition géographique des suffrages. — Bruxelles: Institut de Sociologie Solvay, 1956. — 154 p.

466. *Soares G., Terron S.* Dois Lulas: A geografia eleitoral da reeleição (explorando conceitos, métodos e técnicas de análise geoespacial) // Opinião Pública. — Vol. 14. No. 2. — P. 269–301.
467. *Srivastava M.K.* Electoral Geography of an Indian State: Space-Time Sociological Models of Congress Support in Uttar Pradesh. — Allahabad: Atul Dissertations, 1982. — 172 p.
468. *Stephanopoulos N.O., McGhee E.M.* Partisan Gerrymandering and the Efficiency Gap // The University of Chicago Law Review. — 2015. — Vol. 82. No. 2. — P. 831–900.
469. *Stockemer D., Blais A., Kostelka F., Chhim C.* Voting in the Eurovision Song Contest // Politics. — 2018. — Vol. 38. No. 4. — P. 428–442.
470. *Sui D.Z.* Tobler's First Law of Geography: A Big Idea for a Small World? // Annals of the Association of American Geographers. — 2004. — Vol. 94. No. 2. — P. 269–277.
471. *Taagepera R., Grofman B.* Mapping the Indices of Seats-Votes Disproportionality and Inter-Election Volatility // Party Politics. — 2003. — Vol. 9. No. 6. — P. 659–677.
472. *Taagepera R., Shugart M.* Seats and Votes: The Effects and Determinants of Electoral Systems. — New Haven: Yale University Press, 1989. — 288 p.
473. *Takagi A.* Recent Trends in Electoral Geography // Japanese Journal of Human Geography. — 1986. — No. 38. — P. 26–40. [in Japanese].
474. *Tapp K.* Measuring Political Gerrymandering // The American Mathematical Monthly. — 2019. — Vol. 126. No. 7. — P. 593–609.
475. *Tatalovich R.* «Friends and Neighbors» Voting: Mississippi, 1943–73 // The Journal of Politics. — 1975. — Vol. 37. No. 3. — P. 807–814.
476. *Taylor A.H.* The Electoral Geography of Welsh and Scottish Nationalism // Scottish Geographical Magazine. — 1979. — Vol. 89. No. 1. — P. 44–52.
477. *Taylor P.J.* Quantitative Methods in Geography: An Introduction to Spatial Analysis. — Boston and London: Houghton Mifflin, 1977. — 386 p.
478. *Taylor P.J.* Some Implications of the Spatial Organization of Elections // Transactions of the Institute of British Geographers. — 1973. — No. 60. — P. 121–136.

479. *Taylor P.J., Gudgin G.* The Myth of Non-partisan Cartography: A Study of Electoral Biases in the English Boundary Commission's Redistribution for 1955–1970 // Urban Studies. — 1976. — Vol. 13. No. 1. — P. 13–25.
480. *Taylor P.J., Johnston R.J.* Geography of Elections. — London: Penguin Books, 1979. — 528 p.
481. *Terron S.* Geografia Eleitoral Em Foco // Revista Em Debate. — 2012. — Vol. 4. No. 2. — P. 8–18.
482. *Thiervoz R.* L'Industrie en Valdaine et ses Répercussions Démographiques, Sociales et Électorales // Revue de Géographie Alpine. — 1954. — Vol. 42. Part 1. — P. 81–105.
483. *Tilly C.* (ed.) The Formation of National States in Western Europe. — Princeton: Princeton University Press, 1975. — 711 p.
484. *Tingsten H.* Political Behavior: Studies in Election Statistics. — London: P.S. King and Son, 1937. — 231 p.
485. *Tučas R.* Rinkimų geografija: mokomoji knyga. — Vilnius: Vilniaus universiteto leidykla, 2016. — 200 p.
486. *Unwin A., Unwin D.* Exploratory Spatial Data Analysis with Local Statistics // The Statistician. — 1998. — Vol. 47. No. 3. — P. 415–421.
487. *Van Wingen J.R., Parker J.B.* Measuring Friends-and-Neighbors Voting // American Politics Quarterly. — 1979. — No. 7 (3). — P. 367–383.
488. *Venyet-Verner G.* Quelques Réflexions sur la Géographie Politique des Alpes du Nord et leur Avant-pays // Revue de Géographie Alpine. — 1954. — Vol. 42. Part 1. — P. 107–110.
489. *Verma A.* Electoral Geography: Approaches to Study Voting Behavior // Research Review. — 2022. — Vol. 7. No. 3. — P. 68–73.
490. *Vickrey W.* On the Prevention of Gerrymandering // Political Science Quarterly. — 1961. — Vol. 76. No. 1. — P. 105–110.
491. *Voeten E.* Data and Analyses of Voting in the UN General Assembly / Reinalda B. (ed.) Routledge Handbook of International Organization. — London: Routledge, 2013. — P. 54–66.
492. *Vos de S.* De omgeving telt. Compositionele effecten in de sociale geografie: PhD thesis. — Amsterdam: Instituut voor Sociale Geografie UvA, 1997. — 221 p.

493. *Walter D.O.* Reapportionment and Urban Representation // The Annals of the American Academy of Political and Social Science. — 1938. — Vol. 195. No. 1. — P. 11–20.

494. *Warf B., Leib J.* Revitalizing Electoral Geography. — Farnham: Ashgate Publishing, 2011. — 238 p.

495. *Weaver J.B.* Fair and Equal Districts. — NY: National Municipal League, 1970. — 130 p.

496. *Webster G.R.* Representation, Geographic Districting, and Social Justice // Journal of Geography. — 2004. — Vol. 103. No. 3. — P. 111–126.

497. *Wildgen J.K.* The Measurement of Hyperfractionalization // Comparative Political Studies. — 1971. — Vol. 4. No. 2. — P. 233–243.

498. *Woolstencroft R.P.,* Electoral Geography: Retrospect and Prospect // International Political Science Review. — 1980. — Vol. 1. No. 4. — P. 540–560.

499. *Wright J.K.* Voting Habits in the United States: A Note on Two Maps // The Geographical Review. — 1932. — Vol. 22. No. 4. — P. 666–672.

500. *Wusten van der H., Mamadouh V.* «It is the Context, Stupid!». Or is It? British-American Contributions to Electoral Geography since the 1960s // L'Espace Politique. — 2014. — Vol. 23. No. 2.

501. *Wuthrich F.M.* An Essential Center-Periphery Electoral Cleavage and the Turkish Party System // International Journal of Middle East Studies. — 2013. — Vol. 45. No. 4. — P. 751–773.

502. *Zavala R.G.* Génesis de la geografía electoral // Revista Espacialidades. — 2012. — Vol. 2. No. 1. — P. 80–95.

503. *Zolnerkevic A., Raffo J.G.* Geografia Eleitoral: Representação Espacial da Volatilidade do Voto // GEOUSP: Espaço e Tempo. — 2013. — Vol. 17. No. 1. — P. 221–228.

504. *Zuiderveen Borgesius F., Möller J., Kruikemeier S., Ó Fathaigh R., Irion K., Dobber T., de Vreese C.H.* Online Political Microtargeting: Promises and Threats for Democracy // Utrecht Law Review. — 2018. — Vol. 14. No. 1. — P. 82–96.

505. *Zvada L., Petlach M., Ondruška M.* Where Were the Voters? A Spatial Analysis of the 2019 Slovak Presidential Election // Slovak Journal of Political Sciences. — 2020. — Vol. 20. No. 2. — P. 176–205.

List of Tables and Figures

Table 1.1.	Types of electoral space	13
Table 1.2.	Elected bodies of power in different forms of government	26
Table 1.3.	Types of political culture	32
Table 1.4.	Relationship between electoral and political systems	34
Table 1.5.	Electoral statistics and electoral sociology	36
Table 1.6.	Political technology strategies in elections	39
Table 2.1.	Type of majority	50
Table 2.2.	Types of electoral system	55
Table 2.3.	Electoral systems used in parliamentary elections around the world (as of 2022)	56
Table 2.4.	Electoral systems for heads of state around the world (as of 2022)	62
Table 2.5.	Electoral thresholds in parliamentary elections around the world (as of 2022)	70
Table 2.6.	The majoritarian component of non-compensatory mixed electoral systems	72
Table 2.7.	Types of mixed electoral systems	73
Table 2.8.	Types of quotas for distributing seats in parliament	77
Table 2.9.	Calculation of the distribution of seats using the quota method	78
Table 2.10.	Distribution of seats using the divisor method	79
Table 2.11.	Calculation of the distribution of seats using the divisor method	80
Table 2.12.	Calculation of the distribution of influence according to the Banzhaf index	83
Table 2.13.	Calculation of the distribution of influence according to the Shapley–Shubik power index	84
Table 2.14.	Degree of geographic favouritism of electoral systems	87
Table 2.15.	Influence of elements and properties of electoral systems on geographic favouritism	88
Table 3.1.	Territorial differentiation of types of political party	99

Table 3.2.   Types of party systems by level of nationalization 111
Table 3.3.   Cleavages in the ideological and political space 121
Table 4.1.   Projections of spatial models of voting 146
Table 5.1.   The scalar and vector spatial effects of voting 152
Table 5.2.   The neighbourhood effect 156
Table 5.3.   Wasted votes and the efficiency gap 166
Table 5.4.   Wasted votes and the weighted efficiency gap 168
Table 6.1.   Levels of determination of correlational links 189
Table 6.2.   Basic indicators in spatial statistics 196
Table 6.3.   Comparison of adjacency neighbourhood matrices ... 208
Table 6.4.   Comparison of metric neighbourhood matrices 211
Table 6.5.   Calculation of spatial lag indicators 214
Table 6.6.   Comparison of types of correlation 219
Table 6.7.   Moran's *I* for different neighbourhood matrices 225
Table 6.8.   Levels of spatial autocorrelation links 227
Table 6.9.   Comparison of types of spatial autocorrelation 229
Table 6.10. Comparison of spatial autocorrelation types 235
Table 6.11. Spatial regression models 241
Table 6.12. Diagnostics for multicollinearity 246

Fig. 1.1.   André Siegfried – the founder of electoral geography 19
Fig. 1.2.   Kevin R. Cox, Peter J. Taylor and Ron Johnston
           (left to right) – founders of the theory of electoral
           geography 21
Fig. 1.3.   Patrick Moran, Keith Ord and Luc Anselin (left to
           right) – founders of spatial statistical analysis 24
Fig. 1.4.   Chorochromatic map of the foreign policy similarity
           index (based on voting in the UNGA) 43
Fig. 2.1.   Electoral systems used to elect candidates to the
           lower (or only) house of parliament 58
Fig. 2.2.   Electoral systems used to elect candidates to the
           upper house of parliament 58
Fig. 2.3.   Pie chart of electoral systems used in parliamentary
           elections around the world 59
Fig. 3.1.   Groups of influence within political parties 102
Fig. 3.2.   The ideological spectrum of the electoral field 116

List of Tables and Figures 461

Fig. 3.3.	Spatial model of ideological and political cleavages ..	119
Fig. 3.4.	Values map of the world	126
Fig. 3.5.	The cyclical nature of ideological and political cleavages	127
Fig. 4.1.	Hotelling's linear city model	135
Fig. 4.2.	Downs' spatial model of voting	136
Fig. 4.3.	Two-dimensional Downs' spatial model of voting	138
Fig. 4.4.	The Enelow–Hinich perceptual voting model	140
Fig. 4.5.	The three-dimensional Enelow–Hinich voting model	141
Fig. 4.6.	The Granberg–Brown parabolic model of voting	143
Fig. 4.7.	The Granberg–Brown three-dimensional parabolic model of voting	144
Fig. 4.8.	The Rabinowitz–MacDonald three-dimensional parabolic model of voting	145
Fig. 5.1.	Miller's environmental (neighbourhood) effect	158
Fig. 5.2.	Gerrymandering	163
Fig. 5.3.	Efficiency gap	168
Fig. 5.4.	Main compactness tests for districts	172
Fig. 6.1.	Chorochromatic map of five equal classes (quintiles)	182
Fig. 6.2.	Chorochromatic map of five equal intervals	183
Fig. 6.3.	Histogram of five equal intervals	183
Fig. 6.4.	Chorochromatic map of five natural intervals	184
Fig. 6.5.	Chorochromatic map of specified intervals	185
Fig. 6.6.	Gauss–Laplace normal distribution	187
Fig. 6.7.	Box-and-whiskers plot with a standard deviation of 3σ	188
Fig. 6.8.	Scatterplot and regression line	190
Fig. 6.9.	The modifiable areal unit problem	194
Fig. 6.10.	The boundary problem	195
Fig. 6.11.	Algorithm for calculating the geographical average	197
Fig. 6.12.	Algorithm for calculating standard distance	199
Fig. 6.13.	Algorithm for calculating the ellipse of the geographical average	200
Fig. 6.14.	Ellipse of the geographical average	201

Fig. 6.15.	Types of neighbourhood by adjacency	206
Fig. 6.16.	Types of neighbour by adjacency order.	207
Fig. 6.17.	Queen's adjacency graph for U.S. counties	208
Fig. 6.18.	Adjacency graph for U.S. counties using the k-nearest neighbours method	211
Fig. 6.19.	Spatial lag chorochromatic map.	214
Fig. 6.20.	Diagram of spatial lag changes by neighbourhood order	215
Fig. 6.21.	Ratio chorochromatic map	216
Fig. 6.22.	Spatially weighted ratio chorochromatic map	217
Fig. 6.23.	Examples of spatial autocorrelation	221
Fig. 6.24.	Moran scatterplot for spatial autocorrelation	222
Fig. 6.25.	Quadrants of the Moran scatterplot	223
Fig. 6.26.	Significance test for Moran's I	224
Fig. 6.27.	Moran scatterplot for bivariate spatial correlation	226
Fig. 6.28.	Local Moran chorochromatic map	231
Fig. 6.29.	Local Moran significance chorochromatic map	231
Fig. 6.30.	Bivariate Local Moran's I chorochromatic map	232
Fig. 6.31.	Local Geary's C chorochromatic map	233
Fig. 6.32.	Getis–Ord general G chorochromatic map	234
Fig. 6.33.	Scatterplot for temporal autocorrelation	239
Fig. 6.34.	Natural breaks map marked with a zone of anomalous temporal autocorrelation	239
Fig. 6.35.	Differential Local Moran's I chorochromatic map	240
Fig. 6.36.	Spatial regression models	242

Index of Terms

absenteeism 30
alliance 103
anarchism 118
axis (ideological spectrum) 114

bearer of power 25
bloc 103

chorochromatic map 178
  equal class 179
  equal interval 179
  natural interval (breaks) 180
caucus 29
centrography 192, 196
cleavage 118
coalition 103
communism 118
compactness test 169
consensus 48
conservatism 114
contamination 72
correlation (covariation) 188
  bivariate spatial 218
  two spatially weighted
    variables 218
cracking/fracturing
  (gerrymandering) 164

deviation 187
  standard 187
diffusion of the electoral field 111
divisor 78

effect
  campaign 154
  distance decay 196
  fabricated majority 85
  friends-and-neighbours 152
  issue voting 155
  neighbourhood 156
  spoiler 61
efficiency gap 165
elections
  by-elections 29
  direct 28
  early 29
  general 29
  indirect 28
  mid-term 29
  primary 29
  regular 29
  snap 29
electoral cycle 29
electoral district 52
  multi-member 53
  single 53
  single-member 53
electoral field 13, 114
electoral-geographic
  engineering 16
electoral-geographic field trips 17
electoral-geographic
  mapping 17, 177
electoral-geographic modelling
  17

electoral-geographic zoning 17
electoral geography
  critical 15
  international 15
  party and party system 15
  political campaigns 15
  representation 15
  social cleavages 15
  urban 15
  voting 15
electoral landscape 13
electoral limology 15
electoral preferences
  multi-peaked 134
  single-peaked 134
electoral space 13
electoral system
  compensatory 71
  dual-vote 71
  hybrid 71
  with seat linkage 71
  with vote linkage 71
  with no linkage 71
  majoritarian 54. 60
  mixed 71
  multi-component 71
  non-compensatory 71
  non-parallel 72
  parallel 71
  plural 54
  plurinominal 55
  proportional 54, 66
  semi-proportional 64
  single-vote 71
  uninominal 55
  weighted 81
electoral threshold 69
electoral worldview 13
electorate
  changeable 38

core 38
opposition 38
exit poll 37

fascism 118
faction 103
funnel of causality 132

geofencing 40
geoframing 40
geographic determinism 19
geographic favouritism 85
geographic midpoint 197
geographic nihilism 20
geographic possibilism 20
geotargeting 40
gerrymandering 162
  discriminatory 165
  incumbent 164
  inter-party 164
  legitimate 165
government
  majority 103
  minority 103

heteroskedasticity 246
hijacking (gerrymandering) 164

ideological spectrum 114
ideology 117
incumbent 29
index
  Geary 226
  Getis–Ord 234
  Moran 219
inversion
  electoral 86
  spatial 196
isolate 205

kidnapping (gerrymandering) 164

liberalism 114
libertarianism 117
lists
  closed 67
  free 68
  open 67
  regional 67
Local Indicator of Spatial
  Association (LISA) 228, 361

majority
  absolute 49
  constitutional 49
  mixed 50
  qualified 49
  relative 50
  simple 49
majority bonus 68
malapportionment 160
mean
  arithmetic 186
  arithmetic weighted 186
  geographical 197
  geographical weighted 197
mobilization of the periphery 122
median 186
  centre 198
median voter 117, 137
model
  Spatial Autoregressive 241
  Spatial Durbin 241
  Spatial Error 241
monism 33
multicollinearity 245

nationalism 117
neighbourhood effect 156
  consensual 157

no effect 157
  Przeworski 157
  reactive 157
normal distribution 187
notables 98
null hypothesis 191

outlier 181
  spatial 201

packing (gerrymandering) 164
panachage 68
party 98
  block 64
  branch-type 99
  catch-all 101
  cadre 98
  cell-type 100
  centrist 101, 117
  clientelist 99
  committee-type
  congress-type 100
  electoralist 101
  elite 98
  identity 101
  mass-based 100
  militia-type 100
  moderate 117
  movement 102
  personalist 102
  pluralist 100
  programme 101
  proto-hegemonic 100
  radical 118
  regional 104
  regionalist 105
  ruling 103
party system 105
  bivalent 106
  dominant 107

hegemonic 107
monovalent 106
multi-party 107
nationalized 111
non-party 98
one-party 107
polyvalent 106
regionalized 111
segmented 111
mono-party 107
territorialized 111
two-party 107
Pearson coefficient 189
Plebiscite 26
pluralism 33
political culture 31
  parochial 32
  subject 32
  participant 32
polygon 165
primary 29
problem
  boundary problem 195
  modifiable areal unit
    problem 194
  multiple comparisons
    problem 195
psephology 12

quantile 179, 291
quota 69

redistricting 164
referendum 29

scalar 152
shortest split-line algorithm 165
significance level
  (p-value) 288, 404

social democracy 117
socialism 17
source of power 25
spatial analysis 18
spatial autocorrelation 217
  global 217, 228
  local 228
spatial lag 218
spatial neighbourhood weights 218
spatial regression 240
spatial voting model
  vector 144
  dynamic 147
  linear 139
  one-dimensional 133
  parabolic 142
  static 146
standard
  error 191
  distance 198
standard deviation ellipse 199
status quo (gerrymandering) 164
suffrage 30

transitivity 51

validity 36
variance 186
variation 186
vector 152
veto 50
vote
  excess 86
  lost 85
  wasted 85
voting
  absentee 30
  approval 54
  categorical 54

compulsory 30
conditional 64
cumulative 54
disapproval 54
early 30
preferential 54
proxy 30
remote 30
strategic 61
voting district 52

www.ingramcontent.com/pod-product-compliance
Ingram Content Group UK Ltd.
Pitfield, Milton Keynes, MK11 3LW, UK
UKHW021324180426
11947UKWH00017B/1432